# Shipboard Propulsion, Power Electronics, and Ocean Energy

# Shipboard Propulsion, Power Electronics, and Ocean Energy

## Mukund R. Patel

**CRC Press**
Taylor & Francis Group
Boca Raton London New York

CRC Press is an imprint of the
Taylor & Francis Group, an **informa** business

Cover illustration courtesy of Azipod® Propulsion System (With permission from ABB Marine.).

CRC Press
Taylor & Francis Group
6000 Broken Sound Parkway NW, Suite 300
Boca Raton, FL 33487-2742

First issued in paperback 2017

© 2012 by Taylor & Francis Group, LLC
CRC Press is an imprint of Taylor & Francis Group, an Informa business

No claim to original U.S. Government works

Version Date: 20120106

ISBN 13: 978-1-4398-8850-6 (hbk)
ISBN 13: 978-1-138-07544-3 (pbk)

| Library of Congress Cataloging-in-Publication Data |
| --- |

Patel, Mukund R., 1942-
    Shipboard propulsion, power electronics, and ocean energy / Mukund Patel.
       p. cm.
    ISBN 978-1-4398-8850-6 (alk. paper)
    1. Ship propulsion, Electric. 2. Power electronics. 3. Electric power-plants, Offshore.
4. Ocean wave power. I. Title.

VM773.P37 2012
623.87--dc23
                                        2011033522

**Visit the Taylor & Francis Web site at**
**http://www.taylorandfrancis.com**

**and the CRC Press Web site at**
**http://www.crcpress.com**

*To Anthony, Patricia, and Malena*

*for enhancing our family*

# Contents

## PART A   Power Electronics and Motor Drives

# *PART B*   *Electric Propulsion Technologies*

## PART C   *Emerging Ocean Energy Technologies*

# PART D   *System Integration Aspects*

# Preface

The marine industry demand for larger cargo and cruise ships with higher speed, lower life-cycle cost, lower environmental impact, and greater maneuverability, reliability, and safety has been rapidly growing. The conventional ship today can deliver transatlantic cargo in 2 to 3 weeks. Fast cargo ships being designed today would deliver the same freight in a week at about one-fifth the cost of airfreight. Moreover, high emissions from the land and air transportation on congested routes are now favoring marine transportation, which is mostly away from the population. Among large passenger cruise ships built today, the 339-m long, 38.6-m wide Freedom-class cruise of the Royal Caribbean International takes 4375 passengers and 1365 crew members with a diesel-electric power plant approaching 100 MW$_e$ rating that compares in size with some small power plants on land.

No other technology than power electronics has brought a greater change in the electrical power industry—on ship or on land—and still holds the potential to bring future improvements. The power electronics equipment prices have declined to approximately one-tenth in cost since the 1990s, fueling rapid growth in their applications throughout the power industry. Most cruise ships and many icebreakers today use electric propulsion, which uses power electronics in abundance. The navies of the world have undertaken electric propulsion with power electronics as their next goal for numerous benefits it offers.

Shipbuilders around the world have added requirements to minimize noise and vibration and maximize usable space. This allows larger combat weapons in navy warships and more paying passengers in passenger cruise ships, where the premium on space is the greatest. The integrated electrical power system makes the navy ship reconfigurable for greater survivability. The Office of Naval Research has been providing research funding for developing electric propulsion with an integrated electrical power system. The research is conducted by the Electric Ship Research and Development Consortium, which includes Florida State University, Massachusetts Institute of Technology, Mississippi State University, Purdue University, the U.S. Naval Academy, University of South Carolina, and the University of Texas–Austin.

For these reasons, shipboard power systems have undergone significant new developments since 2000 and will continue to do so at an even faster rate in the current decade. Today's shipbuilders of both commercial and navy ships, along with their support industries, are now taking an active part in research and development to advance shipboard electric propulsion and power electronics systems, both of which are covered at length in this book.

New high-power density permanent magnet and superconducting motors developed to fit the confines of propulsion pods are discussed in this book. Potential applications of the recently developed high-temperature superconductors in degaussing the ship and in energy storage to provide pulse power on combat ships are also discussed.

It is in this light that modern commercial and military shipbuilders are looking for electrical power and power electronics engineers to meet their rapidly growing need for large, fast, efficient, reconfigurable, and economical ships to compete in the growing international trade and national defense. Until now, there was no single book available that covered the entire scope of shipboard power and propulsion systems. The industry professionals had been relying on limited publications presented at various conferences and a few books with short sections with sketchy coverage of this vast subject.

This book is a companion to *Shipboard Electrical Power Systems* by the same author published by CRC Press in 2011. It is the first comprehensive volume of its kind that focuses on the shipboard electric propulsion and power electronics systems, in addition to the renewable ocean energy technologies, in which marine engineers will gradually become involved. As such, the electrical power system for extracting ocean energy is the exact reverse of the variable-frequency motor drive already in wide use for ship propulsion.

It is hoped that this book is a timely addition to the literature and a one-volume resource for students at various marine and naval academies around the world and a range of industry professionals.

**Mukund R. Patel**
*Kings Point, New York*

# Acknowledgments

A book of this nature incorporating new technologies for shipboard propulsion, power electronics, and renewable ocean energy extraction cannot possibly be written without help from many sources. I have been extremely fortunate to receive full support from many organizations and individuals in the field. They not only encouraged me to write the book on this timely subject, but also provided valuable suggestions and comments during the development of the book.

At the U.S. Merchant Marine Academy, Kings Point, New York, I am grateful to Professor Jose Femenia, director of the graduate program; Dr. David Palmer, engineering department head; and Dr. Shashi Kumar, academic dean, for supporting my research and publications that led to writing this book. I have benefited from many midshipmen at the academy, both the undergraduate seniors and the graduate students, who contributed to my learning by pointed questions and discussions based on their professional experience.

Several shipbuilders and organizations worldwide provided current data and reports on shipboard power technologies. They are ABB Marine, Rolls Royce Marine, Converteam Incorporated, and the Office of Naval Research. Expert individuals at these organizations gladly provided all the help I requested. Dr. Kaushik Rajashekara, chief technologist, Propulsion & Power Systems Engineering, Rolls-Royce Corporation, provided valuable help. In addition, I have benefited from many graduate students at the U.S. Merchant Marine Academy who provided some data used in this book. They are David Condron, Ted Diehl, Chad Fuhrmann, Bill Frost, James Hogan, Derrick Kirsch, Enrique Melendez, Dana Walker, Bill Veit, Raul Osigian, Edward Woida, Edgar Torres, and others.

The patience of Mr. Jonathan Plant, executive editor at CRC Press/Taylor & Francis Group, kindly allowed me to reconfigure and complete the book even with many interruptions. My gratitude also goes to Lt. Anthony J. Indelicato Jr., U.S. Navy Reserve (Ret.), Professor John Hennings of the Webb Institute of Naval Architecture, and Professor Ehsan Mesbahi of Newcastle University in the United Kingdom for reviewing the book proposal and providing valuable comments.

My wife Sarla and grandchildren Rayna, Dhruv, and Naiya cheerfully contributed the time they would have otherwise spent with me.

I offer my heartfelt acknowledgment for the valuable support and encouragement from all.

**Mukund R. Patel**

# About the Author

**Mukund R. Patel, PhD, PE,** is a professor of engineering at the U.S. Merchant Marine Academy in Kings Point, New York. He has over 45 years of hands-on involvement in research, development, and design of state-of-the-art electrical power equipment and systems. He has served as a principal engineer at the General Electric Company in Valley Forge, Pennsylvania; fellow engineer at the Westinghouse Research and Development Center in Pittsburgh, Pennsylvania; senior staff engineer at Lockheed Martin Corporation in Princeton, New Jersey; development manager at Bharat Bijlee (Siemens) Limited, Bombay, India; and 3M McKnight Distinguished Visiting Professor at the University of Minnesota, Duluth.

Dr. Patel obtained his PhD degree in electric power engineering from Rensselaer Polytechnic Institute, Troy, New York; MS in engineering management from the University of Pittsburgh; ME in electrical machine design from Gujarat University; and BE from Sardar University, India. He is a fellow of the U.K. Institution of Mechanical Engineers, associate fellow of the American Institute of Aeronautics and Astronautics, senior life member of the IEEE (Institute of Electrical and Electronics Engineers), registered professional engineer in Pennsylvania, chartered mechanical engineer in the United Kingdom, and a member of Eta Kappa Nu, Tau Beta Pi, Sigma Xi, and Omega Rho.

Dr. Patel is an associate editor of *Solar Energy*, the journal of the International Solar Energy Society, and a member of the review panels for the government-funded research projects on renewable energy in the state of California and the emirate of Qatar. He has authored five books, two of which have been translated into Chinese and Korean, and major chapters in two international handbooks. He has taught 3-day courses to practicing engineers in the electrical power industry for over 15 years, has presented and published over 50 papers at national and international conferences and in journals, holds several patents, and has earned NASA recognition for exceptional contribution to the power system design for the Upper Atmosphere Research Satellite. He can be reached at patelm@usmma.edu.

# About This Book

This book evolved from the author's 30 years of work experience at General Electric, Lockheed Martin, and Westinghouse Electric Corporations and 15 years of teaching at the U.S. Merchant Marine Academy in Kings Point, New York. The book has 16 chapters, divided in four parts

**Part A:** Power Electronics and Motor Drives
**Part B:** Electric Propulsion Technologies
**Part C:** Emerging Ocean Energy Technologies
**Part D:** System Integration Aspects

Since power electronics and variable-frequency motor drives dominate the electric propulsion in modern ships, they are covered before the discussion of ship propulsion, although the book title suggests otherwise. Part A coverage is needed to cover the electric propulsion of ships discussed in Part B and the renewable ocean energy that is the topic in Part C. Moreover, Part A topics on power electronics are treated broadly for applications in all power systems—on land, on ships, or at renewable power sites in an ocean. The analytical treatment is kept simpler than that found in specialized books on power electronics, keeping in mind that most marine engineers are mechanical majors with limited background and interest in detailed analyses of power electronics equipment.

As a textbook, the discussion can be covered in one elective course in marine and naval academies offering engineering programs. The book has many examples integrated within the text, exercise problems at the end that can help students calibrate their learning, and concept questions to engage students in group discussions on the topics covered in the chapter. The rationale for grouping these four parts together in one book is as follows:

Part A covers basic power electronics and variable-frequency motor drives that convert fixed-frequency power to a variable-frequency power at the motor terminals as needed during various phases of ship operations, such as approaching and leaving the port and while cruising at various speeds. These drives are widely spread in the power industry at present. The shipboard power systems are now catching up with the trend on land by their increasing use on ships. This part dominates the book, centering on ship propulsion and ocean energy.

Part B covers at length the electric propulsion systems and various propulsion motors, including the superconducting motor recently developed by the U.S. Navy for future ships. New high-power-density permanent magnet and superconducting motors developed to fit the confines of the propulsion pod are covered. Potential applications of the high-temperature superconductors in degaussing the ship and in energy storage to provide pulse power on combat ships are also covered.

Part C at first glance may look out of place with the other two parts. However, the ocean is truly an infinite source of energy, which is being developed for electric power generation around the world. The construction, operation, and logistic support from the shore to the renewable energy site in the ocean will surely be provided by marine engineers. Moreover, the ocean power generated at randomly varying ocean waves, marine currents, and wind speed in offshore and far-shore energy farms—spreading rapidly in Europe and many countries—will generate power at variable frequency, which needs to be converted into fixed-frequency power for local use or for feeding to the grid. Here also, power electronics are involved as much as in ship propulsion. As such, the electrical power system for extracting ocean energy will be the exact reverse of the variable-frequency drive for ship propulsion. In electrical propulsion, the power electronics interface the fixed-frequency power generator and the variable-frequency motor load. In ocean energy extraction, the power electronics interface the variable-frequency power generator and the fixed-frequency loads or the grid.

Part D covers two aspects—energy storage and system reliability—necessary for integrating a large system, electrical or mechanical. For example, in a practical ocean power system design for local use, such as for the local shore communities or for offshore oil-drilling platforms, some form of medium- to large-scale energy storage is necessary. And, the reliability in the system design is important since the electrical power is now taken for granted by most users. For this reason, the reliability fundamentals and estimation methods are presented in this part. The reliability considerations are important in any system, electrical or mechanical, but more so in power electronics systems, for which there are numerous parts. Engineers know that *the system is weaker than its weakest link*.

Thus, all four parts logically fit together in one book that should be of interest to undergraduate and graduate students in marine engineering. It is also a reference book for a range of commercial and military shipbuilders, ship users, port operators, renewable ocean energy developers, classification societies, machinery and equipment manufacturers, research institutes, universities, and others.

Both systems of units—international and British—are used in the book to present data as they came from various sources. An extensive conversion table connecting the two systems of units is therefore included in this book.

# Systems of Units and Conversion Factors

Both international units (SI or MKS [meter-kilogram-second] system) and British (foot-pound-second) units are used in this book. This table relates the international units with the British units commonly used in the United States.

| Category | Value in SI Unit = | Factor Below × | Value in British Unit |
|---|---|---|---|
| Length | meter | 0.3048 | foot |
| | mm | 25.4 | inch |
| | micron | 25.4 | mil |
| | km | 1.6093 | mile |
| | km | 1.852 | nautical mile |
| Area | $m^2$ | 0.0929 | square feet |
| | $mm^2$ | 506.7 | circular mil |
| Volume | liter ($dm^3$) | 28.3168 | cubic foot |
| | liter | 0.01639 | cubic inch |
| | $cm^3$ | 16.3871 | cubic inch |
| | $m^3/s$ | 0.02831 | cubic foot/hr |
| | liter | 3.7853 | gallon (U.S.) |
| | liter/sec | 0.06309 | gallon/minute |
| Mass | kg | 0.45359 | pound mass |
| | kg | 14.5939 | slug mass |
| Density | $kg/m^3$ | 16.020 | pound mass/$ft^3$ |
| | $kg/cm^3$ | 0.02768 | pound mass/$in^3$ |
| Force | N | 4.4482 | pound force |
| Pressure | kPa | 6.8948 | pound/$in^2$ (psi) |
| | kPa | 100.0 | bar |
| | kPa | 101.325 | std atm (760 torr) |
| | kPa | 0.13284 | 1 mm Hg at 20°C |
| Torque | Nm | 1.3558 | pound-force foot |
| Power | W | 1.3558 | foot pound/sec |
| | W | 745.7 | horsepower |
| Energy | J | 1.3558 | foot pound-force |
| | kJ | 1.0551 | Btu International |
| | kWh | 3412 | Btu International |
| | MJ | 2.6845 | horsepower hour |
| | MJ | 105.506 | therm |
| Temperature | °C | (°F - 32)·5/9 | °F |
| | K | (°F + 459.67) × 5/9 | °R |
| Heat | W | 0.2931 | Btu (Int.)/hour |
| | kW | 3.517 | ton refrigeration |

|           |                       |              |                              |
|-----------|-----------------------|--------------|------------------------------|
|           | $W/m^2$               | 3.1546       | $Btu/(ft^2\ hr)$             |
|           | $W/(m^2°C)$           | 5.6783       | $Btu/(ft^2\ hr\ °F)$         |
|           | $MJ/(m^3°C)$          | 0.0671       | $Btu/(ft^3\ °F)$             |
|           | $W/(m\ °C)$           | 0.1442       | $Btu\ inch\ /(ft^2\ hr\ °F)$ |
|           | $W/(m\ °C)$           | 1.7304       | $Btu\ ft/(ft^2\ hr\ °F)$     |
|           | $J/kg$                | 2.326        | Btu/pound                    |
|           | $MJ/m^3$              | 0.0373       | $Btu/ft^3$                   |
|           | $J/(kg\ °C)$          | 4.1868       | $Btu/(pound\ °F)$            |
| Velocity  | m/s                   | 0.3048       | foot/sec                     |
|           | m/s                   | 0.44704      | mile/hour                    |
|           | m/s                   | 0.51446      | knot                         |
| Magnetics | weber                 | $10^{-8}$    | line                         |
|           | $wb/m^2$ (tesla)      | 0.0155       | $kiloline/inch^2$            |

## PREFIXES TO UNITS

| m | micro | $10^{-6}$ | m | milli | $10^{-3}$  |
|---|-------|-----------|---|-------|------------|
| k | kilo  | $10^{3}$  | M | mega  | $10^{6}$   |
| G | giga  | $10^{9}$  | T | tera  | $10^{12}$  |

## OTHER CONVERSIONS

| | |
|---|---|
| 1 nautical mile | = 1.15081 mile |
| 1 bar pressure | = 14.50 psi = 100 kpascals |
| | = 29.53 in Hg = 10.20 m water = 33.46 ft water |
| 1 calorie (CGS unit) | = 4.1868 J |
| 1 kg cal (SI unit) | = 4.1868 kJ |
| 1 horsepower | = 550 ft-lb/s |
| 1 tesla magnetic flux density | = 1 $Wb/m^2$ = 10,000 gauss (lines/$cm^2$) |
| Absolute zero temperature | = 273.16 °C = 459.67° F |
| Acceleration due to Earth's gravity | = 9.8067 $m/s^2$ (32.173 5 $ft/s^2$) |
| Permeability of free space $\mu_o$ | = $4\pi \times 10^{-7}$ henry/m |
| Permittivity of free space $\varepsilon_o$ | = $8.85 \times 10^{-12}$ farad/m |

## ENERGY CONTENT OF FUELS

| | | |
|---|---|---|
| 1 tip of matchstick | = | 1 Btu (heats 1 lb water by 1°F) |
| 1 therm | = | 100,000 Btu (105.5 MJ = 29.3 kWh) |
| 1 quad | = | $10^{15}$ Btu |
| 1 $ft^3$ of natural gas | = | 1000 Btu (1055 kJ) |
| 1 gallon of LP gas | = | 95,000 Btu |
| 1 gallon of gasoline | = | 125,000 Btu |
| 1 gallon of no. 2 oil | = | 140,000 Btu |
| 1 gallon of oil (U.S.) | = | 42 kWh |

| | | |
|---|---|---|
| 1 barrel | = | 42 gallons (U.S.) |
| 1 barrel of refined oil | = | $6 \times 10^6$ Btu |
| 1 barrel of crude oil | = | $5.1 \times 10^6$ Btu |
| 1 ton of coal | = | $25 \times 10^6$ Btu |
| 1 barrel of crude oil | = | $5.1 \times 10^6$ Btu |
| 1 ton of coal | = | $25 \times 10^6$ Btu |
| 1 cord of wood | = | $30 \times 10^6$ Btu |
| 1 million Btu | = | 90 lb coal, or 8 gallons gasoline, or 11 gallons propane |
| 1 quad ($10^{15}$ Btu) | = | 45 million tons coal, or $10^{12}$ cubic feet natural gas, or 170 million barrels oil |
| 1 lb of hydrogen | = | 52,000 Btu = 15.24 kWh primary energy |

World's total primary energy demand in 2010 was about 1 quad ($10^{15}$ Btu) per day = $110 \times 10^{16}$ J/day = $305 \times 10^{12}$ kWh/day.

About 10,000 Btu of primary thermal energy input at the power-generating plant produces 1 kWh of electrical energy at the user's outlet.

The information provided herein does not necessarily represent the view of the U.S. Merchant Marine Academy or the U.S. Department of Transportation.

Although reasonable care was taken in preparing this book, the author and/or the publisher assumes no responsibility for any consequences resulting from the use of this information. The text, diagrams, technical data, and trade names presented herein are for illustration purposes only and may be covered under patents. For shipboard power and propulsion system design and analysis, the equipment manufacturers should be consulted for the current exact data.

# Part A

---

# Power Electronics and
# Motor Drives

No other technology has brought a greater change in the electrical power industry, and still holds the potential to bring future improvements, than power electronics. Power electronics equipment prices have declined to about 1/10 of their price since the 1990s, fueling an exponential growth in their applications throughout the power industry.

Since power electronics dominates electric propulsion in modern ships, it is discussed here before ship propulsion, although the book title suggests otherwise. Moreover, Part A topics on power electronics are treated broadly for applications in all power systems on land or on ships or for renewable power sites in an ocean. In that sense, Part A coverage is needed for the discussion of electric propulsion of ships in Part B and renewable ocean energy in Part C. The analytical treatment is simpler than that found in specialized books on power electronics, keeping in mind that most marine engineers are mechanical majors with limited background and interest in detailed analyses of power electronics equipment.

The subject of *electronics* is broad, covering the following five classes of functionally different electronics

1. Power electronics, for which solid-state semiconducting devices are used primarily as *on* or *off* switches to process and condition power—change voltage, current, or frequency—to increase the energy efficiency of the system.
2. Audio and video electronics in radio and TV circuits, for which three-terminal semiconducting devices (transistors) are primarily used to amplify weak signals from the airwaves before feeding to the speaker coil, TV, or the display monitor screen.

3. Operational amplifiers, which are used extensively in feedback control systems (servo controls).
4. Digital electronics in computer-type equipment, for which semiconducting devices are used to register the presence or absence of signals (1 or 0 digit).
5. Programmable logic controller (PLC), which uses digital electronics and relays to program discrete steps of operation in a logical sequence that is set in the program.

These five branches of electronics are so different in operation and analysis that a course focusing on one branch would generally not cover the other four branches. Part A of the book covers only power electronics.

Power electronics saves energy while offering great flexibility in electrical equipment operation. This is achieved using semiconductor junction devices with two terminals (diode) or three terminals (controlled diode and transistor) as switches to turn the power on and off in the desired direction and at the desired time. The term *switch* in Part A of this book means a power electronics semiconducting device periodically switched at line frequency (60 or 50 Hz) or at high frequency, up to hundreds of kilohertz. In theory—and only in theory—one can achieve the same results using ordinary wall switches if one can switch them on and off by hand at the desired frequency for a long time. Even then, the wall switches would wear out (erode) under unavoidable sparking at the moving metal contacts. The power electronics switches, on the other hand, can switch power on and off at hundreds of kilohertz frequency for decades without wearing out. Because the semiconducting switches are solid in form and were invented to replace hollow vacuum tubes, they are often called solid-state switching devices.

As for ships, electric propulsion has been selectively used in special applications since the 1950s and widely used in cruise ships since the 1980s after significantly improved power electronics variable-frequency motor drives became available at low prices. Navies worldwide are moving toward electric propulsion, primarily due to the continuing progress being made in power electronics. Part A of the book, therefore, is devoted to prepare the student in power electronics commonly used on land and on ships. It then feeds into Part B, which covers electric propulsion of ships, and Part C, which covers renewable ocean energy technologies, both of which widely use power electronics.

# 1  Power Electronics Devices

Power electronics devices in electrical power systems basically perform the following functions by periodically switching the current on and off at a desired frequency: (a) convert ac into dc, (b) convert dc into ac, (c) convert frequency, (d) control ac and dc voltages, and (e) control power without converting voltage or frequency.

The direct or indirect benefit of a power electronics controller is system efficiency improvement. Figure 1.1 is a simple example in a process heat control, which can be achieved two ways: (a) by inserting a control resistance (rheostat) or (b) by a control switch that is repetitively turned on for duration $T_{on}$ and off for duration $T_{off}$. In the rheostat control, the average power going in the load is fraction $R_L \div (R_L + R_c)$ of that coming out of the source, and the remaining power is wasted in $R_c$, giving the energy efficiency of $R_L \div (R_L + R_c)$. In switch control with an ideal switch, all energy coming from the source goes to the load, and the energy efficiency is 100%, although the average power going in the load is lower by the duty ratio $D = T_{on} \div (T_{on} + T_{off})$. Thus, the power electronics switch controls the load power without wasting energy in the control (dummy) resistor.

## Example 1.1

A 240-V dc generator is powering a chemical process heater that has 20-$\Omega$ load resistance and produces 2880 watts heat when fully on. A certain operation requires a partial heat of only 2160 watts on average. Determine (a) the required series resistance (rheostat) value and the process energy efficiency, and (b) the switching duty ratio D and the energy efficiency if an ideal power electronics control is employed to reduce the heat.

### SOLUTION

(a) With rheostat control of power, the ratio of the reduced to full power gives $2160 \div 2880 = R_L \div (R_L + R_c) = 20 \div (20 + R_c)$, which leads to $R_c = 6.667\ \Omega$. Then, the energy efficiency $= 20 \div (20 + 6.667) = 0.75$, which means 25% power is wasted in the rheostat.

(b) With power electronics control of power, the power ratio $2160 \div 2880 =$ Duty ratio $D = T_{on} \div (T_{on} + T_{off})$. The actual values of $T_{on}$ and $T_{off}$ can depend on the mass (thermal inertia) of the load. For a small mass to be heated uniformly in time, $T_{on}$ and $T_{off}$ could be seconds. For a large mass, they can be longer. The energy efficiency of the switch control is always 100% regardless of the switching times, meaning no energy is wasted in the switch or anywhere else.

Obviously, power electronics control is energy efficient, saving 25% energy in this example.

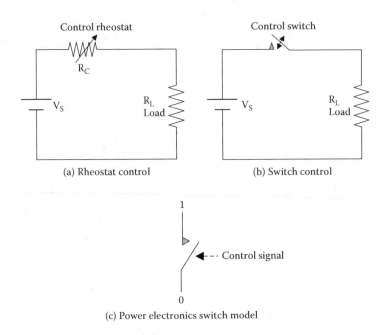

(a) Rheostat control                    (b) Switch control

(c) Power electronics switch model

**FIGURE 1.1**   Power control by variable series resistor versus on-off switch.

The U.S. Department of Energy (DoE) estimated that over 100 billion kWh can be saved every year in the United States by using power electronics motor drives. At $0.10/kWh energy cost, that would save about $10 billion per year and would eliminate about 11 1000-MW power plants in the United States. A significant fraction of electrical power in the United States is already supplied through power electronics, and this value is perhaps higher in other countries where the energy cost is higher.

Semiconducting devices have discrete conducting and nonconducting states that depend on the direction of the prospective current flow and the bias signal—voltage or current—at its control terminal. They are made with one or more junctions of two differently doped semiconducting materials, known as p-type and n-type. The semiconducting devices, when used to process high power (as opposed to signals), are called *power electronics devices*. In such applications, they are generally used only in two operating states, either on with near-zero resistance or off with a very large resistance across the main power terminals.

The diode is a two-terminal device, which automatically turns on (virtually offering zero resistance) when the p-terminal voltage is positive (higher) over the n-terminal voltage and is off (virtually offering infinite resistance) under the negative (lower) voltage at the p-terminal. The switchover from the on to off state is automatic and is not controllable by any signal. Other devices covered in this chapter use the diode as the building block, in which a third terminal is added to control (trigger) the on and off states by injecting a control signal at the control terminal. Thus, they virtually work as variable resistance that can be controlled ideally between zero (on) and infinity (off) by a small control signal applied at the third terminal. For this reason, the power electronics device in the electrical circuit diagram is often shown by a simple switch symbol between two power terminals and the third control terminal as shown in Figure 1.1c. A variety of switching devices is available for use as controlled switches. However, a few basic switching devices are as follows:

- Thyristor or silicon-controlled rectifier (SCR)
- Gate turnoff (GTO) thyristor
- Bipolar junction transistor (BJT)
- Insulated gate bipolar transistor (IGBT)
- Metal-oxide semiconducting field effect transistor (MOSFET)
- Junction field effect transistor (JFET)

The power electronics switching device of any type is triggered periodically on and off by a train of control signals of suitable frequency. The control signal may be of rectangular, triangular, or other wave shape and is generated by a separate triggering circuit, which is often called the firing circuit. Although it has a distinct identity with many different design features, it is generally incorporated in the main power electronics assembly.

Figure 1.2 shows the external construction features of these devices: (a) the diode with two power terminals; (b) the thyristor with three terminals (p-terminal with large eye, n-terminal with threads to go on a grounded assembly plate, and a control [gate] terminal with small eye); (c) the IGBT with three terminals; and (d) the gate triggering integrated circuit (IC) chip. The generally used circuit symbols of the six

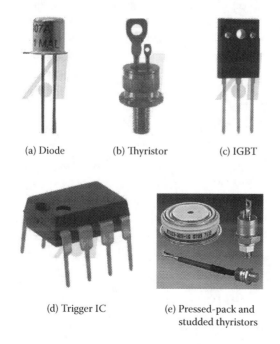

(a) Diode      (b) Thyristor      (c) IGBT

(d) Trigger IC      (e) Pressed-pack and studded thyristors

**FIGURE 1.2** Power electronics devices external construction features.

switching devices covered in this chapter are shown in Figure 1.3. A common feature among these devices is that all are 3-terminal devices. The power terminals 1 and 0 are connected in the main power circuit. The control gate terminal C is connected to the auxiliary control circuit called the firing circuit. In a normal conducting operation, terminal 1 is generally at higher voltage than terminal 0. Since the device is primarily used for switching power on and off as required, it is functionally represented by a gate-controlled switch as shown in Figure 1.4a. In absence of the gate control signal, the device resistance between the power terminals is large—the functional equivalence of an open switch. When the control signal is applied at the gate, the device resistance approaches zero, making the device function like a closed switch. The device in this state lets the current flow freely through its body. With the control signal on and off periodically, the device conduction can be turned on and off at high frequency as shown in Figure 1.4b. With no moving contacts, such a virtual switch can be operated on and off trillions of times per day for decades with no wear at all, requiring no maintenance over the entire operating life.

The term *voltage-versus-current (v-i) characteristic* of the device is often used to describe the device performance at its own two power terminals, although the device makes only a part of a long circuit loop. To be more specific, it is the plot of voltage drop versus current at the device power terminals measured locally as shown in Figure 1.5. The current would flow only if the power loop is complete (not shown here) and the device is conductive; otherwise, it will have zero current even in a closed loop. The current in any closed loop is calculated by setting up Kirchhoff's voltage law (KVL) equation and solving it for the loop current. The KVL states the following: In any closed loop, the sum of source voltages equals the sum of voltage drops in the loads. The voltage drop $v$ across the device at a given current $i$ in formulating the KVL equation is that given by the $v$-$i$ characteristic relation of the

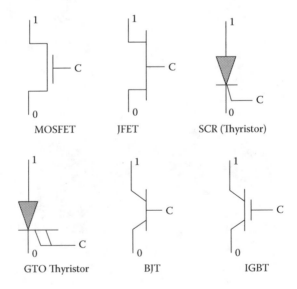

**FIGURE 1.3** Circuit symbols of power electronics devices used as controlled switches.

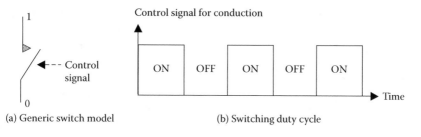

(a) Generic switch model                          (b) Switching duty cycle

**FIGURE 1.4**  Power electronics device operation as periodic on-off switch.

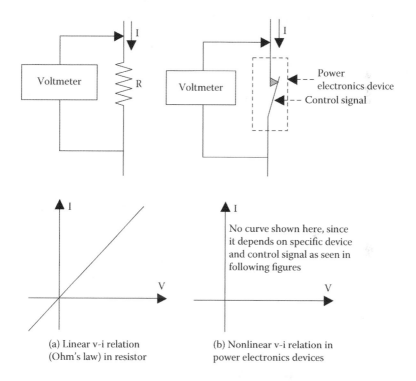

(a) Linear v-i relation                    (b) Nonlinear v-i relation in
(Ohm's law) in resistor                    power electronics devices

**FIGURE 1.5**  Local voltage drop versus current characteristic at device terminals.

device, which is linear (Figure 1.5a) for the resistor and nonlinear (Figure 1.5b) for the power electronics devices. For this reason, the power electronics devices are called nonlinear devices, and the loads with power electronics devices are called nonlinear loads.

## 1.1  DIODE

A diode is often used to rectify (convert) ac into dc; hence, it is also known as the *rectifier*. Although not a controllable switching device, the diode is the most basic semiconducting device that forms the building block for other devices. Made of p- and n-type semiconductors forming a p-n junction, it conducts current from P (anode) to N (cathode) terminals, but not in the reverse direction. Figure 1.6 depicts the circuit symbol and the *v-i* characteristics of the ideal and practical diodes. It shows that the voltage drop in the ideal diode conducting current in the forward direction is zero regardless of the current magnitude. The practical diode, however, has a small voltage drop, typically less than 1 V. If the voltage of negative polarity is applied at the power terminals (i.e., if the diode is reverse biased), the current in the reverse direction is negligible until the junction breaks down at what is known as the *reverse breakdown voltage*, which is typically hundreds of volts. After the breakdown, a large current flows, with the reverse voltage drop approximately constant.

The diode is rated in terms of the reverse voltage it can withstand without breakdown and the forward current it can carry without overheating. The diode is also

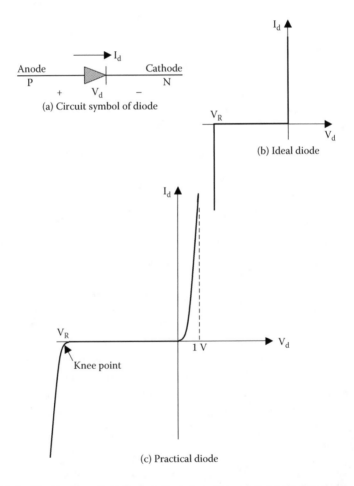

**FIGURE 1.6**    Ideal and practical diode voltage drop versus current characteristics.

rated by the switching time from the off state to the on state and vice versa. Large power diode ratings can be as high as 5000 V, 5000 A, with small leakage current in the off state. However, high-power diodes with large charge stored at the junction take a longer time to switch and hence are limited to relatively low-frequency applications, such as for rectifying up to 400-Hz ac into dc. Diodes in power electronics converters operating at a switching frequency of 10 to 20 kHz require special high-speed diodes.

*Fast-switching diode*: For fast-switching operations, the reverse recovery time of the diode must be short to minimize the power loss. Fast recovery diodes for high-frequency applications are available, but their power-handling capability is lower due to higher power loss at higher switching frequency. The power frequency (60 or 50 Hz) diodes are available up to 7 kV and 10 kA ratings, a few kilohertz diodes up 3000 V and 1500 A, a few hundreds of kilohertz diodes up to 1000 V and 100 A, and Schottky diodes for high-frequency applications up to 1 MHz in ratings below 150 V and a few amperes.

*Zener diode*: This special diode permits *current* in the forward direction like a normal diode, but also in the reverse direction if the voltage is greater than the reverse *breakdown voltage,* which is called the *Zener knee voltage* or *Zener voltage* $V_Z$. The device is named after Clarence Zener, who first discovered it. The conventional diode draws negligible current in reverse bias below the reverse breakdown voltage $V_R$. However, when the applied reverse bias voltage exceeds $V_R$, the conventional diode draws high current due to avalanche breakdown and generates large power loss equal to $V_R \times I$. The resulting heat would permanently damage the diode junction unless the avalanche current is limited by external circuit elements. The Zener diode exhibits almost the same characteristic, except the device is specially designed to have much reduced breakdown voltage (Zener voltage). Moreover, it exhibits a controlled breakdown and allows the current just enough to maintain a constant voltage drop across the terminals equal to the Zener voltage $V_z$ without thermal damage. For example, a Zener diode with $V_z = 3.2$ V in Figure 1.7 will maintain a constant voltage of 3.2 V at the output terminals even if the input voltage (reverse-bias voltage applied across the Zener) varies over a wide range above the Zener voltage. In this way, the Zener diode is typically used to generate a constant reference voltage for an amplifier stage or as a voltage stabilizer for low-current applications. The Zener breakdown voltage $V_z$ can be controlled accurately in the semiconductor doping process. The most widely used voltage tolerance is 5%, although close tolerances within 0.05% are available for precise applications.

*Trigger (PNPN) diode*: It blocks current in both directions but can conduct with about a 1-V drop in the forward direction if the applied forward voltage exceeds the forward breakdown value $V_{BO}$ as shown in Figure 1.8. It turns off when the diode current falls below the holding current $I_H$, which is typically in several milliamperes.

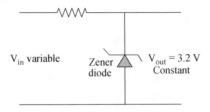

**FIGURE 1.7**   Zener diode circuit to derive constant 3.2-V reference voltage.

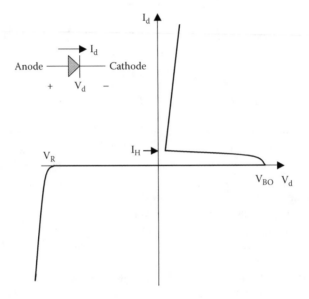

**FIGURE 1.8**   Trigger diode voltage drop versus current characteristic.

*Diac*: The diac is made of two trigger diodes placed back to back. It conducts when the voltage exceeds the breakdown voltage in either direction as shown in Figure 1.9, and remains conducting with a small voltage drop across it until the current falls below the holding current $I_H$.

## Example 1.2

Determine the current, voltage, and power loss in the diodes of circuits (a), (b), and (c) shown in Figure E1.2, where all diodes have a forward voltage drop of 1 V and reverse breakdown voltage of 100 V.

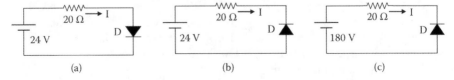

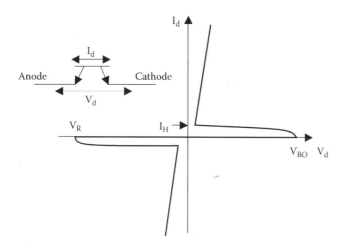

**FIGURE 1.9** Diac voltage drop versus current characteristic.

## SOLUTION

We first recall that (1) the voltage drop in the diode when it conducts in the forward or reverse direction is independent of the current magnitude, and (2) the power loss in the diode equals the voltage drop in the diode times the current through the diode.

In circuit (a), the diode is forward biased (in the conducting direction); hence, it will conduct since the applied voltage exceeds the prospective voltage drop of 1 V. Using KVL, we derive the current $I = (24 - 1)$ V $\div$ 20 $\Omega = 1.15$ A. The voltage across the diode = 1 V (regardless of the current magnitude), and power loss in the diode = 1 × 1.15 = 1.15 W.

In circuit (b), the diode is reverse biased with the applied voltage less than the reverse breakdown voltage. Hence, the diode will not conduct ($I = 0$), the voltage across the diode = 24 V (full-source voltage since the voltage drop across the resistor is zero), and the power loss in the diode = 0 × 24 = 0.

In circuit (c), the diode is reverse biased with the applied voltage greater than the reverse breakdown voltage. Hence, the diode will break down and conduct current in the reverse direction (against the symbolic arrow of the diode) with a voltage drop of 100 V (again, regardless of the current magnitude). That gives $I = (180 - 100)$ V $\div$ 20 $\Omega = 4$ A, voltage drop across the diode = 100 V, and the power loss in the diode = 4 × 100 = 400 W. Such high power loss in the diode conducting in the reverse direction—as opposed to 1.15 W in the forward direction—will burn the junction almost instantly. Therefore, the circuit diodes are normally selected with sufficient margin to avoid reverse breakdown under normal operation and transient conditions.

## Example 1.3

For the circuit shown in Figure E1.3 with ideal diodes (zero forward voltage drop and high reverse breakdown voltage), determine the average power absorbed by the resistor.

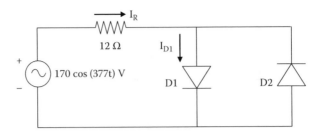

## SOLUTION

When the ac voltage has positive polarity at the top as shown, diode D1 is forward biased and conducts, but D2 in the reverse bias does not conduct. With zero voltage drop in the ideal diode D1, the current in the resistor will be 170 cos(377*t*) ÷ 12 A, left- to right-hand side.

In the following half cycle, the ac voltage will have positive polarity at the bottom, and diode D2 will conduct, but diode D1 will not. The current in the resistor will be −170 cos(377*t*) ÷ 12 A, now right- to left-hand side.

Therefore, for the whole cycle, the resistor current will be a complete ac sine wave, as if two diodes do not exist in this circuit. The average power absorbed by the resistors is then the same as that in any ac circuit, that is,

$$P_{avg} = I_{rms}{}^2 R = \left( \frac{170}{\sqrt{2} \, x \, 12} \right)^2 12 = 1204 \; watts$$

## 1.2   THYRISTOR

The thyristor (SCR) is a three-terminal device derived from the diode by adding a control terminal. It is made of four PNPN layers with two power terminals—A (anode) and C (cathode)—and a control terminal G (gate). Figure 1.10 shows its circuit symbol and the terminal *v-i* characteristic. The thyristor is normally nonconducting in both directions, but can be made to conduct in the forward direction by applying a gate current. A larger gate current starts the conduction sooner. Once it is triggered in the forward conduction mode, its voltage drop is low as in the diode (<1 V), and it remains conducting until the current falls below the holding current $I_H$. In the reverse direction, it remains nonconducting until the reverse breakdown voltage is reached and then conducts current that is determined by the entire circuit loop. The thyristor is available in ratings from a few amperes to about 3000 A.

The thyristor is also known as the silicon controlled rectifier (SCR) since it is usually made using a silicon diode (rectifier) and can control the conduction by a gate signal. The third terminal offers the control on making the device conducting or nonconducting or delaying the start of conduction until reaching a desired phase angle (firing angle) on a sinusoidal voltage wave. A few other variations to the thyristor are available with somewhat varying operating characteristics as described next:

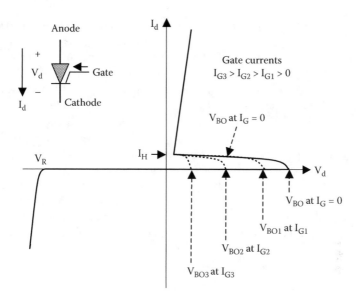

**FIGURE 1.10**   Thyristor (SCR) voltage drop versus current characteristic.

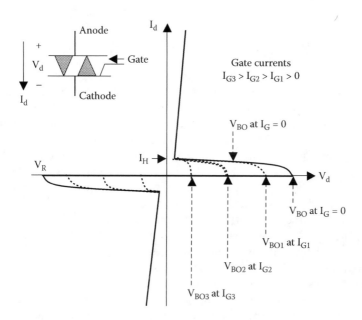

**FIGURE 1.11**   Triac voltage drop versus current characteristic.

*Triac:* It has five semiconducting layers (PNPNP) and behaves like two
   PNPN thyristors placed back to back. Its circuit symbol and the *v-i* characteristic are shown in Figure 1.11. Because a single triac can conduct
   in both directions, it can replace a pair of thyristors in many ac voltage

control circuits and is widely used to step-down the ac voltage by delaying the firing angle. However, since its switching speed is slower than the thyristor, it is limited in use in low-to-medium power at 60 or 50 Hz, such as in lighting circuits.

*Gate turnoff (GTO) thyristor:* The GTO thyristor is a recent introduction in high-power applications. The thyristor, once triggered into the conduction, remains conducting until the control voltage is removed and the current is brought back below the holding current $I_H$ (i.e., when the thyristor is commutated or quenched). That requires some special circuit—called the commutating circuit—which adds complexity and cost. The GTO thyristor eliminates the need for the commutating circuit. It can be turned off even from current exceeding $I_H$ by applying a large enough negative pulse (about 20% of the load current) at the gate terminals for about 20–30 µs. Many large GTO thyristors have a hockey puck (pressed pack) shape and are water cooled to remove the large power loss produced internally while processing high power.

## 1.3 POWER TRANSISTOR

The power transistor (PTR) also has three terminals, C (collector), E (emitter), and B (base). The collector and emitter terminal names are continuations from the vacuum tubes that were replaced by the transistor. Among a variety of transistors available in the market, some commonly used devices are as follows:

*Bipolar junction transistor (BJT):* The circuit symbol and the *v-i* characteristic of the BJT are shown in Figure 1.12. Even in the forward direction, the transistor remains nonconducting if the base current is zero. A base current triggers the transistor into the conduction mode and builds up the current until a certain value, beyond which the current remains constant (called the saturation current limit). A larger base current drives the transistor into a higher saturation current limit. Since a small base current signal can produce a large collector current, the transistor is widely used as an amplifier. However, in the power industry, it is used as a controlled switch that can be effectively turned on or off by the base current signal.

The BJT comes with NPN or PNP junctions. The NPN transistor has higher voltage and current ratings and hence is better suited for power electronics. It has low gain ($Ic/I_B$ ratio less than 10), meaning that the base current must be more than 10% of the current to be switched, which may require a large base circuit. Bipolar transistors have lower power-handling capabilities than thyristors in general but have good control characteristics and high-frequency switching capabilities. Turning on and off the current in the BJT is inherently slow due to its own internal stray inductance. Since BJTs are difficult to operate in parallel at high frequency above 10 kHz, they have been traditionally used in low-power applications in the past. However, the BJT has been largely replaced by MOSFET in low-voltage and IGBT in high-voltage applications.

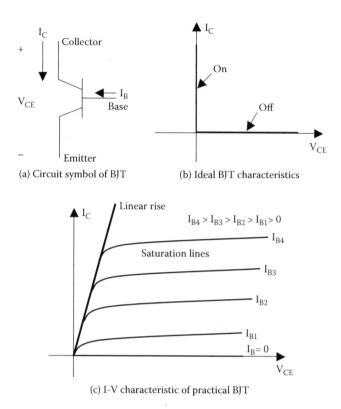

(a) Circuit symbol of BJT    (b) Ideal BJT characteristics

(c) I-V characteristic of practical BJT

**FIGURE 1.12**    Bipolar junction transistor voltage drop versus current characteristic.

*Metal-oxide semiconducting field effect transistor (MOSFET):* This device is controlled by voltage (<15 V in power MOSFET) rather than current, as shown in Figure 1.13. Because of the high gate impedance, the gate current is low, making the drive circuit simple. It can operate above 100 kHz. The power capability is low (a few kilowatts), but it can be easily paralleled for greater power capability. MOSFETs are available in ratings up to 600 V with lower current or up to 200 A with lower voltage.

*Insulated gate bipolar transistor (IGBT):* The IGBT combines the good characteristics of MOSFET and BJT. It is a relatively recent device that works on the voltage signal applied at the base and does not require the commutating circuit. Its circuit symbol and *v-i* characteristics are shown in Figure 1.14. Since turning on and off the voltage signals in the IGBT is faster, it has been used in high-power, high-frequency applications in motor drives up to several hundred horsepower. IGBTs are available from small to high ratings up to 1200 A at 330 V, 900 A at 4500 V, and some even higher ratings. Like GTO thyristors, large IGBTs have a hockey puck (pressed pack) shape and are water cooled to remove large internal power loss produced while processing high power (Figure 1.15).

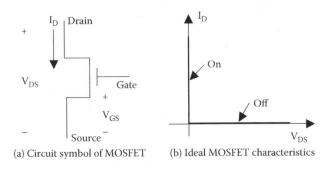

(a) Circuit symbol of MOSFET          (b) Ideal MOSFET characteristics

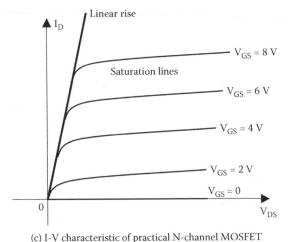

(c) I-V characteristic of practical N-channel MOSFET

**FIGURE 1.13**   MOSFET voltage drop versus current characteristic.

## 1.4   HYBRID DEVICES

Thyristor technology has advanced into a variety of hybrid devices. The GTO thyristor and the static induction thyristor (SITH) turn on by a positive pulse to the thyristor gate and turn off by applying a negative pulse at the gate. Thus, they offer good forced-commutation techniques, but require high firing power. The GTO thyristors are available up to 4500 V/3000 A, which makes multimegawatt inverters possible for motor drives. SITHs are available up to 1200 V/300 A. Both GTO thyristors and SITHs have high flexibility in design and can be easily controlled. SITHs have high-frequency switching capability, way above GTO thyristors, which are limited to about 10 kHz. The SCRs and IGBTs, on the other hand, are limited to 100 Hz at present.

The gate-commutated thyristor (GCT) and the integrated gate-commutated thyristor (IGCT) are based on the GTO thyristor. They conduct like a GTO thyristor but turn off like a transistor, which reduces switching power loss to about one-half compared to that in the GTO thyristor. Commercially developed around 2000, the GCT has started to replace the GTO thyristor in high-power applications.

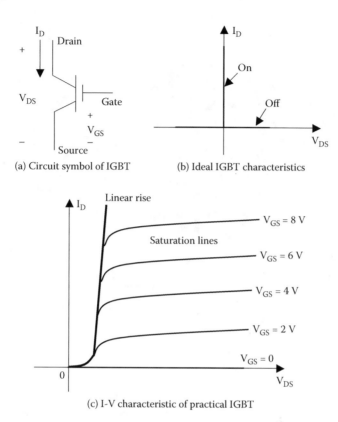

(a) Circuit symbol of IGBT

(b) Ideal IGBT characteristics

(c) I-V characteristic of practical IGBT

**FIGURE 1.14**  IGBT voltage drop versus current characteristic.

The MOS-controlled thyristor (MCT), reverse-conducting thyristor (RCT), gate-assisted turnoff thyristor (GATT), and light-activated silicon-controlled rectifier (LASCR) are other examples of specialty devices, each having niche advantage and application.

Figure 1.16 shows the power-handling versus switching frequency capabilities of the most commonly used power electronics devices available to the design engineer at present, and higher-rated devices are emerging in an evolutionary manner.

## 1.5  *DI/DT* AND *DV/DT* SNUBBER CIRCUITS

The voltage drop across two terminals of a current-carrying inductor $L$ is $v = L \times di/dt$. Therefore, switching the current on and off in a short time in a high-frequency switching circuit gives high $di/dt$, which causes high-voltage stress on the junction, which may damage the device unless suppressed by snubber circuit.

The voltage drop across two terminals of a current-carrying capacitor $C$ is $v = \int i \cdot dt/C$, the differentiation of which leads to $i = C \times dv/dt$. Therefore, switching the voltage on and off in a short time in a high-frequency switching circuit gives high $dv/dt$, which causes high current stress on the junction, possibly damaging the device.

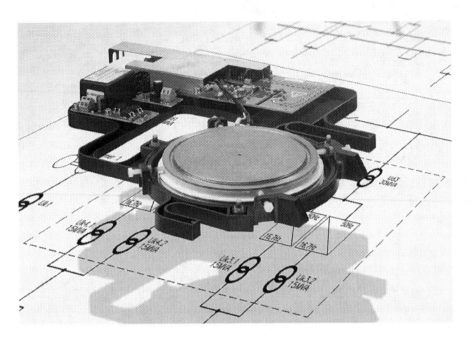

**FIGURE 1.15**  Large IGBT in hockey puck (*pressed pack*) shape assembled in base frame. (With permission from Converteam, Inc.)

Moreover, both high *di/dt* and high *dv/dt* can cause high electromagnetic interference (EMI) in the neighboring circuits.

When a firing signal is applied at the gate terminal of the thyristor and GTO-type devices under forward voltage bias, the conduction across the junction starts in the immediate neighborhood of the gate connection and spreads from there across the whole area of the junction. A high rate of rise in anode current (i.e., high *di/dt*) results in high current density around the junction area. This may cause a local hot spot, thermally damaging the junction and making the thyristor inoperative. The maximum *di/dt* a thyristor can withstand without damage is specified by the manufacturer. The converter design engineer must limit the *di/dt* on the anode current by inserting a small inductor in series with the thyristor as shown in Figure 1.17. Recall that the inductor slows the *di/dt* to keep its magnetic energy constant.

If a sinusoidal voltage is applied to the thyristor, the minimum required inductor to limit the *di/dt* below the maximum permissible value can be derived from the *v* = *L* × *di/dt* relation, which gives

$$L_{min} = \frac{V_{peak(anode)}}{(di/dt)_{max(allowed)}} \text{henrys} \qquad (1.1)$$

The thyristor *dv/dt* rate is controlled by placing a capacitor in parallel with the thyristor as shown in Figure 1.17, where a small series resistance is to protect the capacitor from high inrush charging current. Under high *dv/dt*, which is virtually a

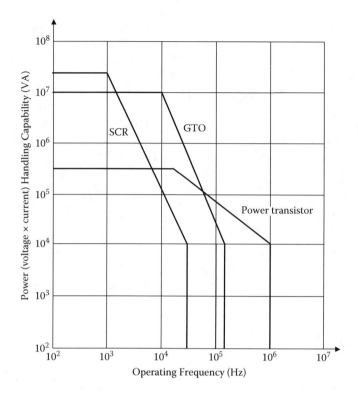

**FIGURE 1.16**   Power-handling capability versus switching frequency of various power electronics devices.

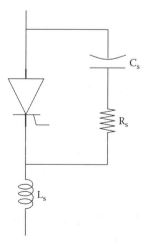

**FIGURE 1.17**   Snubber circuit for *di/dt* and *dv/dt* suppression as part of thyristor power circuit.

high-frequency voltage, the snubber capacitor branch offers much lower impedance, which draws most of the inrush current, diverting it away from the thyristor. The required value of such a capacitor is determined from the $i = C \times dv/dt$ relation. It is also related to $L_s$.

Figure 1.17 is a typical snubber circuit for the thyristor or a transistor. Its main purpose is to protect the device from a large rate of change in voltage and current. The series inductor limits $di/dt$, and the parallel capacitor branch limits $dv/dt$. Typical values of $L_s$, $R_s$, and $C_s$ are a few millihenrys, few ohms, and a few tenths of a microfarad, respectively. The power loss in $R_s$ decreases the efficiency of the converter; hence, the efficiency is traded with the level of $di/dt$ and $dv/dt$ protection desired.

Another problem associated with the thyristor and GTO-type devices is a high rate of rise in the forward voltage. High $dv/dt$ applied at the anode terminal turns on the thyristor even in absence of the gate firing signal. The possibility of such anomalous conduction due to stray $dv/dt$ signals from the radiated EMI of neighboring power equipment is one reason why thyristors are not used in spacecraft with extremely high-reliability requirements, for which the probability of malfunction in each device must be less than one in billions.

### Example 1.4

A thyristor in a certain 440-V converter design can see a maximum forward voltage of 254 $V_{LNrms}$. Its data sheet specifies the maximum allowable $di/dt$ of 100 A/μs. Determine the snubber inductor you must include in the converter design to limit the $di/dt$ stress to 70 A/μs to allow for a desired margin.

#### SOLUTION

To limit the $di/dt$ to 70 A/μs, which is $70 \times 10^6$ A/sec, we use Equation (1.1) to obtain
$L_{min} = 254 \times \sqrt{2} \div (70 \times 10^6) = 5.13 \times 10^{-6}$ H or 5.13 μH

### Example 1.5

A power electronics device turns off 50 A in 200 μs and builds up the voltage from 0 to 110 V in 220 μs. Determine the $di/dt$ and $dv/dt$ stresses on the device.

#### SOLUTION

We first must be clear about the definition of change, which is defined as:

Change $\Delta$ = (New value − Old value)

50 A turnoff means $\Delta i = (0 - 50) = -50$ A in $\Delta t = 200$ μs

100 V buildup means $\Delta v = (110 - 0) = +110$ V in $\Delta t = 220$ μs

∴ $di/dt = -50/200 = -0.25$ A/μs    and    $dv/dt = +110/220 = +0.50$ V/μs

## 1.6 SWITCHING POWER LOSS

The power electronics switch, when turned on or off, encounters some power loss. The converter design engineer chooses the device with low power loss and provides adequate cooling to maintain the junction temperature below the allowable limit. The power loss in any switch can be estimated from a generic diagram shown in Figure 1.18, where the switch is turned on by the control signal in time $T_{swon}$ and off in time $T_{swoff}$. On the on signal, the switch current takes time $t_{ir}$ to rise to full value, and the voltage takes time $t_{vf}$ to fall to zero, making $T_{swon} = t_{ir} + t_{vf}$. On the off signal, the voltage rises to full value in time $t_{vr}$, and current falls to zero value in time $t_{if}$, making $T_{swoff} = t_{vr} + t_{if}$. The power loss in the switch is the product of the voltage across the switch and the current through the switch, and the energy loss = power × time duration, which is shown by the shaded area. To sum up the total energy loss,

Switching energy loss in two hump areas,

$$E_{sw} = \tfrac{1}{2} V_{oc} \times I_{on} \times T_{swon} + \tfrac{1}{2} V_{oc} \times I_{on} \times T_{swoff} \qquad (1.2)$$

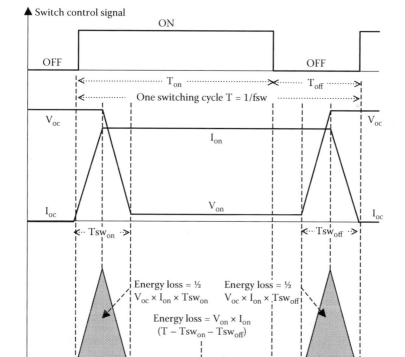

**FIGURE 1.18** Energy losses during rising voltage and falling current in one switching period of fast switch (for worst-case analysis).

Switching repetition period $T = T_{on} + T_{off}$, and switching frequency $f_{sw} = 1/T$ (1.3)

Switching power loss in fast switch,

$$P_{sw} = E_{sw}/T = \frac{1}{2}V_{oc} \times I_{on} \times (T_{swon} + T_{swoff}) \times f_{sw}$$ (1.4)

Other power losses occur during conduction on and off. The power loss during the on state, shown in the narrow, long, shaded area, amounts to the average value of

$$P_{on} = V_{on} \times I_{on} \times (T - T_{swon} - T_{swoff}) \div T = V_{on} \times I_{on} \times D$$ (1.5)

where $D$ = duty ratio, the fraction of cycle time the conduction is on. Here, we have ignored the switching on and off times since they are very short compared to the switching repetition period $T$.

$$\text{Power loss during off state, } P_{off} = V_{oc} \times I_{leak} \times (1 - D)$$ (1.6)

The value of $P_{off}$ is usually negligible and is ignored in most practical calculations.

We note that $P_{sw}$ and $P_{on}$ vary directly with the switching frequency and the switching times. Therefore, the device with a short switching time and low on-state voltage $V_{on}$ minimizes the switching power loss, allowing high-frequency operation.

Equation (1.2) and the subsequent results are for a fast turn-on and turn-off switch with short $T_{swon}$ and $T_{swoff}$ times. In slow switches, the rise time and fall time overlap. The voltage falls as the current builds up simultaneously during the turn-on time, and the current falls as the voltage builds up simultaneously during the turnoff time. The hump area therefore has height only one-half of that in a fast switch, and the shape of one-half sine wave, which has the shaded area equal to Peak $\times 2/\pi$. The net result is reduced switching energy loss, which is given by $E_{sw} = (2/\pi) \{\frac{1}{2}V_{oc} \times \frac{1}{2}I_{on} \times T_{swon} + \frac{1}{2}V_{oc} \times \frac{1}{2}I_{on} \times T_{swoff}\}$, which is about 1/3 of that in a fast switch. Therefore,

$$\text{For slow switch, } P_{sw} = E_{sw}/T = 0.16V_{oc} \times I_{on} \times (T_{swon} + T_{swoff}) \times f_{sw}$$ (1.7)

The switch with a switching speed between the two extreme cases—fast and slow—discussed above would have the factor in the front of the switching power loss formula ranging from a high of 0.50 to a low of 0.16. Equation (1.4) therefore gives a conservative design until the actual switching speed data from the device vendor supports a lower factor. It is important to distinguish the switching speed from the switching frequency.

Switching speed = $T_{swon} + T_{swoff}$ = Time to switch $V$ and $I$ on and off = Duration of the humps (usually in microseconds)

Switching frequency $= f_{sw} = 1/T =$ number of times the voltage and current get switched on and off in 1 s (usually line frequency to 500 kHz in some high-frequency converters).

The switching times come from the device vendor data sheet, whereas the switching frequency is set by the design engineer, and the switching power loss depends on both.

## Example 1.6.

In Figure E1.6 circuit, the power electronics switch is operating at a 1000-Hz switching frequency with a 60% duty ratio. The switch has an on-state voltage drop of 1 V, off-state leakage current of 1.5 mA in reverse bias, switch-on time of 1.2 μs, and switch-off time of 1.5 μs. Determine the average power loss in the switch.

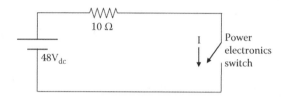

### SOLUTION

We use formulas developed in Section 1.6. One cycle takes 1/1000 = 0.001 s or 1 ms. With 60% duty ratio, the switch conducts for 0.6 ms and is off for 0.4 ms. When it conducts, the current will be (48 − 1) ÷ 10 = 4.7 A with the forward voltage drop of 1 V. In the off state, the voltage across the switch will be equal to the applied voltage of 48 V and the leakage current 1.5 mA.

Since 1000 Hz is a relatively low switching frequency, we use Equation (1.7) to obtain:

Switching power loss $P_{sw} = 0.16 \times 48 \times 4.7 \times (1.2 + 1.5)10^{-6} \times 1000 = 0.0975$ W

Using Equation (1.5), power loss during on time $P_{on} = 1 \times 4.7 \times 0.6 = 2.82$ W

Power loss during off time $P_{off} = 48 \times 0.0015 \times 0.4 = 0.0288$ W (negligible, 1% of $P_{on}$)

Power loss averaged over the cycle = 0.0975 + 2.82 + 0.0288 = 2.95 W

If this switch were operated at very high frequency, say 100 kHz, Equation (1.4) would apply for

$P_{sw} = \frac{1}{2} \times 48 \times 4.7 \times (1.2 + 1.5)10^{-6} \times 100,000 = 30.5$ W

which is about 300 times the 0.0975 W loss at 1000 Hz. This is clearly high, which makes this switch unsuitable for such high switching frequencies.

The on-state power loss in the transistor is always on as long as the power terminals are under voltage, although it becomes conducting only when the base signal is present. It does not multiply by the duty ratio (i.e., conducting/nonconducting time ratio).

## 1.7 DEVICE APPLICATION TRENDS

The device selection for a specific application depends on the voltage, current, and frequency requirements of the system to be designed. The characteristics of the switching devices presently available to the design engineer are listed in Table 1.1. The maximum voltage and current ratings with unique operating features of thyristors and transistors widely used in high-power applications are listed in Table 1.2.

**TABLE 1.1**

**Characteristics of Power Electronics Semiconductor Device**

| Type | Function | Voltage (V) | Current (Å) | Upper Frequency (kHz) | Switching Time (µsec) | On-State Resistance (mΩ) |
|------|----------|-------------|-------------|-----------------------|-----------------------|--------------------------|
| Diode | High power | 6500 | 8000 | 1 | 100 | 0.1–0.2 |
| | High speed (50 kHz) | 3000 | 1500 | 10 | 2–5 | 1 |
| | Schottky (1 MHz) | <150 | <100 | 20 | 0.25 | 10 |
| Forced turned-off thyristor | Reverse blocking | 5000 | 5000 | 1 | 200 | 0.25 |
| | High speed | 1200 | 1500 | 10 | 20 | 0.50 |
| | Reverse blocking | 2500 | 400 | 5 | 40 | 2 |
| | Reverse conducting | 2500 | 1000 | 5 | 40 | 2 |
| | GATT | 1200 | 400 | 20 | 8 | 2 |
| | Light triggered | 6000 | 1500 | 1 | 200–400 | 0.5 |
| TRIAC | Back-to-back thyristors | 1200 | 300 | 1 | 200–400 | 3–4 |
| Self-turned-off thyristor | GTO | 4500 | 3000 | 10 | 15 | 2–3 |
| | SITH | 1200 | 300 | 100 | 1 | 1–2 |
| Power transistor | Single | 400 | 250 | 20 | 10 | 5 |
| | | 400 | 40 | 20 | 5 | 30 |
| | | 600 | 50 | 25 | 2 | 15 |
| | Darling-ton | 1200 | 400 | 10 | 30 | 10 |
| Power MOSFET | Single | 500 | 10 | 100 | 1 | 1 |
| | | 1000 | 5 | 100 | 1 | 2 |
| | | 500 | 50 | 100 | 1 | 0.5 |
| IGBTs | Single | 1200 | 400 | 100 | 2 | 60 |
| MCTs | Single | 600 | 60 | 100 | 2 | 20 |

**TABLE 1.2**

**Maximum Voltage and Current Ratings of Power Electronics Switching Devices**

| Device | Voltage Rating (V) | Current Rating (Å) | Operating Features |
|--------|-----|-----|--------------------|
| BJT | 1500 | 400 | Requires large current signal to turn on |
| IGBT | 1200 | 400 | Combines the advantages of BJT, MOSFET, and GTO |
| MOSFET | 1000 | 100 | Higher switching speed |
| SCR | 6000 | 3000 | Once turned on, requires heavy turnoff circuit |

The thyristor used as the switching device requires the current commutation (turn off the device) when its anode-cathode voltage is reversed. The thyristor current in a line frequency (60 or 50 Hz) load-commutated converter is naturally reverse biased and turns off when the next thyristor is gated on. However, in other—such as a switch-mode pulse width modulated (PWM)—converters, it requires a separate forced-commutation circuit, which adds to the converter cost and power loss. With the availability of high-power BJT, IGBT, and GTO thyristors, which do not require commutation circuits, the switch-mode thyristor converter is seldom used in new designs. The GTO thyristor is a new technology that has commutation capability by a gate signal in the reverse. It has been implemented successfully in some marine propulsion applications. One particular attraction of this technology is the consequent ability to select the induction motor in favor of the synchronous motor if beneficial. GTO thyristors are currently available for power levels over 10 MW and can be connected in parallel for even higher power levels. However, the controlling electronics for device turn on and turn off cost more.

The IGBT is now widely used in new designs due to its faster switching frequency, better controllability, lower power losses, and simpler and compact control circuit than is possible with GTO thyristors. However, most are available for relatively small prepackaged drive applications (up to a few megawatts), with their principal limitation in the voltage rating up to about 1500 V rms.

## 1.8   DEVICE COOLING AND RERATING

The device junction temperature must be below the allowable limit to avoid junction failure. This is difficult considering that the operating heat is generated in a small wafer-thin volume around the junction way inside the device. The design power rating of any device is limited by the permissible temperature rise of the junction, which depends on the cooling method and the temperature of the ambient air where all the switching power loss is ultimately dissipated. Lower ambient air temperature therefore allows greater power-handling capability and vice versa. The device vendor provides the nominal rating at the industry standard ambient air temperature of 40° C. If the device is operating a a higher ambient temperature, it cannot handle the vendor-rated power and must be derated accordingly. On the other hand, it can handle higher than the vendor-rated power in ambient air cooler than 40°C. The device rerating (derating or uprating) factor for the ambient air deviating from 40°C is shown in Figure 1.19.

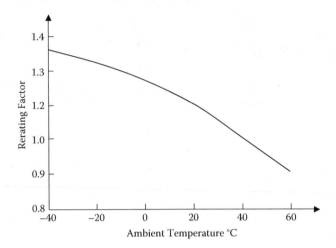

**FIGURE 1.19**   Device rerating factor versus ambient air temperature.

## PROBLEMS

*Problem 1.1*: An industrial process heater delivers power from a 480-V dc generator to a 30-Ω load resistance that produces 7680 watts of heat when fully on. A reduced operation requires only 4224-watts average heat generation rate. Determine (a) the required series rheostat resistance and the process energy efficiency and (b) the switching duty ratio $D$ and the energy efficiency if an ideal power electronics control is employed.

*Problem 1.2*: Determine the current, voltage, and power loss in the diodes of circuits (a), (b), and (c) shown in Figure P1.2, where all diodes have the forward voltage drop of 0.8 V and reverse breakdown voltage of 50 V.

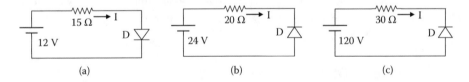

*Problem 1.3*: With ideal diodes (zero forward voltage drop and high reverse breakdown voltage) in Figure P1.3, determine the average power absorbed by the resistor.

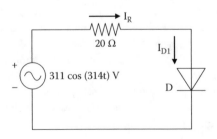

*Problem 1.4*: A thyristor in a certain 240-V converter design can see a maximum forward voltage of 120 $V_{LNrms}$. Its data sheet specifies the maximum allowable *di/dt* of 80 A/μs. Determine the snubber inductor you must include in the converter design to limit *di/dt* to 60 A/μs to allow for a desired margin.

*Problem 1.5*: A power electronics device builds up the voltage from 0 to 120 V in 180 μs and turns off 100 A in 250 μs. Determine the *dv/dt* and *di/dt* stresses on the device.

*Problem 1.6*: In Figure P1.6 circuit, the power electronics switch is operating at 2000 Hz switching frequency with a 50% duty ratio. The switch has an on-state voltage drop of 0.8 V, off-state leakage current of 2 mA in reverse bias, a switch-on time of 1.5 μs, and a switch-off time 2.0 μs. Determine the average power loss in the switch.

## QUESTIONS

*Question 1.1*: Explain the difference between the diode and the thyristor in construction and in operation.

*Question 1.2*: In which two different ways can the transistor be used in electronic circuits?

*Question 1.3*: Differentiate the terms *switching speed* and *switching frequency*.

*Question 1.4*: High *di/dt* causes which: high voltage stress or high current stress? Why?

*Question 1.5*: High *dv/dt* causes which: high voltage stress or high current stress? Why?

*Question 1.6*: When would you need to rerate (derate or uprate) the device and for what reason?

## FURTHER READING

Ahmed, A. 1999. *Power Electronics for Technology.* Upper Saddle River, NJ: Prentice Hall.

Ang, S., and A. Oliva. 2010. *Power Switching Converters.* Boca Raton, FL: CRC Press/Taylor & Francis.

Hart, D. 2011. *Power Electronics.* New York: McGraw-Hill.

Mohan, N., T.M. Undeland, and W. Robbins. 2003. *Power Electronics Converters, Applications, and Design.* New York: Wiley/IEEE Press.

# 2 DC-DC Converters

The power electronics devices covered in Chapter 1 make the building blocks of a variety of power electronics converters for changing the ac and dc voltage level or changing the frequency at the interface of the power mains and the load. This chapter covers dc-dc converters; ac-dc-ac converters are covered in the next chapter. The dc-dc converter is needed to change voltage for load equipment or for charging and discharging a battery in the dc power distribution system. Since such a converter achieves the desired function by switching the dc power on and off (chopping the power) at high frequency, it is also known as the *dc chopper* or *switch-mode power converter*. The following sections describe the circuit topology and the operational characteristics of various types of dc-dc converter.

## 2.1 BUCK CONVERTER

The buck converter steps down the input voltage; hence, it is also known as the *step-down converter*. Figure 2.1a is the most widely used buck converter circuit. The switching device used in this converter may be the bipolar junction transistor (BJT), metal-oxide semiconducting field effect transistor (MOSFET), or insulated gate bipolar transistor (IGBT). The switch is turned on and off periodically at high frequency, typically in tens of kilohertz, as shown in Figure 2.1b. The duty ratio $D$ of the switch is defined as

$$Duty\ ratio\ D = \frac{Time\ on}{Period} = \frac{T_{on}}{T} = T_{on} \cdot Switching\ frequency \qquad (2.1)$$

Since the buck converter is widely used to charge the battery, it has yet another name, the *battery charge converter*. It is required to buck (step down) the dc bus voltage to the battery voltage during charging. Its operation during the triggering signal of one on-and-off period is shown in Figure 2.2. During the on time, the switch is closed, and the circuit operates as in Figure 2.2a. The dc source charges the inductor and capacitor, in addition to supplying power to the load. During the off time, the switch is open, and the circuit operates as in Figure 2.2b. The power drawn from the dc source is zero. However, full load power is supplied by the energy stored in the inductor and the capacitor, with the diode carrying the return current. Thus, the inductor and the capacitor provide short-time energy storage to ride through the off period of the switch. The load current during this period is known as the freewheeling current, and the diode is known as a freewheeling diode. The voltage and current waveforms over one complete cycle are displayed in Figure 2.3. The line voltage is either fully turned on or off. The inductor current decays during off time and rises during on time, as do the switch current and the diode current. Either the switch or the diode carries the inductor current, as shown in the last two waves. A suitable

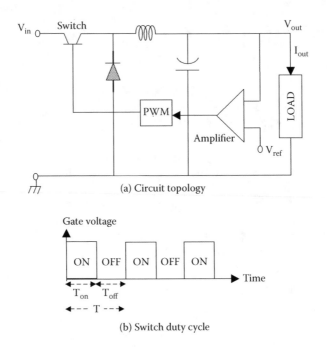

(a) Circuit topology

(b) Switch duty cycle

**FIGURE 2.1**    Buck converter circuit and switch duty ratio.

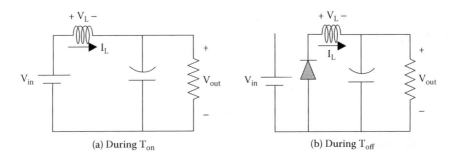

(a) During $T_{on}$                              (b) During $T_{off}$

**FIGURE 2.2**    Buck converter operation during on and off periods of switch.

bleeding (dummy) resistor at the load terminals is sometimes incorporated in the design to keep the converter working without the load.

The analytical principles detailed herein for the buck converter are applicable also for other converters described in this chapter. The analysis is primarily based on the energy balance over one switching period, that is,

- Energy supplied to load over total period $T$ = Energy drawn from source during on time, which is the only energy fed to the circuit, and
- Energy supplied to load during off time = Energy discharged from inductor and capacitor during off time, which equals the energy deposited in the inductor and capacitor during on time.

The freewheeling diode circulates the load current through the inductor. The inductor depletes the energy during the off time, and then refurbishes the energy during the on time. Therefore, the performance analysis is based on the energy balance in the inductor during one cycle since the inductor draws power from the source during on time. The inductor stores energy in the magnetic flux and generally uses a magnetic core with a lumped or distributed air gap. The flux density in the core has to be kept below the magnetic saturation level. For that reason, the net change in flux over one cycle of switching has to be zero. Otherwise, the core would eventually walk away to the magnetic saturation on the high side or become depleted of energy on the low side. The magnitude of the voltage drop $V_L$ across the inductor is given by Faraday's law, that is, $V_L = Nd\phi/dt$, where $N$ = number of inductor turns, $\phi$ = flux in the inductor core, and $t$ = time. It can also be written as $V_L dt = Nd\phi$. Since the inductor flux must maintain balance over one cycle in the steady state, we must have $d\phi = 0$ over one complete cycle, or

$$\int_o^T V_L \cdot dt = 0 \tag{2.2}$$

This requires that the (voltage × time) products during on and off periods must be equal and algebraically opposite to maintain the energy balance over one complete switching cycle. Therefore, equating the two is called the volt-second balance method of analysis. Since the voltage is a measure of energy perunit charge, the $v \times dt$ product balance during on and off periods essentially gives the energy balance in the inductor over one steady-state cycle. Since $v = L \times di/dt$ or $v \times dt = L \times di$, the $v \times dt$ balance also requires the $L \times di$ balance. The steady-state current in the inductor may rise and fall during one cycle, but the net change in one cycle has to be zero for it to be in a steady-state operation.

We now apply the volt-second or the $L \times di$ balance in the voltage and current waveforms shown in Figure 2.3, where $I_L$ = inductor current,

During on time,

$$L \times \Delta I_L = \text{Voltage across inductor} \times T_{on} = (V_{in} - V_{out}) \times T_{on} \tag{2.3}$$

During off time,

$$L \times \Delta I_L = \text{Voltage across inductor} \times T_{off} = V_{out} \times T_{off} \tag{2.4}$$

Equating (2.3) with (2.4) and rearranging the terms result in the following output-to-input voltage relation, where $D$ = duty ratio,

$$\frac{V_{out}}{V_{in}} = \frac{T_{on}}{T_{on} + T_{off}} = \frac{T_{on}}{T} = D \tag{2.5}$$

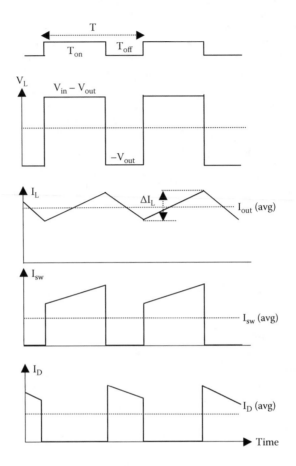

**FIGURE 2.3**  Current and voltage waveforms in buck converter operation.

Ignoring the internal power loss, the balance of input power and output power gives $V_{in} \times I_{in} = V_{out} \times I_{out}$, which leads to the current relation,

$$\frac{I_{out}}{I_{in}} = \frac{1}{D} \tag{2.6}$$

Since the duty ratio $D$ is always less than unity, the output voltage of the buck converter is always less than the input voltage. Thus, the buck converter can only step down the voltage, hence the name.

### Example 2.1

A buck converter operating at 50-Hz switching frequency has $T_{on} = 5$ ms. Determine the average source current if the load current is 40 A.

## SOLUTION

In the buck converter, we expect the source side at higher voltage to draw lower current than the load side at lower voltage.

Switching period $T = 1/50 = 0.02$ s $= 20$ ms          Duty ratio $D = 5/20 = 0.25$

Average source current $= 40 \times 0.25 = 10$ A

The inductor current $I_L$ rises during the switch on time (charging) and decays during the off time (discharging), with an average value equal to the output current $I_{out}$. The capacitor charging and discharging does not affect the average value of $I_{out}$. If the inductor is large enough, as is usually the case in practical designs, the change in the inductor current is small, and the peak value of the inductor current is given by

$$I_{peak} = I_{out} + \tfrac{1}{2}\Delta I_L \qquad (2.7)$$

where the average load current $I_{out} = V_{out}/R_{load} =$ average value of the inductor current, and $\Delta I_L =$ ripple current in the inductor as shown in Figure 2.3.

This converter can charge the battery only if $V_{out}$ is greater than the battery voltage, which rises as the battery approaches full charge. Therefore, $V_{out}$ must also rise with the battery state of charge. It is seen from Equation (2.5) that varying the converter duty ratio controls the output voltage, which is done in a feedback control loop with the battery voltage or charge current as the feedback signal. Since the duty ratio is controlled by modulating the pulse width of $T_{on}$, such a converter has another name, the *pulse width modulated (PWM) converter*.

Relations (2.5) and (2.6) between the output and input voltages and currents (all dc) are analogous to the ac transformer relations. The duty ratio $D$ in the dc-dc converter effectively works like the turn ratio in an ac transformer. This converter is therefore often incorrectly called the dc transformer, although it works on a fundamentally different principle than the ac transformer as we know it.

## Example 2.2

A buck converter is operating at 1-kHz switching frequency from a 120-V dc source. The inductance is 50 mH. If the output voltage is 60 V and the load resistance is 12 $\Omega$, we determine the following:

Duty ratio $D = 60$ V $\div 120$ V $= 0.50$          Period $T = 1/1000 = 0.001$ s $=1$ ms

$T_{on} = 0.50 \times 1 = 0.5$ ms          $T_{off} = 1 - 0.5 = 0.5$ ms

Average load current $= 60$ V $\div 12 \ \Omega = 5$ A          Output power $= 60 \times 5 = 300$ W

Average source current = 5 × 0.50 = 2.5 A        Input power = 120 × 2.5 = 300 W

Equation (2.4) gives peak-to-peak ripple current:

$$\Delta I_L = V_{out} \times T_{off} \div L = 60 \times 0.0005 \div 0.050 = 0.6 \text{ A}$$

If the switching frequency were 10 kHz, $T$ = 1/10,000 = 0.1 ms and $T_{off}$ = 0.05 ms. The inductance required for the same ripple current = 60 V × 0.05 ms/0.6 A = 5 mH, which is 1/10 the size of that required at 1 kHz. This illustrates the benefit of a high switching frequency.

We have assumed in the preceding analysis that the buck converter is operating in the continuous conduction mode, which requires a minimum output current. If $I_{out} < \frac{1}{2}\Delta I_L$, the load current would fall to zero and remain there until the switch is turned on again. This is known as the discontinuous mode of operation. To establish the boundary between the continuous and discontinuous conduction modes of the buck converter, we again use the basic relation $v = Ldi/dt$ or $di = v \times dt \div L$, which leads to

$$I_{out.bounday} = \frac{1}{2} \frac{(V_{in} - V_{out})T_{on}}{L} \tag{2.8}$$

The converter operates in continuous conduction mode if the load current $I_{out} > I_{out.boundary}$.
   Another design approach is to establish the minimum required inductance that will store the energy to keep the output current continuous even during the off time. That requires $I_{pk\text{-}pk} = 2 \times I_{out} = T_{off}V_{out} \div L$, which gives $L_{min} = T_{off}V_{out} \div (2 \times I_{out})$. Since $V_{out}/I_{out} = R_L$, we obtain

$$L_{min} = \frac{T_{off} R_L}{2} \tag{2.9}$$

With a capacitor filter, in addition to the inductor, the output voltage ripple magnitude is

$$\Delta V_{out} = \frac{\Delta Q}{C} = \frac{1}{C}\frac{1}{2}\frac{\Delta I_{out}}{2}\frac{T}{2} = \frac{T}{8C}\frac{V_{out}}{L}(1-D)T = \frac{T^2 V_{out}}{8\,CL}(1-D) \tag{2.10}$$

### Example 2.3

A buck converter has 120-V input voltage, $R_L$ = 12 Ω, switching frequency is 1 kHz, and on time is 0.5 ms. If the average source current is 2 A, we determine the following:

For 1-kHz switching frequency, T = 1/1000 sec = 1 ms, duty ratio = 0.5/1 = 0.5, $T_{on}$ = 0.5 ms, and $T_{off}$ = 1 − 0.5 = 0.5 ms.

Average output voltage = $0.5 \times 120 = 60$ V

Average output current = $2/0.5 = 4$ A

Average output power = $V_{out} \times I_{out} = 60 \times 4 = 240$ W

For continuous conduction, using Equation (2.9), we have $L_{min} = 0.5$ ms $\times$ 12 $\div$ 2 = 3 mH.

## Example 2.4

For a buck converter with $V_{out} = 5$ V, $f_s = 20$ kHz, $L = 1$ mH, $C = 470$ μF, $V_{in} = 12.6$ V, $I_{out} = 0.2$ A, determine the peak-to-peak ripple $\Delta V_{out}$ in the output voltage.

### SOLUTION

Switching period $T = 1/20,000 = 0.00005$ sec, and $D = V_{out}/V_{in} = 5/12.5 = 0.4$. Using Equation (2.10), we obtain

$$\Delta V_{out} = \frac{0.00005^2}{8 \times 470 \times 10^{-6}} \times \frac{5}{0.001} \times (1 - 0.40) = 0.002 \text{ V}$$

This is quite a low ripple voltage, giving a smooth dc voltage output in this converter.

The converter efficiency is calculated as follows: During the on time, the input voltage less the voltage drop in the switch equals the circuit voltage. The energy transfer efficiency is therefore $(V_{in} - V_{sw})/V_{in}$, where $V_{in}$ = input voltage, and $V_{sw}$ = voltage drop in the switch. During the off period, the circuit voltage equals the output voltage plus the diode drop. The efficiency during this period is therefore $V_{out} \div (V_{out} + V_{diode})$, where $V_{out}$ = output voltage, and $V_{diode}$ = voltage drop in the diode. In addition, there is some loss in the inductor, the efficiency of which we denote by $\eta_{ind}$. The losses in the capacitor and wires are relatively small and can be ignored for simplicity. The overall efficiency of the buck converter is therefore the product of the previous three efficiencies, namely,

$$\eta_{conv} = \left( \frac{V_{in} - V_{sw}}{V_{in}} \right) \left( \frac{V_{out}}{V_{out} + V_{diode}} \right) \eta_{ind} \tag{2.11}$$

With commonly used devices in such converters, the transistor switch and diode voltage drops are typically 0.6 V each, and the inductor efficiency can approach 0.99. Therefore, the overall efficiency of a typical 28-V output converter is around 95%. The losses in the wires, capacitors, and magnetic core may further reduce the efficiency by a few percent, making it in the range of 92–94%. Equation (2.11) indicates that the buck converter efficiency is a strong function of the output voltage. It decreases with decreasing output voltage since the switch and diode voltage drops remain constant. For example, a 5-V output converter would have efficiency in the

75–80% range, and a 3-V converter would have even lower efficiency, around 65%, unless special low-voltage devices are used in the design.

## 2.2 BOOST CONVERTER

The boost converter steps up the input voltage to a higher output voltage. The converter circuit is shown in Figure 2.4. When the transistor switch is on, the inductor is connected to the input voltage source, the inductor current increases linearly while the diode is reverse biased, and the capacitor supplies the load current. When the switch is off, the diode becomes forward biased, the inductor current flows through the diode and the load, and the inductor voltage adds to the source voltage to increase the output voltage. The output voltage of the boost converter is derived again from the volt-second balance principle in the inductor. With duty ratio $D$ of the switch, it can be shown, in a manner similar to that in the buck converter analysis, that the output voltage and current are given by the following expressions:

$$\frac{V_{out}}{V_{in}} = \frac{T_{off} + T_{on}}{T_{off}} = \frac{1}{1-D} \qquad (2.12)$$

$$\frac{I_{out}}{I_{in}} = 1 - D \qquad (2.13)$$

Since $D$ is always a positive number less than one ($0 < D < 1$), the output voltage is always greater than the input voltage. Therefore, the boost converter can only step up the voltage; hence, it is also called a *step-up converter.* It is widely used for discharging the battery to feed loads at constant voltage, providing yet another name: the *battery discharge converter.* Since the battery voltage under discharge sags with increasing depth of discharge, a voltage converter with a feedback-controlled duty ratio is required to continuously boost the battery voltage to a regulated output voltage for the loads.

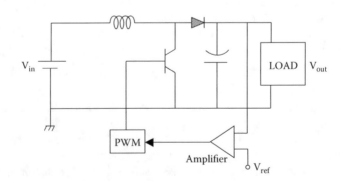

**FIGURE 2.4**   Boost converter circuit.

The efficiency of the boost converter can be shown to be

$$\eta_{conv} = \left(\frac{V_{in} - V_{sw}}{V_{in}}\right)\left(\frac{V_{out}}{V_{out} + V_{diode}}\right)\eta_{ind} \tag{2.14}$$

which is generally higher than in the buck converter due to higher output voltage.

## Example 2.5

A boost converter powers a 4-$\Omega$ resistor and 1-mH inductor load. The input voltage is 60 V, and the output load voltage is 80 V. If the on time is 2 ms, we determine the following:

Using Equation (2.12), $80 \div 60 = 1 \div (1 - D)$, which gives the duty ratio $D = 0.25$.

$\therefore T_{on} = 0.25\ T$, that is, 2 ms = 0.25 $T$,

which gives $T = 8$ ms, where $T = T_{on} + T_{off}$

Switching frequency = $1/T = 1/0.008 = 125$ Hz

Output current = $80/4 = 20$ A

Input current = $20\ (1 - 0.25) = 15$ A

## Example 2.6

For a boost converter with a 500-Hz switching frequency, 50-V dc input, 75-V dc output, 2-mH inductor, and 2.5-$\Omega$ load resistance, we determine the following:

Period $T = 1/500 = 0.002$ sec = 2 ms

$$V_{out}/V_{in} = 75/50 = 1/(1 - D)$$

which gives $D = 0.333$.

$$T_{on} = 0.333 \times 2\text{ ms} = 0.666\text{ ms}$$

$$T_{off} = 2 - 0.666 = 1.334\text{ ms}$$

$$I_{out} = 75/2.5 = 30\text{ A}$$

$$I_{in} = 30 \div (1 - 0.333) = 45\text{ A}$$

$\Delta I_{Ripple.pk\text{-}pk} = V_{in} \times T_{on} \div L = 50\text{ V} \times 0.666\text{ ms} \div 2\text{ mH} = 16.65\text{ A (a large ripple)}$

## 2.3 BUCK-BOOST CONVERTER

Combining the buck and boost converters in a cascade (Figure 2.5a) gives a buck-boost converter, which can step down or step up the input voltage. A modified buck-boost converter often used for this purpose is shown in Figure 2.5b. The voltage and current relations in either circuit are obtained by cascading the buck and boost converter voltage relations as follows:

$$\frac{V_{out}}{V_{in}} = \frac{D}{1-D} \tag{2.15}$$

$$\frac{I_{out}}{I_{in}} = \frac{1-D}{D} \tag{2.16}$$

Equation (2.15) shows that the output voltage of this converter can be higher or lower than the input voltage depending on the duty ratio $D$ (Figure 2.6), and the voltage ratio equals 1.0 when $D = 0.5$.

The buck-boost converter is capable of a four-quadrant operation (i.e., both $V$ and $I$ can be positive or negative), making it suitable in variable-speed drives for a dc motor with regenerative braking. The converter is in the step-up mode during the generating operation and in the step-down mode during the motoring operation.

## 2.4 FLYBACK CONVERTER (BUCK OR BOOST)

Figure 2.7 shows the topology of the flyback converter, which can buck or boost the voltage depending on the turn ratio $N_{in}/N_{out}$ of the coupled inductor. The energy is stored in the primary side of the inductor during on time. When the switch is turned

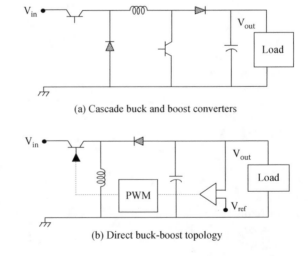

(a) Cascade buck and boost converters

(b) Direct buck-boost topology

**FIGURE 2.5**  Buck-boost converter alternative circuits.

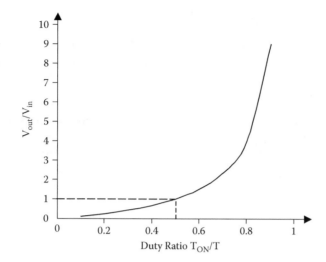

**FIGURE 2.6**  Buck-boost converter voltage ratio versus duty ratio.

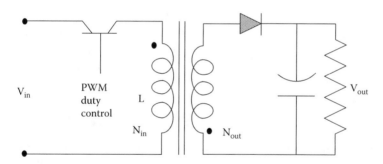

**FIGURE 2.7**  Flyback converter circuit with inductor.

off, the primary side energy is inductively transformed—gets kicked or flyover—to the secondary side and delivered to the load via diode. The polarity marks of the two coils are of essence in this circuit.

The magnetic flux in air only—as opposed to flux in a magnetic core—can store magnetic energy. Therefore, the energy storage inductor must have air in its magnetic flux path. The inductor core in the converter is therefore made with an air gap in one lump or distributed throughout the core, such as in a potted magnetic metal powder core with nonmagnetic binder interleaved that is magnetically equivalent to air. The energy is stored in distributed air gap of the core.

The ideal transformer core cannot store magnetic energy since it has zero air gap and magnetically soft steel. Therefore, the magnetic flux in the core is established with negligible magnetizing current $I_{mag}$ in the inductance $L_{mag}$, and the energy stored in the magnetic flux $= \frac{1}{2}L_{mag} \times I_{mag}^2 = 0$ since $I_{mag} = 0$. The mechanical analogy of this is an infinitely soft spring. When compressed, it cannot store potential energy in the spring compression.

The voltage ratio of the flyback converter is the cascade of the duty ratio and the turn ratio,

$$\frac{V_{out}}{V_{in}} = D \cdot \left( \frac{N_{out}}{N_{in}} \right) \tag{2.17}$$

This gives great flexibility in the output voltage without limitation of the duty ratio. It can buck or boost depending on whether the turn ratio is less than or greater than one.

A flyback converter uses a dual-purpose two-winding inductor with an $N_{in}/N_{out}$ turn ratio to step up or down the voltage like a transformer and to store energy like an inductor.

Denoting the efficiency of coupled inductors by $\eta_{ind}$ (typically 96–98%), the overall converter efficiency derived in a similar manner as previously results in the converter efficiency

$$\eta_{conv} = \left( \frac{V_{in} - V_{sw}}{V_{in}} \right) \left( \frac{V_{out}}{V_{out} + V_{diode}} \right) \eta_{ind} \tag{2.18}$$

An advantage of the flyback converter over the classical buck-boost converter is that it electrically isolates the two sides, thus minimizing the conducted electromagnetic interference (EMI) and enhancing safety. Moreover, it stores the inductive energy and simultaneously operates like a classical forward buck converter with a transformer, stepping up or down the input voltage as needed.

## 2.5   TRANSFORMER-COUPLED FORWARD CONVERTER

The converter circuit shown in Figure 2.8 employs a transformer with suitable turn ratio $N_{in}/N_{out}$ in the conventional buck converter such that the output voltage can be below or above the input voltage. The polarity marks are of essence. A bleeding resistor is required in parallel with the load to keep the converter working with a certain minimum load. Its voltage relation is identical to that for the buck converter modified by the transformer turn ratio. That is,

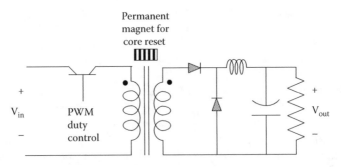

**FIGURE 2.8**   Forward converter circuit with transformer.

$$\frac{V_{out}}{V_{in}} = D \cdot \left( \frac{N_{out}}{N_{in}} \right) \qquad (2.19)$$

And the converter efficiency

$$\eta_{conv} = \left( \frac{V_{in} - V_{sw}}{V_{in}} \right) \left( \frac{V_{out}}{V_{out} + V_{diode}} \right) \eta_{ind} \, \eta_{trfr} \qquad (2.20)$$

where $\eta_{trfr}$ = transformer efficiency, which is typically 96–98%.

## 2.6  PUSH-PULL CONVERTER

The push-pull converter is again a buck-boost converter as shown in Figure 2.9. The converter topology uses a center-tapped transformer, which gets square wave excitation. It is seldom used at low power levels but finds applications in systems dealing with several tens of kilowatts and higher power. It can provide a desired output voltage by setting the required transformer turn ratio. The voltage ratio and the efficiency are the same as those given by Equations (2.19) and (2.20).

## 2.7  INDUCTOR-COUPLED BUCK CONVERTER

The inductor-coupled converter is also known as a *Cuk converter* after its inventor. All converters presented previously need an inductor-capacitor (L-C) filter to control the ripple in the output voltage. The buck converter needs a heavier filter than the boost converter. Since ripples are present in both the input and output sides, it is possible to cancel them by coupling the ripple current slopes of two sides with matched inductors, as shown in Figure 2.10. The ripple magnitudes are matched by using the required turn ratio, and the polarities are inverted by winding the two inductors in magnetically opposite directions. The net ripple on the output side is zero at a certain air gap, below which it changes the polarity. A gap-adjusting screw is needed to tune the gap for a precise match. The output capacitor $C_o$ (not shown) is not really needed in a perfectly coupled design, but a small value of $C_o$ improves the performance. The coupling capacitor $C$ is needed for the performance. The total value of the two

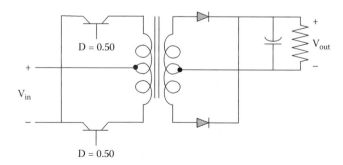

**FIGURE 2.9**  Push-pull converter with center-tapped transformer.

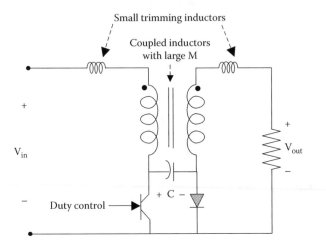

**FIGURE 2.10** Cuk converter with coupled inductors.

capacitors is about the same as that needed in the classical buck converter. On the output side, some ripple may be left due to the winding resistances. The input side ripple is the same as that without the coupled inductor. The load current changes do not affect the ripple cancellation.

## 2.8 DUTY RATIO CONTROL CIRCUIT

The duty ratio shown in Figure 2.1b can be controlled by a feedback control in the triggering circuit as shown in Figure 2.11. In the functional schematic of Figure 2.11a, the actual output voltage of the converter is compared with the desired voltage, and the difference is fed to the operational amplifier. The amplified error signal—called the control voltage—is subtracted from a sawtooth voltage by a comparator, whose output is applied to the gate terminals of the power electronics switch of the converter. If the power electronics device is a transistor, it conducts as long as $(V_{sawtooth} - V_{control})$ is positive, or else it blocks the conduction. The resulting duty ratio is shown in Figure 2.11b. The device turns on when $V_{sawtooth} > V_{control}$, which happens when angle $\alpha = V_{control.dc}/V_{sawtooth.peak} 180°$ and remains on for $(180 - \alpha)°$. Therefore, it yields the duty ratio

$$D = \frac{180 - \alpha}{180} = 1 - \frac{V_{control.dc}}{V_{sawtooth.pk}} \tag{2.21}$$

Obviously, $D = 1$ if $V_{control.dc} = 0$, and $D = 0$ if $V_{control.dc} = V_{sawtooth.pk}$. The value of $D$ can be adjusted to any value between 0 and 1 by varying the value of $V_{control.dc}$.

## 2.9 LOAD POWER CONVERTER

The load power converter (LPC) is connected between the power bus and the load, with its output suitable for a specific load. Its design requirements come primarily from the user requirements on the load and output regulation (such as constant

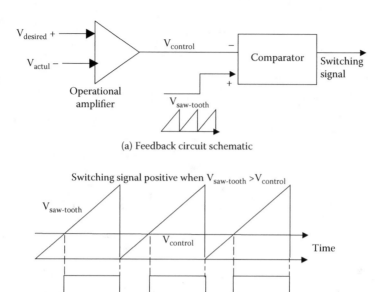

(a) Feedback circuit schematic

(b) Resulting duty ratio control

**FIGURE 2.11** Feedback duty ratio control circuit and operation.

voltage, constant current, constant power, quality of power, etc.). The necessary control circuit to regulate the output is generally incorporated in the LPC design package.

Loads having a stringent ripple requirement may require a prohibitively large filter capacitor at the power supply output. The inductor stores much greater energy per unit mass and therefore significantly reduces the filter size. Alternatively, an L-C filter, shown in Figure 2.12a at the load interface, may be more cost effective. When using such an L-C filter, the transient analysis becomes important on switch on, as both the current and voltage may overshoot to damaging levels. If the LPC is disconnected from the bus using a mechanical relay, the energy stored in the L and C elements of the filter may cause sparks at the relay contacts. Moreover, large negative voltage transients could occur at the user interface when the relay contact is opened. This can result in component damage. A freewheeling diode (Figure 2.12b) can eliminate any such sparking and negative voltage transients by providing a circulating current path when the switch is opened.

The basic converter—called the main power train—in the LPC box constitutes the bulk of the LPC design, around which many bells and whistles are added to make the LPC suitable for the user load. The input voltage may vary over a wide range in some poor-quality power bus. If the LPC output voltage is required to be constant with variable voltage input, a feedback voltage regulator is used in the design with a PWM converter with duty ratio control. If the input voltage variation is small, a

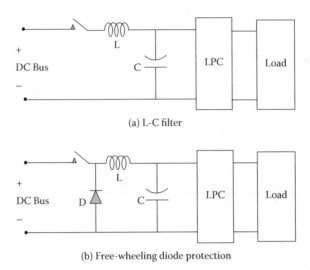

(a) L-C filter

(b) Free-wheeling diode protection

**FIGURE 2.12** L-C ripple filter between power bus and load power converter.

linear dissipative series voltage regulator may be used. The linear resistive control is more reliable than a switching converter because there are fewer parts and no high-voltage switching stress, although it has lower efficiency and runs hotter. The duty ratio-controlled buck converter is simple and dynamically stable in operation. All other converters are relatively complex and may become unstable as they can have poles in the right-hand side of the Laplace plane.

If a current-limiting feature is desired, the converter control circuit regulates and limits the current. In such an LPC design, the current is limited to some high value during overloads or even under a load fault during which the bus voltage is not driven to zero. It just folds back the current to some low value.

Thus, the LPC control loop can be designed to maintain constant voltage, constant current, or constant power. It can also be designed for a combination of two or more performance parameters. For example, the converter may maintain constant voltage up to a certain current limit and then maintain constant current at the set limit or even fold back a little in current. It is all in the control system, rather than in the basic power train design.

## 2.10 POWER SUPPLY

A power supply is a converter having one or more regulated output voltages such that the output is suitable for a range of loads. The output may be electrical power isolated from the input side by a transformer for safety or EMI reasons. The power supply design is focused primarily on the user requirements for the load and the output regulation (such as a constant voltage, constant current, constant power, quality of power, etc.). The necessary control circuit to regulate the output is generally incorporated in the power supply packaging (Figures 2.13 and 2.14). Thus, the main power train of the converter constitutes the bulk of the power supply design around which the controls are added to make it suitable for various loads. The output-to-input relations

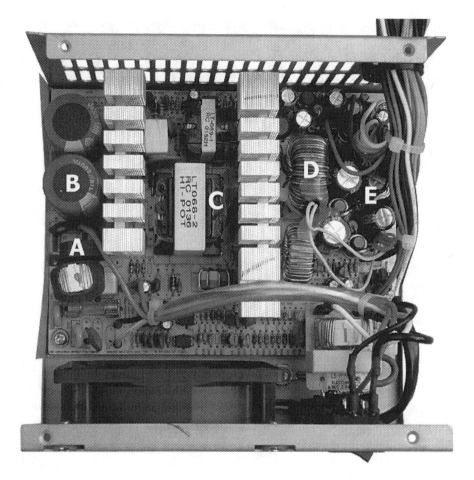

**FIGURE 2.13**   Switch mode power supply interior packaging (ATX).

for the power supply are essentially the same as those for the main power train, but with finer controls. Thus, we can say that power supply is equal to the power train converter in the basic form plus fine controls for the finished output with the desired regulation. Typical power supply requirements are as follows:

- Regulated voltages within a few percent (regardless of wide variations in input voltage, temperature, etc.)
- Ripple limited to low specified value
- Isolations with $n$:1 or 1:1 turn ratio transformer
- Multiple output (using three-winding transformer, if needed)
- Protection (overvoltage, overcurrent, current limiting, etc.)

The power supply uses a buck, boost, or buck-boost converter with a transformer inserted to step up or down the voltage or just for the electrical isolation of the output side from the input side.

**FIGURE 2.14**   Switch mode power supply exterior view (Voltcraft 4005).

In all types of converters covered in this chapter, there is a trend to increase the switching frequency to reduce the size of the power train inductor and the output capacitor filter. Frequencies up to a few hundred kilohertz are common at present but are approaching 1 MHz in some high-density power converters. The design challenge increases as the switching frequency exceeds 500 kHz as the lead wire inductance and capacitance become significant due to stray and parasitic effects, which are difficult to analyze.

## PROBLEMS

*Problem 2.1:* A buck converter operating at 60-Hz switching frequency has $T_{on} = 7$ ms. Determine the average source current if the load current is 12 A.

*Problem 2.2:* A buck converter is operating at 2-kHz switching frequency from a 220-V dc source. The inductance is 30 mH. If the output voltage is 110 V and the load resistance is 15 Ω, determine (a) the switch on time and off time, (b) load current and power output, (c) source current and power input, and (d) peal-to-peak ripple current.

*Problem 2.3:* A buck converter has 220-V dc input voltage, $R_L = 18$ Ω, switching frequency is 5 kHz, and on time is 80 µs. If the average source current is 10 A, determine (a) the average output voltage, current, and power; and (b) the minimum inductance needed to ensure continuous conduction.

*Problem 2.4:* For a low-voltage buck converter with $V_{out} = 3.5$ V, $f_s = 50$ kHz, $L = 200$ µH, $C = 500$ µF, $V_{in} = 12$ V, $I_{out} = 1.5$ A, determine the peak-to-peak ripple $\Delta V_{out}$ in the output voltage.

*Problem 2.5:* A boost converter has input voltage of 115 V and an output voltage of 230 V dc connected to a 10-Ω resistor and 2-mH inductor load. If the on time is 3 ms, determine (a) switching frequency, (b) output current, and (c) input current.

*Problem 2.6:* A boost converter has 60-V dc input, 100-V dc output, 3-mH inductance, and 4-Ω load resistance. If its switching frequency is 2 kHz, determine (a) output and input currents and (b) peak-to-peak value of the ripple current.

## QUESTIONS

*Question 2.1:* Explain the principle on which the buck converter analysis is based.

*Question 2.2:* As the duty ratio changes from 0.1 to 0.9, how does the buck-boost converter voltage ratio change?

*Question 2.3:* At what duty ratio does the buck-boost converter change from buck to boost?

*Question 2.4:* What does the term *electrical isolation* mean? How is it achieved in a converter design.

*Question 2.5:* Identify the similarity and difference between the dc-dc converter and the ac transformer.

*Question 2 6:* Identify the construction and functional differences between the transformer and the inductor.

*Question 2.7:* What flies back in the flyback converter?

*Question 2.8:* Identify and explain the benefit of using high-frequency switching in the switch-mode power converter design.

## FURTHER READING

Ahmed, A. 1999. *Power Electronics for Technology*. Upper Saddle River, NJ: Prentice Hall.

Fang, Lin Luo, and Hong Ye. 2003. *Advanced DC/DC Converters*. Boca Raton, FL: CRC Press/Taylor & Francis.

Fang, Lin Luo, and Hong Ye. 2005. *Essential DC/DC Converters*. Boca Raton, FL: CRC Press/Taylor & Francis.

Kazimierczuk, M. 2008. *Pulse-Width Modulated DC-DC Power Converters*. Hoboken, NJ: John Wiley & Sons.

Wu, K.C. 1997. *Pulse Width Modulated DC/DC Converters*. New York: Chapman and Hall.

# 3 AC-DC-AC Converters

This chapter covers the following power electronics converters widely used in electrical power systems, with their alternative names given in parentheses:

- AC-DC converter (rectifier)
- AC-AC voltage converter (phase-controlled converter)
- DC-AC converter (inverter)
- AC-DC-AC (dc link) frequency converter
- AC-AC direct frequency converter (cycloconverter)

Converters perform their functions using power electronics (solid-state semiconductor) devices periodically switched on and off at a certain frequency. Numerous circuit topologies have been developed for these converters, and new ones are evolving every year. A number of computer software programs, such as PSPICE™ and SABER™, are available to analyze such converters in great detail. However, the following sections present the basic circuits and the voltage, current, and power relationships at the input and output terminals of some widely used converters, leaving details for more advanced books dedicated to the subject of power electronics.

## 3.1 AC-DC RECTIFIER

### 3.1.1 SINGLE-PHASE, FULL-WAVE RECTIFIER

The single-phase, full-wave rectifier shown in Figure 3.1 is built with four thyristors. It is typically used to charge a battery or drive a dc motor, often at variable speed. Its generalized circuit includes the ac side source inductance $L_{ac}$, dc side inductance $L_{dc}$, and back emf $E_{dc}$. The $E_{dc}$ could be the internal emf of a battery or the counter emf of a dc motor. The $R_{dc}$ may be the motor armature resistance or the battery internal resistance.

The inductances $L_{ac}$ may be the cable inductance or the filter inductance purposely placed in the circuit. The cable inductance is due to the leakage flux between the cable conductors, the value of which comes from data sheet of the cable manufacturer. The internal source inductance in the Thevenin source model is generally negligible if the source is much larger than the converter rating, which is often the case in practice. If significant, it is included in $L_{ac}$.

In the converter operation, when the upper terminal of the ac source is positive, the thyristors $T_1$ and $T_4$ are fired to conduct downward current in the dc load. After half a cycle, when the lower ac terminal becomes positive, $T_2$ and $T_3$ are fired, again conducting downward current in the dc load. Thus, two pulses of positive voltage are applied to the load over one full ac wave (hence the name), and the load current is always positive downward (rectified dc). The dc side current would have some ripples but can be made smooth using a large dc side filter inductor $L_{dc}$, as is usually done

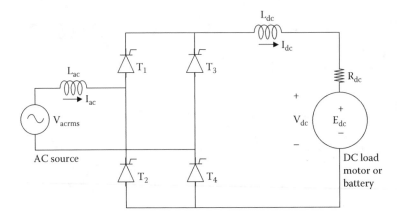

**FIGURE 3.1**   Single-phase full-wave rectifier circuit for dc motor drive or battery charging.

in practical rectifier designs. The dc side voltage can be controlled by delaying the firing angle $\alpha$ of the thyristors as shown in Figure 3.2.

When no dc side load is connected, the no-load dc voltage $V_{dc0}$ at the converter terminals is given by the average value of the dc side voltage over the conducting pulse of the rectified voltage shown in Figure 3.2. Mathematically, it is the average value of the sine function from $\alpha$ to 180° over the 180° span. With $v(t) = \sqrt{2}V_{rms}$ sin $\omega t$ for ac voltage and the angle measured in radians, we have

$$V_{dc0} = \frac{\frac{1}{\omega}\int_{\alpha}^{\pi} v(t) \cdot d(\omega t)}{\pi} = \frac{\sqrt{2}}{\pi} V_{rms}(1+\cos\alpha) = 0.45 V_{rms}(1+\cos\alpha) \qquad (3.1a)$$

However, with inductive load, the output current continues in the negative side even after the converter voltage has come to zero, posing a negative voltage on the load that subtracts from the integral in Equation (3.1a). This results in a slightly reduced output voltage at the converter terminals,

$$V_{dc0} = \frac{\frac{1}{\omega}\int_{\alpha}^{\pi} v(t) \cdot d(\omega t)}{\pi} = \frac{2\sqrt{2}}{\pi} V_{rms}\cos\alpha = 0.90 V_{rms}\cos\alpha \qquad (3.1b)$$

The average value of dc is not a smooth constant value but is superimposed with high-frequency ripples. When the dc side delivers a constant load current $I_{dc}$ via a large ripple filter inductor $L_{dc}$, the smooth dc voltage at the load terminals is given by

$$V_{dc} = 0.90 V_{rms}\cos\alpha - \frac{2\omega L_{ac}}{\pi} I_{dc} = E_{dc} + I_{dc}R_{dc} \qquad (3.2)$$

The $L_{dc}$ has no effects on $V_{dc}$; it merely smoothes the ripples from $I_{dc}$, as is assumed in the analysis.

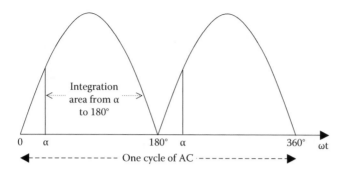

**FIGURE 3.2** Single-phase full-wave rectifier output with firing delay angle α.

The dc side voltage can be controlled by controlling the firing angle α as desired. If the diodes were used instead of thyristors, then α is always zero as the diodes automatically conduct immediately on becoming positively biased. Such a diode rectifier is called the *uncontrolled rectifier*, for which Equation (3.2) holds true with α = 0, giving the maximum dc side voltage of 0.90 $V_{rms}$.

Figure 3.3 shows how the thyristor gate-firing (triggering) signal pulses are generated by comparing a reference sawtooth wave voltage with a dc control voltage. When $V_{sawtooth} > V_{control}$, a positive voltage results on the thyristor gate terminal, starting the conduction (i.e., firing the thyristor). The firing angle is controlled from 0° to 180° by varying the value of $V_{control}$. A higher $V_{control}$ results in later firing. Thus, the firing delay angle α is proportional to $V_{control}$, as given by

$$\alpha = \left( \frac{V_{control(dc)}}{V_{sawtooth.peak}} \right) 180° \tag{3.3}$$

Varying the $V_{control}$ signal varies the ac output voltage. A rectifier with 0 < α < 90° can be easily made an inverter by increasing the firing delay angle α above 90° (i.e., 90° < α < 180°), when cos α becomes negative. Then, Equation (3.2) would give negative $I_{dc}$, taking power from the dc side to the ac side (i.e., inverting the dc into ac).

The voltage across the thyristor is near zero when it is conducting. The reverse voltage equals the full source voltage $V_s$ when the thyristor is not conducting as there are no voltage drops anywhere else in the circuit due to zero current. With an ac source, the peak value of the reverse voltage is 1.414 $V_{ac.rms}$. This can be visualized by thinking of a circuit with only a thyristor and a resistor in the loop.

The output voltage of the rectifier is not a pure smooth dc, as is the characteristic of all other rectifiers covered in this book. For comparing the output performance of various converters, we develop in the following two sections a method of taking into account the amount of deviation from the ideal smooth dc or sinusoidal ac.

## Example 3.1

A single-phase, full-wave rectifier shown in Figure 3.1 is built using diodes and a large inductor. Determine the average and rms values of the current in each diode.

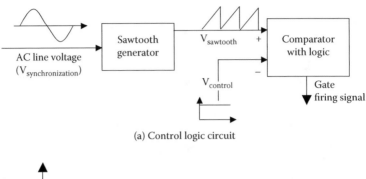

(a) Control logic circuit

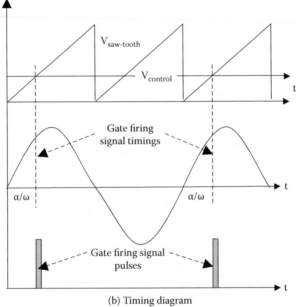

(b) Timing diagram

**FIGURE 3.3**  Thyristor firing angle control logic and operation.

## SOLUTION

The current in each diode is the load side current $I_{dc}$, which is constant as long as that particular diode is conducting and is zero all other times. Since each diode conducts only for half a cycle or $\frac{1}{2}T$, where $T$ = period of the cycle, we have

Diode current = $I_{dc}$ for $\frac{1}{2}T$

= 0 for the rest of the cycle

$$\therefore\ I_{diode(avg)} = \frac{I_{dc} \times \dfrac{T}{2} + 0 \times \dfrac{T}{2}}{T} = \frac{I_{dc}}{2}$$

and

$$I_{diode(rms)} = \sqrt{\frac{I_{dc}^2 \frac{T}{2} + 0^2 \frac{T}{2}}{T}} = \sqrt{\frac{I_{dc}^2 \cdot \frac{T}{2}}{T}} = \frac{I_{dc}}{\sqrt{2}}$$

The quantity in the first square root sign above is the mean of the square over period $T$ of the cycle.

## Example 3.2

A single-phase, full-wave bridge rectifier using phase-controlled thyristors with a large $L_{dc}$ converts 60-Hz, 120-V rms to dc for a dc side load resistance of 10 $\Omega$. For the firing delay angle $\alpha = 60°$, determine the dc side voltage and the power absorbed by the load, ignoring the cable inductances and ripples.

### SOLUTION

We use Equation (3.2) to obtain $V_{dc} = 0.9 \times 120 \times \cos 60° = 54$ V.
   Then,

$I_{dc} = 54/10 = 5.4$ A

and

$P_{dc} = 54 \times 5.4 = 291.6$ W

The dc side ripple current has zero average value, but its rms value will add into the power absorbed by the 10-$\Omega$ resistor. However, we ignore the ripple effect here.

## Example 3.3

The rectifier of Example 3.2 with the ac side cable inductance of 2 mH and the dc side cable resistance of 0.5 $\Omega$ is charging a battery at 5.4-A charge rate instead of powering the 10-$\Omega$ resistance. Determine the voltage available at the battery terminal.

### SOLUTION

From the dc voltage of 54 V calculated in Example 3.2, we deduct the voltage drops due to the ac side inductance and the dc side resistance. With $\omega = 2\pi \times 60 = 377$ rad/sec for 60 Hz, using Equation (3.2) leads to

$E_{dc} = 54 - 2 \times 377 \times 0.002 \times 5.4 \div \pi - 5.4 \times 0.5 = 54 - 2.6 - 2.7 = 48.7$ V

The dc side voltage has now dropped from 54 V in Example 3.2 to 48.7 V due to the inductance and resistance of the cables on both sides.

### 3.1.2 Ripples in DC and Ripple Factor

The output voltage of the rectifier discussed is not a smooth dc. The term *ripple* means periodic rise and fall in dc voltage or current output of the rectifier circuit. Ripples can be filtered out by capacitor, inductor, or both in some combinations. Even then, small rises and falls—exponential in theory but practically linear over a small range—remain superimposed on the average dc output voltage or current. The ripple factor is a measure of the deviation from the smooth dc output of a dc converter and is defined as

$$\text{Ripple factor} \qquad r = \frac{V_{ripple.rms}}{V_{dc}} \tag{3.4}$$

The dc output with ripples has an rms value greater than its average dc value, as given by

$$V_{out.rms} = \sqrt{V_{dc}^2 + V_{ripple.rms}^2} = V_{dc}\sqrt{1 + \frac{V_{ripple.rms}^2}{V_{dc}^2}} = V_{dc}\sqrt{1 + r^2} \tag{3.5}$$

We can also write Equation (3.5) as

$$\left(\frac{V_{out.rms}}{V_{dc}}\right)^2 = 1 + r^2 \quad or \quad \% ripple\ r = \sqrt{\left(\frac{V_{out.rms}}{V_{dc}}\right)^2 - 1} \times 100 \tag{3.6}$$

### 3.1.3 Harmonics in AC and Root Sum Square

On the ac input side of the converter in Figure 3.1, although the ac voltage input is pure sinusoidal, the ac input current will have high-frequency sine waves superimposed to keep balance with the high-frequency ripples in the dc output current. The term *harmonics* means high-frequency sinusoidal components in ac voltage or current at the input or output of a power electronics converter. Thus, the terms *harmonics* and *ripples* are similar, except that ripples are deviation from the dc, whereas harmonics are deviation from pure sine wave ac. Like ripples, the harmonics can also be filtered out by L-C filters, which can be tuned to specific harmonic frequencies.

Using Fourier series theory, any *periodic* wave can be resolved to a series of orthogonal functions. The sine and cosine functions of different frequencies are orthogonal since the product of any two over the period is zero. Accordingly, a non-sinusoidal voltage (or current) that is not a pure sinusoidal wave can be resolved into a series of high-frequency sinusoidal harmonic functions as follows:

$$V(t) = V_1 \cos \omega t + \sum_{h=2}^{\infty} V_h \cos(h\omega t + \alpha_h) \tag{3.7}$$

The first component on the right-hand side of Equation (3.7) is the fundamental component, whereas all other higher-frequency terms ($h = 2, 3, \dots \infty$) are called harmonics. The nonsinusoidal current $i(t)$ can be represented by a similar series of harmonic currents. Frequency of the $h$th harmonic current is $h \times$ fundamental frequency. For example, the frequency of the seventh harmonic in a 60-Hz power system is $7 \times 60 = 420$ Hz.

It can be shown that the rms value of the total nonsinusoidal ac voltage $v(t)$ with harmonics is given by the *root sum square (rss)* value of the individual harmonic rms values, that is

$$V_{ac.rms} = \sqrt{V_{1rms}^2 + V_{2rms}^2 + V_{3rms}^2 + \ldots\ldots V_{hrms}^2} = \sqrt{V_{1rms}^2 + V_{Hrms}^2} \qquad (3.8a)$$

where $V_{hrms}$=rms harmonic voltage of frequency ($h\times$fundamental frequency) for $h=2,3,$ $\ldots \infty$, and the total harmonic rms voltage $V_{Hrms} = \sqrt{V_{2rms}^2 + V_{3rms}^2 + \ldots\ldots V_{hrms}^2} = \sqrt{\sum V_{hrms}^2}$
Equation (3.8a) can be stated in words as

Rms value of total voltage = Root sum square (rss) value of rms harmonic voltages
$$(3.8b)$$

For the *root mean square* value of the total, we do not take the mean (average) of the rms values of the component squares, but the *root sum of the component squares*. This is analogous to what we know from vector algebra: Any three-dimensional vector V can be resolved into three orthogonal axes $x$-$y$-$z$, and the magnitude of the vector is the rss of the component magnitudes, that is, $V = \sqrt{V_x^2 + V_y^2 + V_z^2}$. The same applies to the harmonics, which are mathematically orthogonal functions.

If a harmonic is not present, then its value is taken as zero in the formula given. Practical voltage and current waves are symmetrical about the time axis, for which the Fourier series always results in all even harmonics equal to zero.

The performance of the system with harmonic voltages can be calculated by calculating the system performance under each harmonic voltage separately and then superimposing the individual performances to obtain the total performance of the system.

### 3.1.4 HARMONIC DISTORTION FACTOR

Power engineers express ac voltage and current in customarily implied rms values, generally omitting the suffix rms. The total harmonic content normalized with the fundamental value as the base is called the *total harmonic distortion* (THD) factor. The THD in voltage is then defined with implied rms values as $THD_v$=Total harmonic voltage rms / Fundamental voltage $V_1$ rms , that is,

$$THD_v = \frac{\sqrt{V_{3rms}^2 + V_{5rms}^2 + \ldots\ldots V_{hrms}^2}}{V_{1rms}} \qquad (3.9)$$

Usually, the fundamental voltage $V_{1rms}$ is taken as the base for $THD_v$. Recalling that $V_H$ represents the rss of all harmonics, Equation (3.8a) can then be rewritten as

$$V_{ac.rms} = \sqrt{V_{1rms}^2 + V_{Hrms}^2} = V_{1rms}\sqrt{1 + \frac{V_{Hrms}^2}{V_{1rms}^2}} = V_{1rms}\sqrt{1 + THD_v^2} \quad (3.10)$$

We can also write Equation (3.10) as

$$\left(\frac{V_{ac.rms}}{V_{1rms}}\right)^2 = 1 + THD_v^2 \quad or \quad \%THD_v = \sqrt{\left(\frac{V_{ac.rms}}{V_{1rms}}\right)^2 - 1} \; x \; 100 \quad (3.11)$$

Similarly, the total harmonic rms current $I_{Hrms} = \sqrt{I_{3rms}^2 + I_{5rms}^2 + I_{7rms}^2 + ....}$ , and the THD factor in current with no even harmonic is defined as

$$THD_i = \frac{\sqrt{I_{3rms}^2 + I_{5rms}^2 + .......I_{hrms}^2}}{I_{1rms}} = \sqrt{\left(\frac{I_{3rms}}{I_{1rms}}\right)^2 + \left(\frac{I_{5rms}}{I_{1rms}}\right)^2 + \left(\frac{I_{7rms}}{I_{1rms}}\right)^2 ...} \quad (3.12)$$

The $THD_i$ can be derived from the harmonic current ratios often given for the equipment. Alternatively, we can write the total ac rms current

$$I_{ac.rms} = \sqrt{I_{1rms}^2 + I_{Hrms}^2} = I_{1rms}\sqrt{1 + \left(\frac{I_{Hrms}}{I_{1rms}}\right)^2} = I_{1rms}\sqrt{1 + THD_i^2} \quad (3.13)$$

or,

$$\%THD_i = \sqrt{\left(\frac{I_{ac.rms}}{I_{1rms}}\right)^2 - 1} \; x \; 100 \quad (3.14)$$

### Example 3.4

A single-phase half-wave diode rectifier draws power from a single-phase, 240-V, 60-Hz source and delivers to a 2-kW load on the dc side at constant voltage $V_{dc}$ with a large filter capacitor as shown in Figure E3.4. The ac side harmonics to fundamental current ratios $I_h/I_1$ in percentages are 73.2, 36.6, 8.1, 5.7, 4.1, 2.9, 0.8, and 0.4% for $h$ = 3, 5, 7, 9, 11, 13, 15, and 17, respectively. Determine the rms ripple current in the filter capacitor. Ignore all losses.

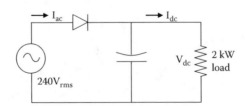

## SOLUTION

With large filter capacitor, the dc bus voltage $V_{dc}$ = peak of the ac input voltage.

$$\therefore V_{dc} = \sqrt{2} \times 240 = 339 \text{ V}$$

For 2-kW dc output,

$$I_{dc(avg)} = 2000 \text{ W}/339 \text{ V} = 5.9 \text{ A}$$

With no firing delay angle in the diode rectifier, and ignoring all losses, we must have $P_{ac} = P_{dc}$.

$\therefore$ Fundamental ac side rms current $I_{s1(rms)} = 2000 \text{ W}/240 \text{ V} = 8.33 \text{ A}$

Since there is no transformer in this circuit, we must have total rms current continuity on both sides, that is

$$I_{ac(rms)} = I_{dc(rms,\,total)} = \sqrt{I_{dc(avg)}^2 + I_{dc(ripple,rms)}^2}$$

$$\therefore I_{dc(ripple,rms)} = \sqrt{I_{ac(rms)}^2 - I_{dc(avg)}^2}$$

From the given ac side harmonics, Equation (3.12) gives

$$THD_i = \sqrt{0.732^2 + 0.366^2 + \ldots + .004^2} = 0.82 \text{ or } 82\%$$

Equation (3.13) gives the total

$$I_{ac(rms)} = I_{ac1} \times \sqrt{1^2 + 0.82^2} = 8.33 \times 1.2932 = 10.8\,A$$

$$\therefore I_{dc(ripple)} = \sqrt{10.8^2 - 5.9^2} = 9A$$

As seen here, the ac side THD and dc side ripples are rather large in the half-wave rectifier. That is why it is generally not used in practical applications.

### 3.1.5    THREE-PHASE, SIX-PULSE RECTIFIER

The circuit diagram of the full wave, three-phase ac-to-dc rectifier is shown in Figure 3.4 with thyristors (silicon-controlled rectifiers) as the power switches. It has two thyristors ($T_5$ and $T_6$) added in the single-phase, full-wave rectifier of Figure 3.1. Keeping the same notations for the ac and dc side parameters, its operation over one ac cycle is as follows: At any instant of time, one of the $T_1$, $T_3$, and $T_5$ thyristors and one of the $T_2$, $T_4$, and $T_6$ thyristors are fired in proper sequence as the ac phase voltage goes from the positive to the negative cycle. For example, when the phase voltage $V_a$ is most positive and the phase voltage $V_b$ is most negative, thyristors $T_1$ and $T_4$ are fired to conduct $I_{dc}$. This lasts for 60° of the full 360° cycle. For the next 60°, when $V_a$ is most positive and $V_c$ is most negative, thyristor $T_6$ is fired, and the $I_{dc}$ flows through

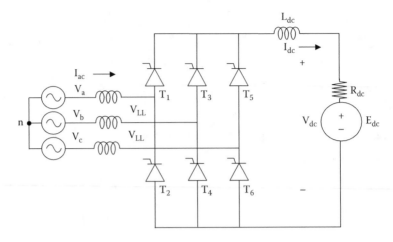

**FIGURE 3.4**   Three-phase, full-wave, six-pulse rectifier circuit.

$T_1$ and $T_6$. The current switchover from $T_4$ to $T_6$ is called the *commutation*, which may be delayed by a few electrical degrees due to the circuit inductance. For the next four 60° intervals, $T_3 + T_6$, $T_3 + T_2$, $T_5 + T_2$, and $T_5 + T_4$ in pairs conduct $I_{dc}$. The operation is like an orchestra of thyristors conducted by the triggering circuit. Each of the three phases contributes two pulses, making a total of six pulses in the output side (hence the name). Each thyristor conducts for 120° every cycle, so it carries the average current of $I_{dc}/3$ and the rms current of $I_{dc}/\sqrt{3}$.

### Example 3.5

A three-phase, full-bridge, six-pulse rectifier shown in Figure 3.4 is built with diodes and a large inductor. Determine the average and rms values of the current in each diode.

### SOLUTION

The current in each diode is the load side current $I_{dc}$, which is constant as long as that diode is conducting and is zero at all other times. Since each diode conducts for only one-third cycle or ⅓$T$, where $T$ = period of the cycle, we have

Diode current = $I_{dc}$ for ⅓$T$

= 0 for the rest of the cycle

$$\therefore \; I_{diode(avg)} = \frac{I_{dc} \times \dfrac{T}{3} + 0 \times \dfrac{2T}{3}}{T} = \frac{I_{dc}}{3} \qquad \text{and}$$

$$I_{diode(rms)} = \sqrt{\frac{I_{dc}^2 \frac{T}{3} + 0^2 \left(\frac{2T}{3}\right)}{T}} = \sqrt{\frac{I_{dc}^2 \cdot \frac{T}{3}}{T}} = \frac{I_{dc}}{\sqrt{3}}$$

The quantity in the first square root sign is the mean of the square over period $T$ of the cycle.

The rectified output voltage has six pulses as shown in Figure 3.5. With a firing delay angle of $\alpha$, the average $V_{dc}$ value is given by

$$V_{dc} = \frac{3\sqrt{2}}{\pi} V_{LLrms} \cos\alpha - \frac{3\omega L_{ac}}{\pi} I_{dc} = E_{dc} + I_{dc} R_{dc} \tag{3.15}$$

Again, $V_{dc}$ is controlled by controlling $\alpha$, and if $\alpha > 90°$, cos $\alpha$ becomes negative, making $I_{dc}$ negative. The three-phase rectifier then becomes a three-phase inverter, taking power from the dc side and delivering it to the ac side. With no dc side load in the rectifier mode,

$$V_{dco} = \frac{3\sqrt{2}}{\pi} V_{LLrms} \cos\alpha = 1.35 V_{LLrms} \cos\alpha \tag{3.16}$$

If diodes were used, $\alpha = 0$, and the maximum dc side voltage with no load is 1.35 $V_{LLrms}$.

If $L_{ac}$ is negligible, then it can be shown that the total and fundamental component of the ac side current have the following relations:

$$I_{rms} = \sqrt{\frac{2}{3}} I_{dc} \qquad I_{1rms} = \frac{3}{\pi}\sqrt{\frac{2}{3}} I_{dc} \qquad and \qquad \frac{I_{1rms}}{I_{rms}} = \frac{3}{\pi} = 0.955 \tag{3.17}$$

Total harmonic current is therefore

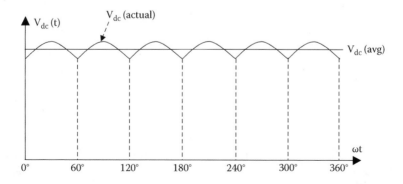

**FIGURE 3.5** Three-phase, full-wave, six-pulse rectifier output voltage.

$$I_{Hrms} = \sqrt{I_{rms}^2 - I_{1rms}^2} = 0.2966 I_{rms}$$

and

$$THD_i = \frac{I_{Hrms}}{I_{1rms}} = \frac{0.2966}{0.955} = 0.31 \; perunit(pu) \; or \; 31\% \qquad (3.18)$$

With line-frequency ac side voltage, only the fundamental ac side current can produce average real power. The average power is therefore reduced by the current ratio $I_{1rms}/I_{rms}$, called the *harmonic power factor*, which is derived from Equation (3.17) to be equal to $3/\pi$ for this rectifier. The firing delay angle $\alpha$ introduces the phase difference $\theta_1 = \alpha$ between the voltage and the fundamental component of the ac side current. This is known as the *displacement power factor*. When $L_{ac}$ is not negligible, it causes further delay in transferring (commutating) the conduction from one thyristor to another by additional angle $u$, called the *commutating angle*, given by the following relation:

$$\cos(\alpha + u) = \cos\alpha - \frac{2\omega L_{ac}}{\sqrt{2} V_{LLrms}} I_{dc} \qquad (3.19)$$

This equation gives the commutating angle $u$ for given $L_{ac}$ and $I_{dc}$.

### Example 3.6

A three-phase, six-pulse thyristor rectifier with a large dc side filter inductor delivers 500-kW power at 525-V dc and draws 60-Hz power from 460 $V_{LL}$ via a cable with leakage inductance of 25 μH. Determine the commutation angle of the thyristor.

#### SOLUTION

From the given dc side power parameters, we have
$I_{dc} = P_{dc}/V_{dc} = 500,000/525 = 952.4$ A
Using Equation (3.15),

$$V_{dc} = \frac{3\sqrt{2}}{\pi} 460 \cos\alpha - \frac{3 \times 2\pi \times 60 \times 25 \times 10^{-6}}{\pi} 952.4$$

That gives $\cos\alpha = 0.8589$ and $\alpha = 30.8°$.
Using Equation (3.19),

$$\cos(\alpha + u) = 0.8589 - \frac{2 \times 2\pi \times 60 \times 25 \times 10^{-6}}{\sqrt{2} \times 460} 952.4$$

That gives cos $(\alpha + u) = 0.831$ and $\alpha + u = 33.8°$
∴ The commutation angle $u = 33.8 - 30.8 = 3°$

With some approximation, it can be shown that an effective phase difference between the $V_{1rms}$ and $I_{1rms}$ results in a virtual power factor, called the total *displacement power factor*, which is given by

$$DPF = \cos\left(\alpha + \frac{u}{2}\right) \tag{3.20}$$

Ignoring the small commutation angle $u$, we have from Equations (3.17) and (3.20),

Total ac side power factor = Harmonic power factor × Displacement power factor

This equation leads to the following:

$$\text{Total ac side power factor } = \frac{3}{\pi}\cos\alpha \tag{3.21}$$

The harmonics superimposed on the average dc voltage in the output of this converter are of the order

$$h = 6k \pm 1 \text{ where } k = 1, 2, 3, \ldots \tag{3.22}$$

Smooth dc can be obtained by a filter made of an inductor in series or a capacitor in parallel with the rectified output voltage. The load determines the dc side current and power, that is,

$$DC \, side \, load \, current \, I_{dc} = \frac{V_{dc}}{R_{Load}} \quad and \quad DC side \, power \, P_{dc} = V_{dc} I_{dc} \tag{3.23}$$

In steady-state operation, the balance of power must be maintained on both ac and dc sides. That is, the power on the ac side must be equal to the sum of the dc load power and the losses in the rectifier circuit. Moreover, the three-phase ac power is given by $P_{ac} = \sqrt{3}\, V_{LL} I_L \, pf$. The balance of ac and dc powers, therefore, gives the following, from which we obtain the ac side line current $I_L$ for a given $I_{dc}$ load current:

$$\sqrt{3}\, V_{LL} I_L \, pf = \frac{V_{dc} I_{dc}}{Converter \, efficiency} \tag{3.24}$$

## 3.2  AC-AC VOLTAGE CONVERTER

### 3.2.1  Single-Phase Voltage Converter

The single-phase ac voltage converter steps down a fixed line voltage to a variable voltage on the load side. It uses two back-to-back thyristors or a triac in the circuit shown in Figure 3.6a. A firing delay angle $\alpha$ controls the output voltage by effectively chopping out a portion of each half cycle of the line voltage. The thyristor $T_1$ is forward biased during the positive half cycle and is turned on at angle $\alpha$. It then conducts from $\alpha$ to 180° and supplies power to the load. The thyristor $T_2$ is turned on after a half cycle at $\alpha + 180°$, which then conducts up to 360° supplying power to the load. With a purely resistive load, the load current follows the output voltage in phase as shown by the lower wave in Figure 3.6b. Both the voltage and current on the load side are alternating at the same frequency, although chopped off for angle $\alpha$ in both halves of each cycle. The output rms voltage and current can be shown to be

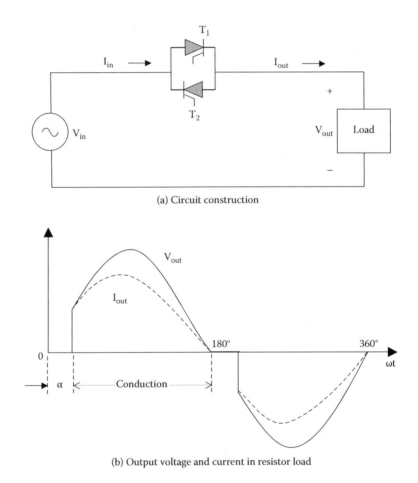

(a) Circuit construction

(b) Output voltage and current in resistor load

**FIGURE 3.6**   Single-phase ac voltage converter.

$$V_{out.rms} = V_{in.rms} \sqrt{1 - \frac{\alpha}{\pi} + \frac{\sin 2\alpha}{2\pi}} \quad and \quad I_{out.rms} = \frac{V_{out.rms}}{R_{Load}} \tag{3.25}$$

Output power $P_{out.avg} = V_{out.rms} \times I_{out.rms} = I_{out.rms}^2 \times R_{Load}$ (3.26)

On the source side, the input rms current is the same as that on the load side, that is, $I_{in.rms} = I_{out.rms}$. The apparent input power $S_{out} = V_{in.rms} \times I_{in.rms}$, and the load power factor due to delayed current conduction even with a resistive load is

$$Load\ power\ factor = \frac{P_{out}}{S_{out}} = \sqrt{1 - \frac{\alpha}{\pi} + \frac{\sin 2\alpha}{2\pi}} \tag{3.27}$$

The output voltage is controlled by varying $\alpha$. With large $\alpha$, the output voltage and effective power factor decrease even for a purely resistive load, and both become zero when $\alpha = \pi$ (only in theory; never done in practice).

Since each thyristor carries $I_{out.rms}$ for a half cycle, its rms current for self-heating calculations is $I_{Thy.rms} = I_{out.rms}/\sqrt{2}$. If a triac were used instead, it would carry current for the whole cycle with $I_{triac.rms} = I_{out.rms}$.

With small load-side inductance with a load power factor around 0.85 (as in most practical loads), the load current not only takes some time to build up initially but also takes about the same time longer to come to zero due to the inductive inertia of the circuit. The net result is that the current conduction time remains essentially the same as in a resistive circuit. The previous equations, therefore, hold approximately true. For purely inductive loads, $\alpha$ must be greater than 90° for the circuit to perform, and the converter operation changes significantly, such that the equations do not hold true even in an approximate sense.

## 3.2.2 THREE-PHASE VOLTAGE CONVERTER

The three-phase voltage converter shown in Figure 3.7 is made of three single-phase converters of Figure 3.6a placed in each phase. It reduces the voltage at the output terminals by delaying the thyristor conduction in each phase. The analysis becomes complex as the operating mode of the converter changes with the value of $\alpha$ in the ranges of 0–30, 30–60, 60–90, 90–120, 120–150, and 150–180°. The power electronics design engineer generally uses simulation software, such as PSPICE, for the performance analysis. However, when the thyristor delay angle $\alpha$ and the load side inductance are small, the single-phase relations presented in Section 3.2.1 hold approximately true with values per phase on the input and output sides. The output voltage will have six pulses, with harmonics of the order $h = 6k \pm 1$, where $k = 1, 2, 3, \dots$.

This converter can be used for controlling the power input to a three-phase static load, which obtains power in proportion to $V_{out}^2$. It can also be used for speed control of a single-phase or three-phase induction motor, particularly driving a pump or a fan. With reduced voltage applied at the motor terminals, the torque-speed curve shrinks vertically in $V_{out}^2$ proportion, keeping the same synchronous speed. The new operating speed is then at the intersection of the motor torque and the load torque characteristics.

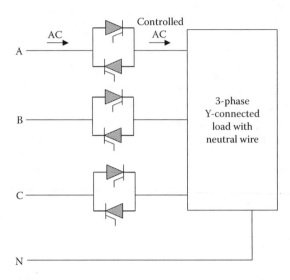

**FIGURE 3.7**   Three-phase ac voltage converter.

## 3.3   DC-AC INVERTER

The power electronics circuit used to convert dc into ac is known as the *inverter*, although the term *converter* is often used to mean either the rectifier or the inverter. The dc input to the inverter can be from a rectifier, battery, photovoltaic panel, or fuel cell.

### 3.3.1   SINGLE-PHASE VOLTAGE SOURCE INVERTER

In its simplest form, the single-phase, full-wave voltage source inverter (VSI) is made using four thyristor switches shown in Figure 3.8a. The four thyristors are fired on and off sequentially in diagonal pairs for equal duration (i.e., each pair for a half cycle). This results in the load being alternatively connected to positive and negative polarities of the dc source, thus making the load voltage alternate between positive and negative polarities as shown by rectangular pulses in Figure 3.8b. For the first half cycle, from time 0 to $\frac{1}{2}T$, thyristors $T_1$ and $T_4$ are fired to conduct the load current downward. For the second half cycle, from time $\frac{1}{2}T$ to $T$, the thyristors $T_3$ and $T_2$ are fired to conduct the load current upward. This way, although the source voltage is pure $V_{dc}$, the load voltage alternates between $+V_{dc}$ and $-V_{dc}$ in a rectangular wave shown in Figure 3.8b, with frequency $f = 1/T$ where $T = $ switching period. This rectangular wave ac output voltage can be resolved into the Fourier series of a fundamental sine wave (dotted curve), plus a third harmonic ac and a series of higher-order harmonics. Such output has $V_{1peak} > V_{dc}$, but $V_{1rms} < V_{dc}$. Since the input side rms value is the same as $V_{dc}$, the total rms values on both sides must be equal, that is.

$$V_{ac.rms} = V_{dc} \tag{3.28}$$

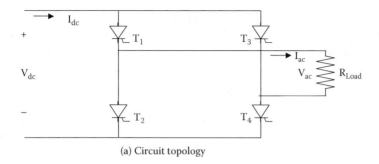

(a) Circuit topology

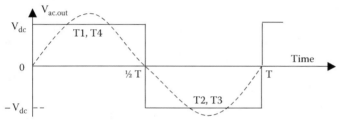

(b) Rectangular AC output voltage

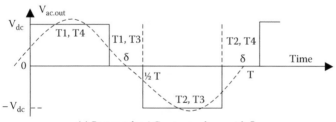

(c) Rectangular AC output voltage with δ

**FIGURE 3.8** Single-phase full-wave inverter with resistive load.

The Fourier series harmonic analysis would give the fundamental and harmonic voltages as

$$V_{1rms} = \frac{4V_{dc}}{\pi\sqrt{2}} \quad and \quad V_{hrms} = \frac{4V_{dc}}{h\pi\sqrt{2}} \quad for \ h = 3, 5, 7, 11, 13,... \quad (3.29)$$

And, the total voltage harmonic distortion factor, which is rather large, is

$$THD_v = \frac{V_{Hrms}}{V_{1rms}} = \frac{\sqrt{V_{acrms}^2 - V_{1rms}^2}}{V_{1rms}} = 0.483 \ pu \ or \ 48.3\% \quad (3.30)$$

The $V_{ac}$ can be controlled either by varying $V_{dc}$ or by introducing an additional thyristor switching state in which the output is shorted ($V_{ac} = 0$) by switching either

$T_1$ and $T_3$ simultaneously or $T_2$ and $T_4$ simultaneously in turn for angle δ in each half cycle as shown in Figure 3.8c. With such an inverter, the total rms value of the ac output voltage is given by

$$V_{ac.rms} = V_{dc}\sqrt{1 - \frac{2\delta}{T}} \qquad (3.31)$$

For inductive loads, we must add freewheeling diodes $D_1$, $D_2$, $D_3$, and $D_4$ to modify the previous circuit in Figure 3.8a to that shown in Figure 3.9a. The function of the diodes is to provide a continuing path for the load current soon after the thyristors are turned off until the current naturally comes back to zero. The diodes are turned on and off automatically as required to carry the freewheeling current. The voltage in the new circuit is the same as before, but the current now becomes continuous due to the inductive inertia, as shown in Figures 3.9b and 3.9c, respectively.

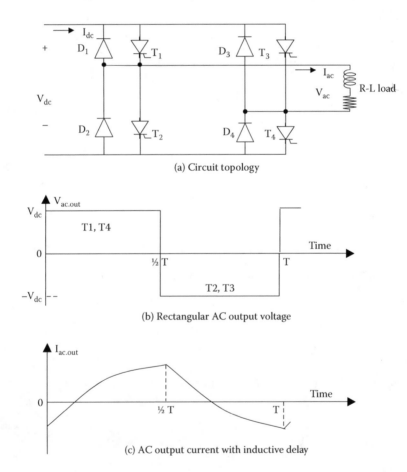

(a) Circuit topology

(b) Rectangular AC output voltage

(c) AC output current with inductive delay

**FIGURE 3.9** Single-phase full-wave voltage source inverter with freewheeling diode for inductive load.

### 3.3.2 SINGLE-PHASE PWM INVERTER

The large 48.3% voltage harmonics distortion factor in Equation (3.30) with the line frequency switching scheme above can be reduced by using a high-frequency switching scheme shown in Figure 3.10, known as the *pulse width modulated (PWM) VSI*. Two VSIs shown in Figure 3.8 are stacked vertically in a bipolar (center tap configuration) to share a common dc voltage source $V_{dc}$, each stack taking $+\frac{1}{2}V_{dc}$ and $-\frac{1}{2}V_{dc}$. The PWM scheme can be implemented by comparing the sinusoidal reference and

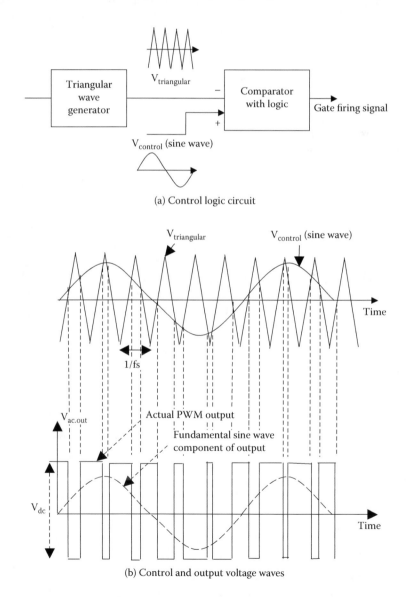

(a) Control logic circuit

(b) Control and output voltage waves

**FIGURE 3.10**   PWM inverter switching scheme and voltage waves.

triangular carrier signal as shown in the middle part of Figure 3.10. The diagonally opposite thyristors $T_1$ and $T_2$ are turned on when $V_{sine.control} > V_{triangular}$, giving $V_o = +\frac{1}{2}V_{dc}$. Thyristors $T_3$ and $T_4$ are turned on when $V_{sine.control} < V_{triangular}$, giving $V_o = -\frac{1}{2}V_{dc}$. Modulating the switch duty ratio this way varies the dc voltage output in a sine wave. The reference frequency of the sine control voltage is the desired output frequency of the inverted voltage, and the triangular carrier frequency is the switching frequency at which the input dc voltage polarity is switched with varying duty ratio. Since the output voltage alternates between $+\frac{1}{2}V_{dc}$ and $-\frac{1}{2}V_{dc}$ as shown in the lower part of Figure 3.10, such inverter is called the bipolar switching inverter. The following important definitions apply to such a PWM switching scheme:

$$\text{Frequency modualtion ratio } M_f = \frac{f_{carrier}}{f_{reference}} = \frac{f_{triangular}}{f_{sinecontrol}} = \frac{f_{switching}}{f_{desired}} \qquad (3.32a)$$

Increasing the carrier frequency increases the switching power loss and the harmonic frequencies. The value of $M_f$ generally ranges from 10 to 100 or even much greater. In the PWM converter, note the following differences carefully:

Frequency of triangular wave = Switching frequency for turning devices on and off

Frequency of sine wave control voltage = Desired ac output frequency          (3.32b)

The switching frequency $f_{sw}$ of the triangular control voltage is 20 kHz or higher in modern converters, and the output power frequency set by the sinusoidal control voltage is typically 60 or 50 Hz or even lower in ac motor drives. The higher switching frequency results in a lower L-C filter energy storage requirement to power the load during off time from the source, leading to lower filter weight and cost. For this reason, the switching frequencies in some modern low-power converters are 200 to 500 kHz and often approach 1000 kHz for achieving high power density in some converters for defense aircrafts and satellites, for which weight has a high premium.

Another important definition in the PWM switching is

$$\text{Amplitude modulation ratio } M_a = \frac{V_{peak\text{-}reference}}{V_{peak\text{-}carrier}} = \frac{V_{peak\text{-}sinecontrol}}{V_{peak\text{-}triangular}} \qquad (3.33)$$

If $M_a < 1$, the peak amplitude of the fundamental frequency output voltage $V_1$ is linearly proportional to $M_a$, that is,

$$V_{1peak} = M_a \cdot V_{dc} \text{ and } V_{1rms} = M_a\, V_{dc}/\sqrt{2} \qquad (3.34)$$

Thus, $V_1$ is linearly controlled by $M_a$. Such control is used for varying $V_1$ under fixed $V_{dc}$ or holding the output $V_1$ constant under fluctuating $V_{dc}$.

If $M_a > 1$, $V_1$ increases with $M_a$, but not linearly, and remains between $V_d$ and $4V_{dc}/\pi$.

If the inverter output load with some inductance draws a sinusoidal current of $I_{out.rms}$, then each thyristor switch must have the peak current rating of $\sqrt{2}I_{out.rms}$ and the peak voltage rating of $V_{dc}$. The inverter kilovolt-ampere output is $V_{1rms} \times I_{out.rms}$

The rectangular pulses of varying width in $V_{ac.out}$ shown in Figure 3.10 can be resolved into the fundamental sine wave and a series of high-frequency harmonics. The harmonic amplitudes in this inverter depend on $M_a$ because each pulse width depends on the relative amplitudes of the sine and triangular waves. The harmonic frequencies are around the integer multiples of the modulation frequency $M_f$ as listed in Table 3.1. For $h = 2M_f \pm$ few and $h = 3M_f \pm$ few, the $V_{h.peak}/V_{dc}$ ratio remains around 0.20 and 0.15, respectively, and 0.100 for all higher-order harmonics up to $9M_f$, beyond which they are not a function of $M_f$.

The PWM switching scheme can also be applied to the three-phase inverter, discussed in the following sections. The harmonics in a PWM inverter modulated with 12 pulses per cycle and 24 pulses per cycle are listed in Table 3.2. The PWM converter has no fifth and seventh harmonics but has much greater higher-order harmonics. It is important to note that, although the lower-order harmonics are eliminated in the PWM inverter, the much-higher-frequency harmonics of much higher amplitudes

### TABLE 3.1
### Harmonic Voltages in PWM Inverter for Large $M_f$ (Normalized to $V_{dc}$)

| Harmonic | $M_a = 0.1$ | $M_a = 0.4$ | $M_a = 0.7$ | $M_a = 1.0$ |
|---|---|---|---|---|
| Fundamental | 0.10 | 0.40 | 0.70 | 1.0 |
| $h = M_f$ | 1.27 | 1.15 | 0.92 | 0.60 |
| $h = M_f \pm 2$ | 0 | 0.06 | 0.17 | 0.32 |
| $h = 2M_f \pm 3$ | 0 | 0.02 | 0.10 | 0.20 |

### TABLE 3.2
### Percentage Harmonics Content of 12-Pulse and 24-Pulse PWM Inverters

| Harmonic Order | 12-Pulse per Cycle PWM | 24-Pulse per Cycle PWM |
|---|---|---|
| 1 | 100 | 100 |
| 5 | 0 | 0 |
| 7 | 0 | 0 |
| 11 | 40 | 0 |
| 13 | 40 | 0 |
| 17 | 5 | 0 |
| 19 | 11 | 0 |
| 23 | 13 | 40 |
| 25 | 13 | 40 |
| 29 | 18 | 0 |
| 31 | 9 | 0 |

are present as seen in Figure 3.11. This can be damaging to motor performance and can increase the motor noise in the audible range.

### 3.3.3 THREE-PHASE, SIX-PULSE VOLTAGE SOURCE INVERTER

Figure 3.12a depicts a six-pulse VSI fed with constant voltage $V_{dc}$ with a large capacitor. The thyristors are fired in square-wave mode for 180° at a time as shown in Figure 3.12b, 120° out of phase in each phase. The resulting step-wave alternating line voltage has fundamental and harmonic voltages as follows:

$$V_{LL1rms} = \frac{\sqrt{3}}{\sqrt{2}} \frac{4}{\pi} \frac{V_{dc}}{2} = \frac{\sqrt{6}}{\pi} V_{dc} = 0.78 V_{dc} \quad (3.35)$$

The output ac voltage magnitude can be controlled only by $V_{dc}$, and the frequency can be controlled by the switching frequency. The voltage is independent of the load and has harmonics on the order $h = 6k \pm 1$ ($k = 1, 2, 3, \ldots$) with line-to-line rms magnitudes given by

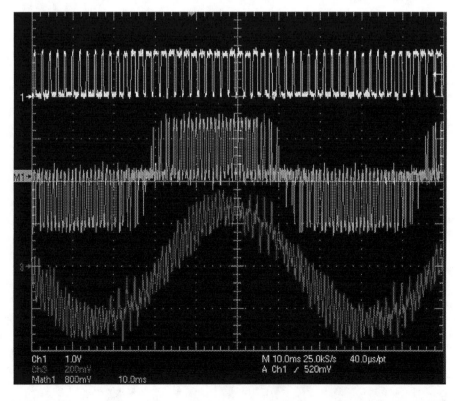

**FIGURE 3.11** High-frequency harmonics in PWM converter. (From Bill Veit, U.S. Merchant Marine Academy.)

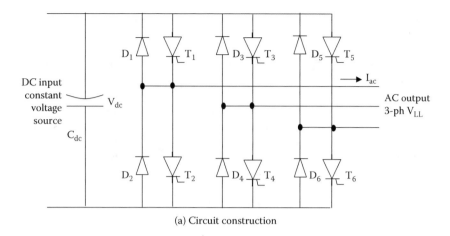

(a) Circuit construction

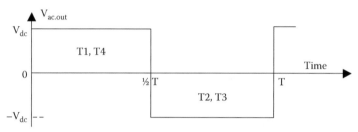

(b) Rectangular AC output voltage

**FIGURE 3.12** Three-phase voltage source inverter.

$$V_{LLhrms} = \frac{\sqrt{6}}{\pi h}V_{dc} = \frac{0.78V_{dc}}{h} \qquad (3.36)$$

## Example 3.7

A three-phase, six-pulse VSI is fed with 240-V dc. Determine the ac side rms fundamental line-to-line voltage and the first four harmonic voltages.

### SOLUTION

Using Equation (3.35),

$V_{LL1rms} = 0.78 \times 240 = 187.2$ V

Harmonics in the six-pulse VSI are on the order $h = 6k \pm 1$, where $k = 1, 2, 3, \dots$. Therefore the first four harmonics are 5, 7, 11, and 13, with the rms voltage magnitudes 187.2/h, which gives $V_5 = 187.2/5 = 37.44$ V, $V_7 = 187.2/7 = 26.74$ V, $V_{11} = 187.2/11 = 17.02$ V, and $V_{13} = 187.2/13 = 14.4$ V, all rms values because 187.7 V is an rms value.

### 3.3.4  THREE-PHASE, SIX-PULSE CURRENT SOURCE INVERTER

The output of a photovoltaic panel at a constant solar intensity is a constant dc current proportional to the panel area. Therefore, the solar panel works like a constant current source, which must be inverted to 60 or 50 Hz ac for local use or for feeding back to the grid. In many variable-frequency motor drives operating with a large dc filter in a dc link, the inverter power source is also a constant-current source. We therefore study the three-phase current source inverter (CSI) circuit shown in Figure 3.13 that is used in a solar photovoltaic power plant and in variable-frequency motor drives. The circuit is essentially the same as in Figure 3.12 except that a large input side capacitor is replaced with a large series inductor, which maintains a constant current at the input terminal of the inverter. At any given instant, only two thyristors, as shown in Figure 3.13b by their numbers, are turned on in sequence. Each thyristor conducts for 120°. The three-phase output current is a 120° wide rectangular step wave of magnitude $I_{dc}$, and the output frequency equals the switching frequency of the thyristors. The current and voltage relations are as follows:

In each phase (leg) of the inverter,

$$I_{out.rms(phase)} = \frac{\sqrt{2}}{3} I_{dc} \tag{3.37}$$

In each output line of the inverter,

$$I_{out.rms(line)} = \frac{\sqrt{2}}{\sqrt{3}} I_{dc} \tag{3.38}$$

Output voltage per phase (line to ground),

$$V_{out.rms} = \frac{V_{dc}}{\sqrt{6}\cos\theta} \tag{3.39}$$

where $\cos\theta$ = load power factor, that is, $\theta = \tan^{-1}(X_{Load}/R_{Load})$

### Example 3.8

A three-phase, six-pulse CSI is fed from a 240-V dc bus with a dc link current of 80 A. The ac side load has an 0.85 power factor lagging. Determine the ac side rms phase and line currents and output phase and line voltages.

#### SOLUTION

Using Equations (3.37) and (3.38),

$$I_{ac.rms.phase} = 80\sqrt{2} \div 3 = 37.71 \text{ A}$$

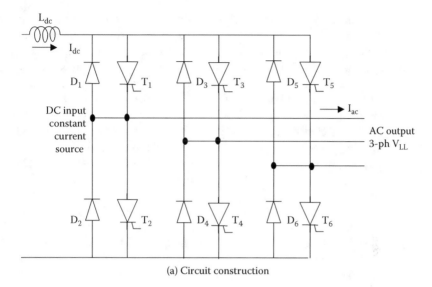

(a) Circuit construction

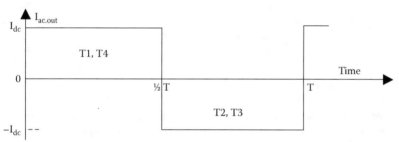

(b) Rectangular AC output current

**FIGURE 3.13**  Three-phase current source inverter.

and

$$I_{ac.rms.phase} = \sqrt{3} \times 37.71 = 65.32 \text{ A}$$

Using Equation (3.39), the ac side phase voltage is

$$V_{ac.rms.phase} = 240 \div (\sqrt{6} \times 0.85) = 115.27 \text{ V line to ground}$$

$$V_{ac.rms.line} = \sqrt{3} \times 115.27 = 200 \text{ V line to line}$$

Unlike in BJTs, MOSFETs, and IGBTs, the thyristor current, once switched on, must be forcefully switched off (commutated) to cease conduction. Therefore, a converter using thyristors as the switching devices must incorporate an additional commutating circuit—a significant part of the inverter—for converter performance. There are two main types of inverters: the line commutated and the forced commutated.

*Line-commutated inverter*: This type of inverter must be connected to the ac system into which it feeds power. The design method is mature and has been extensively used in high-voltage dc transmission line inverters. The inverter is simple and inexpensive and can be designed in any size. The disadvantage is that it acts as a sink of reactive power and generates many high-frequency harmonics. Therefore, its output needs a heavy harmonic filter to improve the quality of power at the ac output. This is done by an inductor connected in series and a capacitor in parallel with the inverted output voltage, just like in the rectifier discussed previously.

The poor power factor and high harmonic content of the line-commutated inverter significantly degrade the quality of power at the grid interface. This problem has been recently addressed by a series of design changes. Among them is the 12-pulse inverter circuit and increased harmonic filtering. These new design features have resulted in today's inverter operating near unity power factor and less than 3–5% THD. The quality of power at the utility interface at many modern solar power sites now exceeds that of the grid they interface.

*Force-commutated inverter*: This inverter type does not have to be supplying load and can be free running as an independent voltage source. The design is relatively complex and expensive. The advantages are that it can be a source of reactive power, and the harmonics content is low.

The power electronics device prices dictate the most economical design for a given system. Present inverter prices are about $1,500/kW for ratings below 1 kW, $1000/kW for 1–10 kW, $600/kW for 10–100 kW, and $400/kW for ratings approaching 1000 kW. The efficiency of an dc-ac converter with a transformer at full load is typically 85–90% in small ratings and 92–95% in large ratings.

In addition to the high efficiency in dc-ac conversion, it must have low harmonic distortion, low electromagnetic interference (EMI), and high power factor. The inverter performance and testing standards are covered by IEEE (Institute of Electrical and Electronics Engineers) 929-2000 and UL (Underwriters Laboratories) 1741 in the United States, EN 61727 in the European Union, and IEC (International Electrotechnical Commission) 60364-7-712 for the international standards. The THD generated by the inverter is regulated by international standard IEC 61000-3-2. At full load, it requires $THD_i < 5\%$ and the voltage $THD_v < 2\%$ for harmonics spectra up to the 49th harmonic. At partial loads, the harmonic distortions are usually much higher.

## 3.4 FREQUENCY CONVERTER

Two main types of frequency converter are (a) an dc link converter, which first converts ac of one frequency into dc voltage or current using a rectifier and then inverts dc into another frequency ac using the voltage source or current source inverter, and (b) a cycloconverter, which converts one frequency ac directly into another frequency ac by ac-ac conversion. In the past, the cycloconverter was more economical than the dc link inverter. With decreasing prices of power semiconductors, it has been gradually replaced by the dc link converter, which gives lower harmonics in the output. However, recent advances in fast switching devices and microprocessor controls are increasing the efficiency and power quality of the cycloconverter, which may come back in large power applications, such as in electric propulsion of ships.

### 3.4.1 DC LINK FREQUENCY CONVERTER

The dc link frequency converter consists of a phase-controlled ($\alpha$-controlled) three-phase rectifier feeding a dc link with a large filter inductor and a three-phase inverter as shown in Figure 3.14. The large inductor in the dc link works like a constant current source for the inverter. The performance of this converter can be analyzed by cascading the performances of the three-phase rectifier and the three-phase CSI covered in Sections 3.1 and 3.3. The advantage of the dc link converter is a common design that can be developed for both the induction motor and the synchronous motor with comparable ratings. For this reason, it is widely used in variable-frequency drives for ac motors, as discussed further in Chapter 4.

The analytical formulation of the instantaneous dc link current is complex. However, it can be simplified by the conservation of power relation, which equates the instantaneous dc power $V_{dc} \times I_{dc}$ with the sum of ac power in three phases, leading to the following simple relation:

$$i_{dc} = \frac{v_a i_a + v_b i_b + v_c i_c}{V_{dc}}$$

(3.40)

In the same manner, $i_{dc.rms} = \sqrt{I_{dc.avg}^2 + I_{ripple.rms}^2}$, which gives the ripple current

$$i_{ripple.rms} = \sqrt{I_{dc.rms}^2 - I_{dc.avg}^2}$$

(3.41)

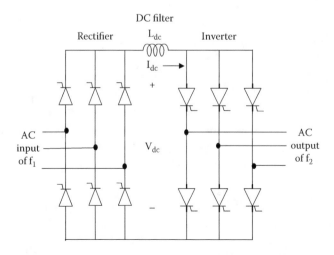

**FIGURE 3.14** Three-phase dc link frequency converter.

The dc link filter inductor $L_{dc}$ is designed to smooth out ripples caused by the source and load side converters while keeping the ripple current low. It should have current rating equal to $I_{dc}$ and low resistance to keep the $I^2R$ loss low. If $I_{ripple(p-p)}$ = permissible ripple current (peak to peak), $V_{ripple.in}$ and $V_{ripple.out}$ = ripple voltage on the line and load sides, respectively, and $\Delta t$ = time increment between the ripple current peaks, then

$$L_{dc} = (V_{ripple.in} - V_{ripple.out})\, \Delta t \div I_{ripple(p-p)} \tag{3.42}$$

A smaller ripple current requirement results in a larger inductor. The ripple current is generally worse at the upper edge of the constant-torque region of the motor drive operation. The size of inductor $L_{dc}$ is usually much larger in the load-commutated inverter than in the high-frequency PWM inverter. In construction, the $L_{dc}$ is usually an iron-core inductor with an air gap to store energy and prevent magnetic saturation.

### Example 3.9

Determine the dc link voltage for a variable frequency drive (VFD) to drive a three-phase, 460-V induction motor.

#### SOLUTION

Using Equation (3.35),

$$V_{LL.rms} = \sqrt{6} \times V_{dc} \div \pi$$

from which we obtain
$$V_{dc} = \pi \times 460 \div \sqrt{6} = 590 \text{ V}$$

### 3.4.2 Single-Phase Cycloconverter

The circuit shown in Figure 3.15a is similar to the three-phase, three-pulse, half-wave, phase-controlled rectifier fed from ac source voltage $V_{rms}$ of frequency $f_1$. Its dc output voltage, as analyzed in Section 3.1.1, is given by Equation (3.1b), that is, $V_{dc0} = 0.90\, V_{rms} \cos\alpha$, where $\alpha$ = conduction delay angle. With fixed $V_{rms}$, the ratio $V_{d0}/V_{rms} = 0.90 \cos\alpha$ changes with the delay angle $\alpha$. It becomes zero at $\alpha = 90°$ and negative for $\alpha > 90°$. Such variations in $V_{dc}$ polarity can be used to change frequency by the conduction delay angle control as follows:

| Conduction delay angle $\alpha°$ | 0 | 30 | 60 | 90 | 120 | 150 | 180 |
|---|---|---|---|---|---|---|---|
| $V_{dc}/V_{rms}$ ratio | 0.90 | 0.78 | 0.45 | 0.00 | −0.45 | −0.78 | −0.90 |

As seen in this table, if the firing delay angle is controlled over a half cycle from 0 to 180°, the $V_{dc}$ as the multiple of $V_{rms}$ varies in discrete steps from a maximum

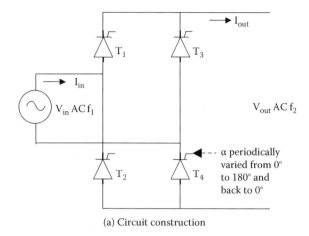

(a) Circuit construction

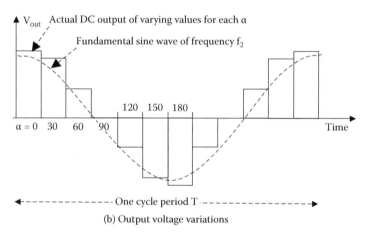

(b) Output voltage variations

**FIGURE 3.15**    Single-phase cycloconverter.

value of 0.90 to 0 to −0.90. Such variations are plotted in Figure 3.15b, which represents one cycle of the ac output made of discrete steps of various $V_{dc}$ values. In such cycling of firing angle $\alpha$, the $V_{dc}$ at any given instant depends on $\alpha$ in force at that instant. If the variations in $\alpha$ continue in small time intervals with a repetition frequency $f_2$, then the converter output is also ac of frequency $f_2$. The superimposed harmonics can be made small by modulating $\alpha$ in small steps.

Notice that we have used the single-phase rectifier of Section 3.1.1 to rectify ac by cycling the firing delay angle $\alpha$ between 0° and 180° to obtain the desired frequency output. This gives an inherent limitation in the cycloconverter operation; its output frequently can only be lower than the input frequency, with the high end of the output frequency about a third of the input line frequency in practical designs. The firing delay angle $\alpha$ varying over a wide range makes it a harmonic-rich converter, requiring a heavy filter.

### 3.4.3  Three-Phase Cycloconverter

The high harmonic content in the single-phase cycloconverter can be reduced by using the three-phase cycloconverter circuit shown in Figure 3.16. Each output phase consists of a single-phase cycloconverter shown in Figure 3.15. The conduction delay angle α in each rectifier and inverter in each phase is cyclically controlled to yield a low-frequency sinusoidal output voltage. Again, due to the harmonic limitations, the maximum output frequency is limited to about a third of the input line frequency. For this reason, the three-phase cycloconverter is generally used with a low-speed, high-power induction motor or synchronous motor typically used for ship propulsion.

The same thyristor commutating techniques are used in the cycloconverter as in the ac-dc rectifier. Some high-power cycloconverters use six thyristors for the positive-wave voltages and another six thyristors for the negative-wave voltage in each phase, thus using a total of 36 thyristors for the three-phase design, many more thyristors than the dc link frequency converter. The conduction angle cycling scheme controls the ac output voltage, and the firing frequency controls the output frequency. Both the α-cycling and frequency controls are complex. The cycloconverter has been used in the past but was generally out of favor for many years due

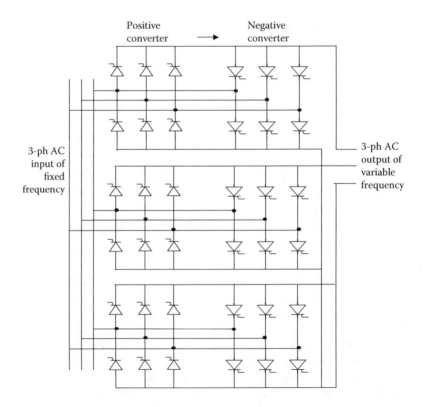

**FIGURE 3.16**  Three-phase cycloconverter.

to complex controls and many more switching devices, which makes it more expensive, less reliable, and less efficient. The new power electronics devices and new cycloconverter designs, however, are gradually bringing this old technology into a new light.

### Example 3.10

A 3-phase, 5000-hp, 12-pole, 60-Hz synchronous motor uses a cycloconverter to vary the speed. Determine the maximum speed it can be operated to limit the harmonics to the generally acceptable level.

#### SOLUTION

The maximum output frequency of the cycloconverter must be limited to a third of the input frequency, that is, $60/3 = 20$ Hz in 60-Hz drives. At this frequency, the 12-pole motor can run at the maximum speed of $120 \times 20 \div 12 = 200$ rpm.

## 3.5   THYRISTOR TURNOFF (COMMUTATION) CIRCUITS

The thyristor can be turned off (commutated) only by reducing the current below the holding current $I_H$ and applying a reverse-bias voltage for a certain minimum duration of time called the turnoff time $t_{off}$. Then only it recovers the forward blocking characteristic for continuing operation. A variety of commutating circuits can be designed for this purpose as discussed next.

### 3.5.1   LINE COMMUTATION

In the line frequency converter, the thyristor current is naturally turned off when the thyristor is reverse biased periodically under a sinusoidal voltage. For example, in a simple circuit shown in Figure 3.17a, the thyristor current naturally comes to zero every half cycle, when it is turned off without an additional commutation circuit. When there are multiple thyristors, as in the full-wave rectifier, the thyristor current is naturally commutated, and the device turns off when the next thyristor in the sequence is fired on. This is not true in the switch-mode converter (SMC), in which the thyristor current needs to be forcefully turned off using an additional commutation circuit.

### 3.5.2   FORCED (CAPACITOR) COMMUTATION

In the SMC, a forced commutating circuit shown in Figure 3.17b is needed to turn off the thyristor. It uses the capacitor energy to send the countercurrent to bring thyristor current to zero. The positive and negative polarities are those of the precharge voltage on the capacitor before the closure of the auxiliary switch to turn off the main thyristor. The inductor controls the $di/dt$ stress on the switch, and the diode provides the freewheeling path for the current when needed.

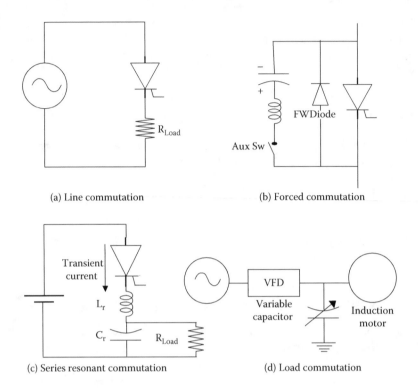

(a) Line commutation

(b) Forced commutation

(c) Series resonant commutation

(d) Load commutation

**FIGURE 3.17** Thyristor turnoff (commutation) circuits.

### 3.5.3 RESONANT COMMUTATION

The resonant converter uses additional inductance $L_r$ and capacitor $C_r$ to cause the converter circuit to resonate locally at the switch as shown in Figure 3.17c. Such a resonant circuit results in transient dc current and voltage oscillating between zero and peak values, and the switching is done at the natural zero of current or voltage. The resonant converter is typically a full-wave converter that includes inductor and capacitor as the resonant elements.

### 3.5.4 LOAD COMMUTATION

For an inductive load, static or dynamic, as shown in Figure 3.17d with induction motor load, the thyristors in the inverter require forced commutation. However, with a parallel capacitor as shown, a resonance is created with the load circuit, and the load current commutation is possible. The inverter frequency should be slightly above the resonance frequency, so that the effective load power factor is leading. The capacitor should be variable to match with the load parameters for satisfactory commutation. This is especially true with the induction motor with variable-frequency drives, for which the motor frequency can vary over a wide range, and the leading power factor can be maintained at all frequencies only with a variable capacitor. The synchronous motor is naturally suitable for the load commutation without the capacitor bank, as the motor itself can be operated at a leading power factor with overexcited rotor field coil. For that reason, the synchronous motor is better suited to work with a VFD.

### 3.5.5 ZERO-CURRENT SWITCHING AND ZERO-VOLTAGE SWITCHING

Some high-power converters with high switching frequency employ zero-current switching (ZCS) or zero-voltage switching (ZVS). The switching is done when the voltage or current is passing through natural zero on a sinusoidal cycle, so that the current or voltage is not abruptly interrupted from a high value to zero in a very short time (in microseconds or even less). Thus, the ZVS and ZCS converters eliminate high $di/dt$ and $dv/dt$ stresses, giving a *soft operation*. They result in higher efficiency, smaller size, lighter weight, and better dynamic response.

## 3.6 OTHER POWER ELECTRONICS APPLICATIONS

### 3.6.1 UNINTERRUPTIBLE POWER SUPPLY

Certain essential loads need uninterruptible power supply (UPS) to continue critical operations even when the main power source is down following a system fault. The UPS comes in two types:

*Type 1*: Provides critical power until a backup generator comes on line in a short time (in minutes). This type has low kilowatt-hour ratings.
*Type 2*: Provides specified power for a specified duration (in hours). This type has large kilowatt-hour ratings (e.g., 10 kW for 1 hour).

In either type, the required backup energy is stored in a battery or in a flywheel in some newly developed UPS units. The stored energy is discharged when needed. The system schematic is shown in Figure 3.18. In normal operation, the line frequency power is rectified to charge the battery, first at full charge rate and then tapered down to the trickle charge rate. In case of an ac line outage, the automatic bus transfer (ABT) switch disconnects the battery from the rectifier and connects it to the inverter. The battery power is then inverted to the line frequency ac and supplied to the critical loads via the same distribution cables. The harmonics in the output power of the rectifier and the inverter are filtered to improve the quality of power.

### 3.6.2 STATIC KVAR CONTROL

As we recall from electrical machines theory, the unloaded synchronous motor with an overexcited rotor field draws the armature current at the leading power factor. Such a capacitor-type machine—called the synchronous condenser—is often used in a large system for power factor improvement. The synchronous condenser can provide variable kVAR to match the system need. Where the vibration and noise coming from a rotating machine are objectionable, large static capacitors are used with a scheme to switch some capacitors on and off. That varies the capacitor kVARs with changing

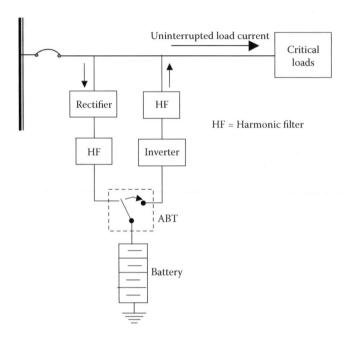

**FIGURE 3.18** Uninterruptible power supply functional diagram.

load power factor to keep the main line power factor near unity at all times. The power electronics kVAR compensation shown in Figure 3.19 is often used to provide variable kVAR to the load. A fixed-base capacitor is connected to provide a fixed kVAR, and the variable kVAR is obtained by the thyristors with variable firing delay angle α necessary to maintain a unity power factor at the load terminals. The capacitors with the thyristors can even be switched off if both thyristors in each phase were not fired at all. A small inductor (not shown) is placed in series with each thyristor or else damaging in-rush current may result when the thyristor is switched directly across the fixed capacitor. The inductor slows the *di/dt* rate in the thyristor to its design limit. If the load power factor becomes leading, it can be corrected to unity by introducing a variable inductor, shown in the left-hand side of Figure 3.19. Such static kVAR control has infinitely fine resolution on kVAR compensation, as opposed to discrete resolution with capacitors switched on and off. Therefore, it is increasingly used in modern power systems.

### 3.6.3 STATIC SWITCH AND RELAY

The power electronics device can be used to turn on or off high power without using mechanical contacts, which may spark and wear out after a certain number of operations. The static switch or static relay—also known as the solid-state switch or solid-state relay (SSR)—has no moving parts; hence, it has no wear-out failure mode and lasts billions of operations over decades. It is made of back-to-back thyristors or a triac in ac power circuits as shown in Figure 3.20. The SSR finds increasing applications for switching a variety of loads, such as motors, transformers, lighting, heating,

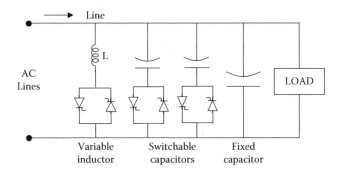

**FIGURE 3.19** Static variable kVAR control.

and so on. It can control large power by a small signal from the control circuit. The light-emitting diode with an optical coupler can provide electrical isolation for safety. The SSR is usually designed for single-pole applications with triac normally off (i.e., normally open contacts).

## 3.7 COMMON CONVERTER TERMS

Widely used terms in discussing power electronics converters are summarized below:

*Switch-mode converter (SMC):* Converter in which the power electronics devices are used as switches (as opposed to amplifiers).

*PWM converter:* Converter in which the output voltage or frequency is controlled by adjusting the pulse width of the gate signal (i.e., on and off time). An SMC can also be a PWM converter.

*Voltage source inverter (VSI):* It maintains a constant voltage on the inverter input side by using a large capacitor. If the dc side capacitor $C_{dc}$ in Figure 3.12 were made infinitely large, it would become a VSI.

*Current source inverter (CSI):* It maintains a constant current on the inverter input side by using a large inductor. If the dc side inductor $L_{dc}$ in Figure 3.13 were made infinitely large, it would become a CSI.

*Commutation:* Refers to turning off one thyristor and shifting the current to another. It can be natural or forced by additional circuits.

*Flyback converter:* The dc-dc converter in which the stored magnetic energy in one inductor coil jumps to another inductor coil—flying from the source side to the load side on the back—when the switch opens. The magnetically coupled flyback inductor coils essentially form an air-core transformer with turn ratio $N_{in}/N_{out}$ that can be less than or greater than one to buck or boost the voltage as required in the design.

*Forward converter:* The dc-dc buck converter with a transformer on the load side (forward in the power flow direction). The transformer turn ratio $N_{in}/N_{out}$ can be less than or greater than one to buck or boost the voltage as required in the design.

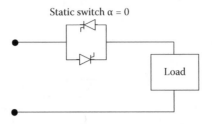

**FIGURE 3.20**    Static (solid-state) switch and relay.

## 3.8   NOTES ON CONVERTER DESIGN

*Fuse protection:* Since the power electronics device has a very small thermal capacity, it must be protected from heavy fault current. This is commonly done in thyristor circuits by using a fast-acting fuse. The thyristor has two types of current ratings: (a) the maximum continuous rms current and (b) the nonrepetitive peak current for half cycle. It also has a subcycle $I^2t$ rating for time duration between 1.5 ms and 8.35 or 10 ms (half cycle of 60 or 50 Hz). As a rule, both the peak current and $I^2t$ ratings of the fuse must be lower than those of the device. A similar protection principle applies to the diode as well. However, since the fast fuse is expensive, a new trend is to design a fuseless thyristor converter. For IGBTs, the fuse protection is not necessary because its gate-blocking capability is lost at high fault current. The same is true for MOSFETs.

*Converter with both sides inductive or capacitive:* In the phase-controlled thyristor converter, both sides can be inductive since the current is commutated from the outgoing to the incoming set of devices. However, the thyristor converter cannot operate if both sides are capacitive due to the capacitor charging in-rush current problem. In general, a PWM converter requires one side to be inductive and the other side capacitive.

*IGBT power rating and heat sink size:* The converter rating is determined by two limiting parameters of the power electronics devices, the peak current $I_{pk}$ and the maximum junction temperature $T_{jmax}$. The converter power losses are primarily made of the ohmic conduction loss and the switching loss. At high power, the losses and the junction temperature increase. At rated power, the steady-state junction temperature $T_j$ must be below $T_{jmax}$. If the cooling is improved, $T_j$ will decrease, and the steady-state power loading can be increased up to the limit of $I_{pk}$ and $T_{jmax}$. Or, with the same cooling (i.e., with the same heat sink), the short-time power rating can be higher (overload) within the peak current limit of the device for a certain time duration within the $T_{jmax}$ limit. Thus, improved cooling increases the converter steady-state rating up to the peak current limit. It also improves the overload capability for a certain time duration limited by the thermal inertia and $T_{jmax}$ of the device.

*PWM switching frequency:* The switching frequency in a motor drive design is selected to minimize the total power loss in the converter and the motor combined. The converter switching loss increases with the switching frequency.

However, the motor harmonic copper loss decreases at higher switching frequency due to lower harmonic current magnitudes, although the harmonic iron loss increases by a relatively small amount. The switching speed of the device also enters in design consideration. Faster devices have low switching loss and vice versa. The total losses in the converter and motor combined are plotted with the switching frequency to determine the optimum switching frequency for the minimum total loss. The following factors may also contribute in determining the optimum switching frequency:

- Switching losses in IGBTs and IGCTs (integrated gate-commutated thyristors) are lower than those in GTO (gate turnoff) thyristors. Therefore, the IGBTs and IGCTs generally operate at higher switching frequencies.
- Harmonic filter weight, volume, and cost become smaller at higher switching frequencies.
- Acoustic noise is another consideration in selecting the switching frequency. In general, a higher switching frequency results in lower acoustic noise.

*Electrolyte capacitor size:* The large dc link filter capacitor size in the VSI is determined by the harmonic current filtering requirement while keeping the temperature rise due to the equivalent series resistance (ESR) of the capacitor within the safe design limit; otherwise, the capacitor may explode. For the three-phase inverter in UPS with a battery, most of the harmonic currents must be filtered out by the capacitor to keep the battery life from degrading due to the harmonic current heating.

*IGBTs versus IGCTs for high-power applications:* IGBT PWM converters have been used in a few megawatt applications and are moving into the higher-power applications. The IGCT—introduced in 1996—is suited to even higher power than the IGBT. The two are similar in construction but have the following differences in operation:

- The IGBT is a voltage-controlled device, whereas the IGCT is a current-controlled device.
- Both devices are suitable for PWM frequency below 1 kHz.
- Both devices can be used in series-parallel combination for high voltage and high current as needed in high-power applications.
- Neither device needs a commutating circuit in operation.
- The IGCT has a lower conduction drop, giving a higher efficiency.

In general, the IGCT gives higher efficiency and needs fewer devices, so it is better suited for high-power applications.

*Double-sided PWM converter versus phase-controlled cycloconverter:* The phase-controlled cycloconverter has been used in tens of megawatt four-quadrant motor drives for icebreaker ships, steel-rolling mills, and mining applications. However, the recent trend is to use a double-sided PWM converter because the cycloconverter has the following disadvantages: (a) the current commutation often fails if the line voltage dips; (b) it generates complex harmonic currents on the line and load sides, which are difficult to filter and may cause excessive losses and magnetic saturation in the motor and line side transformer; and (c) it results in a poor line side displacement power factor.

The cycloconverter, however, being a direct frequency converter without a dc link, does not need an intermediate energy storage element.

*EMI in PWM converter:* Modern PWM converters need many control signal circuits in close proximity with the power circuit. This creates a difficult EMI problem, particularly at high switching frequencies. Analyzing and solving the EMI problem is often difficult as it does not lend itself to a clean analysis. The common-mode EMI generated by an IGBT snubber circuit is particularly difficult to filter. The following design practices generally control the EMI:

- Twisted and shielded signal wires with shield grounded to a low-resistance plane
- Low leakage inductance in the power cables
- Greater physical distance between the power and signal cables
- Using differential and common-mode EMI filters
- Lower switching frequency
- Grounding the metal frame of the transformer, converter, and the motor with a low-resistance ground wire

*VSI versus CSI converter for main frequency power:* The VSI is a mature and less-expensive design, available in 6-pulse, 12-pulse, and 18-pulsle configurations. It provides voltage to the motor in a manner typical of the normal power lines, except at a variable frequency and variable voltage. The CSI, on the other hand, acts as a constant current source to the motor; hence, it is not a popular converter in the power industry. One shortcoming of the CSI is the induction motor crawling at low speed (1/5 of synchronous speed) due to the magnetic phase belt harmonics in space, which rotates at main synchronous speed divided by the space harmonic order. For example, the fifth space harmonic in a two-pole motor flux rotates at $3600 \div 5 = 720$ rpm, causing the motor to run around that speed even at 60 Hz. Since this can be a problem, the range of speed control with CSI is not as large as other drives. The large dc filter required in the dc link can be bulky, often as large as the motor. The advantage of the CSI is that it is a bidirectional converter, which can be used in the regenerative braking scheme to convert the kinetic energy of the load inertia into electrical power and feed back to the source. That can be simply accomplished by having the rectifier delay angle greater than 90°.

*VSI versus CSI converter for 10- to 20-kHz power:* The U.S. Navy often uses high-frequency power converters to meet low-noise standards. For this application, both the CSI and VSI can be considered with the trade-offs listed in Table 3.3.

Power electronics is a wide and developing subject. The intent of this chapter has been to present some basic converters used in electrical power systems. Further details can be obtained from books dedicated to power electronics.

**TABLE 3.3**

**Performance Trade-Off between VSI and CSI for 10 kHz-20 kHz Power**

| Performance Parameter | Voltage Source Inverter | Current Source Inverter |
|---|---|---|
| Circuit configuration | Series resonant | Parallel resonant |
| Inverter frequency | Tracks the load resonance frequency | Tracks the load resonance frequency |
| Load power factor required for soft switching to eliminate switching losses | Lagging | Leading |
| Overall comparison | Lower cost, higher efficiency, and higher frequency possible | Higher cost, lower efficiency, and frequency limitation |

## PROBLEMS

*Problem 3.1:* A single-phase, full-wave rectifier shown in Figure 3.1 is built using a thyristor. Determine the average and rms values of the current in each thyristor when the firing delay angle $\alpha = 0$ and the input voltage is (a) 240-V, 50-Hz ac or (b) 120-V, 60-Hz ac.

*Problem 3.2:* A single-phase, full-wave bridge rectifier using phase-controlled thyristors with a large $L_{dc}$ converts 50-Hz, 240-V rms to dc for a dc side load resistance of 12 $\Omega$. For the firing delay angle $\alpha = 30°$, determine the dc side voltage and the power absorbed by the load, ignoring the cable inductances and ripples.

*Problem 3.3:* The rectifier of Example 3.2 with the ac side cable inductance of 3 mH and the dc side cable resistance of 1.0 $\Omega$ is charging a battery at a 10-A charge rate instead of powering the 10-$\Omega$ resistance. Determine the voltage available at the battery terminal.

*Problem 3.4:* A half-wave diode rectifier draws power from a single-phase, 240-V, 50-Hz source and delivers to a 3-kW load on the dc side at constant voltage $V_{dc}$ with a large filter capacitor as shown in Figure P3.4. The ac side harmonics to fundamental current ratios $I_h/I_1$ in percentage are 80, 40, 10, 6, 4, 3, 1, and 0.5% for the harmonic orders $h = 3, 5, 7, 9, 11, 13, 15,$ and 17, respectively. Determine the rms ripple current in the filter capacitor. Ignore all losses.

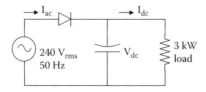

*Problem 3.5:* A 460-V, three-phase, full-wave, six-pulse rectifier shown in Figure 3.4 is built with diodes and a large inductor. Determine the average and rms values of the current in each diode if the output load current is (a) 50-A dc and (b) 135-A dc.

*Problem 3.6:* A three-phase, six-pulse thyristor rectifier with a large dc side filter inductor delivers 200-kW power at 240-V dc and draws 60-Hz power from 220 $V_{LL}$ via a cable with leakage inductance of 30 μH. Determine the firing delay angle of the thyristor and the commutation angle.

*Problem 3.7:* A three-phase, six-pulse VSI is fed with 480-V dc. Determine the ac side rms fundamental line-to-line voltage and the first four harmonic voltages.

*Problem 3.8:* A three-phase, six-pulse CSI is fed from a 120-V dc bus with dc link current of 60 A. The ac side load has 0.90 power factor lagging. Determine the ac side rms (a) phase and line currents and (b) phase and line voltages.

*Problem 3.9:* Determine the dc link voltage for a VFD to drive a three-phase, 6600-V induction motor for propulsion.

*Problem 3.10:* A three-phase, 1000-hp, eight-pole, 50-Hz synchronous motor uses a cycloconverter to vary the speed. Determine the maximum speed it can be operated to limit the harmonics to the generally acceptable level.

## QUESTIONS

*Question 3.1:* In Table Q3.1, fill in the key performance features of controllable power electronics switches covered in this chapter.

---

### TABLE Q3.1
### Key Performance Features of Controllable Power Electronics Switches

| Switch Name | SCR | BJT | MOSFET | GTO | IGBT |
|---|---|---|---|---|---|
| Power terminals name | | | | | |
| Control terminal name | | | | | |
| Latching or nonlatching | | | | | |
| Saturating or nonsaturating | | | | | |
| Control signal, voltage or current | | | | | |
| Control signal, pulse or continuous | | | | | |
| Relative power-handling capability | | | | | |
| Relative switching frequency | | | | | |

*SCR-silicon-controlled rectifier (thyristor).*

---

*Question 3.2:* What is the difference between the terms *ripples* and *harmonics*?

*Question 3.3:* Explain the difference between the firing delay angle α and the commutation delay angle *u*.

*Question 3.4:* Identify the source of the load power factor, displacement power factor, and harmonic power factor.

*Question 3.5:* What is the principal difference between CSI and VSI?

*Question 3.6:* Explain the difference between the snubber circuit and the commutation circuit.

*Question 3.7:* Decades ago, the large UPS design in ac systems used an ac electrical machine coupled to a dc machine connecting to a battery to convert the ac into dc and vice versa. Draw such a electromechanical UPS schematic incorporating the ac bus and the user loads. Today, the electrical machines have been replaced by bidirectional power electronics converters. Discuss the pros and cons of using the electrical machines versus power electronics in UPS.

## FURTHER READING

Ahmed, A. 1999. *Power Electronics for Technology.* Upper Saddle River, NJ: Prentice Hall.

Ang, S., and A. Oliva. 2010. *Power Switching Converters.* Boca Raton, FL: CRC Press/Taylor & Francis.

Bose, B. 1996. *Power Electronics and Variable Frequency Drives.* New York: IEEE Press.

Hart, D. 2011. *Power Electronics.* New York: McGraw-Hill.

Luo, F.L., and H. Ye. 2010. *Power Electronics Advanced Conversion Technologies.* Boca Raton, FL: CRC Press/Taylor & Francis.

Mohan, N., T.M. Undeland, and W. Robbins. 2003. *Power Electronics Converters, Applications, and Design.* New York: Wiley/IEEE Press.

# 4 Variable-Frequency Drives

The power electronics motor drive comes in two major types: (a) the servo motor drive to control the precise speed and position of the rotor and (b) the motor drive for varying motor speed in response to a desired change in the load output, such as the air ventilation, fluid pumping rate, or the propulsion speed of the ship. The response time and precision in position are important in the servo drive but not in the variable-speed motor drive. Our interest in this chapter primarily focuses on the variable-speed motor drives.

The variable-frequency drives (VFDs) for ac motors were initially developed in the 1970s for oil refineries, where numerous motors are used to pump oil all year around. Although costly at the time, the significant energy saving at the end of every month paid back the drive cost within a few years. Since then, for the rest of the 20- to 30-year life of the drive, the VFDs added to corporate profits. Most ships now use VFDs with pumps for water, oil, refrigeration, and with fans for air ventilation. Most cruise ships use VFDs for propulsion motors as well.

The VFD primarily controls the speed, torque, acceleration, or deceleration of an ac motor. Unlike the dual-pole—hence dual-speed—motor, the VFD allows continuous resolution of the speed control within its operating range. The benefit of the infinitely variable speed motor is to increase the energy efficiency—doing the same work at a minimum expenditure of kilowatt-hours—or to optimize the product quality or the production process. The following are a few examples in addition to those in ships:

- The pump supplying water in a high-rise building may run at low speed during nighttime and high speed in the afternoon to provide the required flow rate while maintaining system pressure.
- Lathes or machine tools can run small-diameter jobs at high speed and large-diameter jobs at low speed to increase the production rate.
- A printing press can operate at a speed that optimizes the print quality depending on the type of paper and ink.
- Smoothly controlled acceleration can be provided in a spinning mill to avoid yarn breakage.

The induction motor is the most economical and reliable motor widely used in industry. However, it runs essentially at constant speed with a few percent slip below synchronous speed. The slip—not the speed—is proportional to the load torque. This is a disadvantage in applications where the motor speed needs to change in response to the load requirement. The motor driving (a) an oil pump in a refinery or a cargo ship and (b) a large ventilator fan on a ship or in an industrial plant are just two examples. The ship propulsion motor also falls in this category. The fluid-pumping system can use a throttle valve to reduce the flow rate while the motor runs essentially at a constant speed. Turning the valve partially closed adds resistance to the fluid flow, which lowers the energy efficiency of the system. Reducing the speed to reduce the fluid flow rate is more energy efficient. Oil refineries in the past sometimes used gears or dual-speed motors to change the pump speed. However, the backlash and resulting vibrations following a sudden speed change often broke the shaft and reduced equipment life.

An energy-efficient solution is to reduce the fluid flow rate by reducing the motor speed. We recall that the induction and synchronous motor speeds depend directly on the frequency. Therefore, the motor speed can be changed by varying the frequency of the motor input in response to the flow rate requirement. If a motor larger than 500 hp is running at least 8 hours every day—about 2500 hours per year—the extra investment in the drive cost is generally paid back in a few years by the energy savings every month. The energy-saving potential of the variable-speed drive (VSD) in a fluid-pumping system is first analyzed next.

## 4.1   PUMP PERFORMANCE CHARACTERISTICS

We consider a fluid-pumping system that moves fluid through plumbing that has internal friction in the pipelines. The pump is the pressure source that moves the fluid, and the pipeline friction is the pressure load on the system. Stable system operation is achieved at the flow rate when the pressure rise in the pump is equal to the pressure drop in the pipelines. Figure 4.1 depicts a typical pump pressure versus fluid flow rate (heavy solid line). When there is no resistance to the pump,

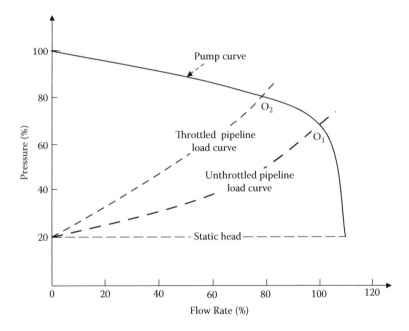

**FIGURE 4.1**   Pump pressure versus fluid flow rate with and without throttle in pipeline.

the flow is maximum and the pump pressure is minimum, which is just equal to the static head value. On the other extreme, if the pump outlet is blocked, the flow is zero, and the pump pressure is maximum (100%). On the load side, the pressure drop in the pipeline varies approximately as a square function of the fluid flow rate, that is

$$\text{Pipeline pressure load} = k \times (\text{Flow rate})^2 \qquad (4.1)$$

where $k$ is the proportionality constant that depends on the friction in the pipeline and throttle valve combined. With no throttle valve or valve fully open (unthrottled system), the load pressure versus flow rate is shown by the heavy dotted curve. The system will operate at point $O_1$ where the pump curve and the load curve intersect. If we throttle the valve, the load curve shifts upward (more pressure drop and less flow), and the operating point moves to $O_2$ and further to the left with increasing throttle. The power required to drive the pump is given by one of the following two equations:

$$\text{Pump motor output power} = (\text{Pressure}) \times (\text{Flow rate}) = k \times (\text{Flow rate})^3 \quad (4.2)$$

$$\text{Propeller or fan motor output power} = k \times \text{Speed}^3 \qquad (4.3)$$

Ignoring the internal losses in the pump and the motor for simplicity here, the motor power input equals the pump power output. The powers at operating points $O_1$ and $O_2$ in Figure 4.1—the products of pressure and flow rate at those points—translate to points $O_1$ and $O_2$ in Figure 4.2, respectively. From point $O_1$ (100% flow rate)

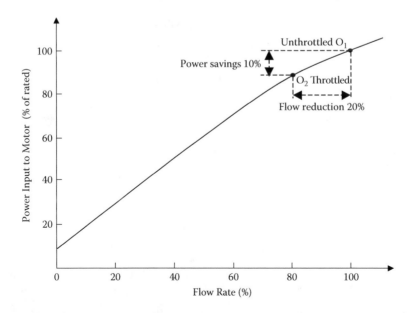

**FIGURE 4.2**   Motor power input versus fluid flow rate with throttle.

to point $O_2$ (throttled to 80% flow rate), we see that the power is reduced by 10% for 20% reduction in flow rate. Thus, the throttled system uses more energy per cubic meter of fluid pumped, which is an uneconomical use of power.

## 4.2   PUMP ENERGY SAVINGS WITH VFD

Now, consider the pump operation with variable-speed motor drive and no throttle valve in the system. In Figure 4.3, the top solid curve at 100% pump speed and dotted unthrottled load curve are the same as in Figure 4.1. At a reduced pump speed, say at 50%, the solid pump curve in Figure 4.3 shifts downward, but the load curve still remains the same—it is still the same pipe and still unthrottled. This results in the operating point moving from point $O_1$ to $O_2$. The power input to the motor is now shown by points $O_1$ and $O_2$ in Figure 4.4. It shows the power dropping to 30% (70% saving) for 50% reduction in the flow rate. This results in less energy required per cubic meter of the fluid pumped. This is expected since the pipeline pressure drop varies with the flow rate squared.

The power savings that can be realized by using a variable-speed motor at various flow rates is shown in Figure 4.5. It shows zero saving at 100% speed (obviously true), the maximum saving around 50% speed, and a small saving at lower speeds (because of some fixed losses regardless of the speed). The actual power savings will be somewhat lower than the chart indicates because of the internal power losses in the VFD, the motor, and the pump.

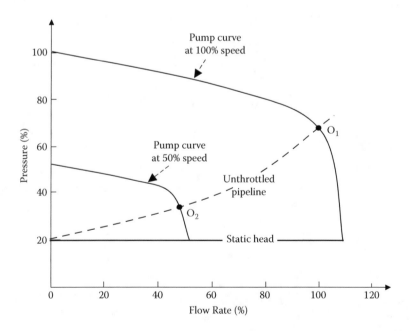

**FIGURE 4.3**   Pump pressure versus fluid flow rate at 100% and 50% motor speed without throttle.

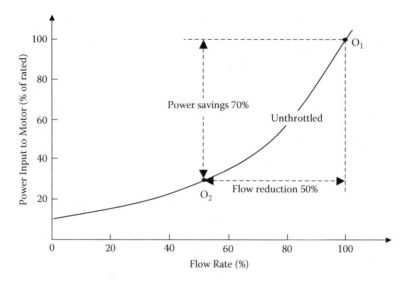

**FIGURE 4.4**  Motor power input versus fluid flow rate without throttle.

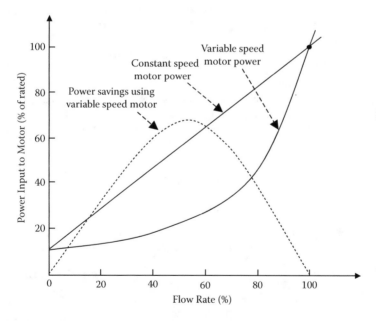

**FIGURE 4.5**  Motor power saving with VFD for fluid flow rate variation from 0% to 100%.

The speed of the pump using an induction motor can be varied by varying the frequency of the motor input. Figure 4.6 depicts the torque-speed characteristics (the humped curves) of the induction motor at various frequencies, where $f_4 < f_3 < f_2 < f_1$ (60 or 50 Hz). The unthrottled pump load torque varies as the speed squared, as plotted by the dotted curve. The operating points (shown by heavy dots at the intersection) shift to lower speeds as the frequency is lowered.

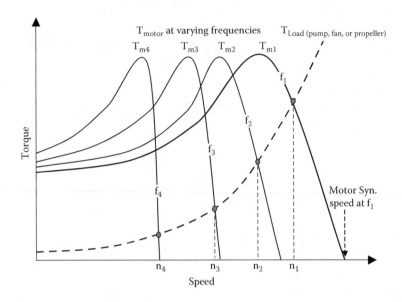

**FIGURE 4.6**  Induction motor torque versus speed characteristics at various frequencies with pump, fan, or propeller load.

### Example 4.1

A water pump driven by a 1000-hp induction motor pumps 100,000 m³/day at the rated speed. If the motor were run at a slower speed using a VFD to reduce the flow rate to 80% over a longer time to pump the same quantity of water per day, determine the percentage saving in kilowatt-hour energy consumed per day.

#### SOLUTION

Using Equation (4.2), power at 80% flow rate = $0.80^3$ × old power.

Hours to pump the same water quantity = Old hours/0.80

∴ New energy consumption to pump the same quantity of water = $(0.80^3$ × Old power) × (Old hours/0.80) = $0.80^2$ × Old kWh = 0.64 × Old kWh per day

This gives the energy saving of 0.36 or 36% of the old energy consumption for pumping the same quantity of water per day. This clearly justifies investing in a VFD for saving energy.

### Example 4.2

An existing cruise ship *New Horizon* was designed to travel at 20 knots with a dedicated 6.6-kV, 46-MW$_e$ propulsion power plant consisting of generators, transformers, VFDs, and motors rated accordingly. A new cruise ship is being designed with essentially the same hull as *New Horizon*, except for a higher speed of 26 knots. Determine the rating of the new propulsion power plant and the recommended voltage level.

## SOLUTION

Using Equation (4.3) in ratio, we have

$$\frac{New\,power}{Old\,power} = \left(\frac{New\,speed}{Old\,speed}\right)^3$$

∴ New propulsion power plant rating = 46 × (26/20)³ = 101 MW

As for the new voltage level, we will discuss in Chapter 7 that

$$\frac{New\,voltage}{Old\,voltage} = \sqrt{\frac{New\,power}{Old\,power}}$$

which gives

$$New\,voltage = \sqrt{\frac{101}{46}} \times 6.6 = 9.78\,kV$$

which can be rounded to the nearest standard voltage of 11 kV.

## Example 4.3

A cargo ship travels 5000 nautical miles transatlantic at 25 knots speed, taking 200 hours and using 520,000 gallons of fuel oil. If the speed were reduced by 10%, determine the percentage change in total gallons of fuel consumption.

### SOLUTION

New speed = 0.9 × 25 knots, and New power = 0.9³ = 0.729 × old power

The fuel consumption rate per hour is proportional to power; hence, it will reduce to 0.729 × Old rate.

New journey time = Old journey time ÷ 0.9 = 200 × 1.111 = 222.2 hours

Gallons of fuel used = (0.729 × Old rate) × (1.111 × Old journey time) = 0.81 × Old gallons

∴ Reduction in fuel consumption = (1 − 0.81) = 0.19 or 19%

We note that gallons of fuel consumption = $k$ × Speed². In this example, the fuel consumption has decreased since the speed has decreased. At a higher speed, the total fuel consumption for the same journey will increase in square proportion of the speed. This is why high-speed ships are not only a design challenge but also consume more fuel for the same journey. On the benefit side, faster speed lowers the manpower and capital costs per journey since the ship can make more trips per year.

## 4.3   SHIPBOARD USE OF VFDS

New modern ships use VFDs widely, although some old ships still have to introduce them onboard. About 80–85% of the installed drives on ships are for auxiliary systems such as fans, pumps, compressors, and so on. Only about 15–20% of the drives are used for ship propulsion. However, the propulsion drives are much larger in ratings, making a significant contribution to energy efficiency in shipboard power utilization. Some common applications of VFD on ships are

- Deck equipment, such as underdeck cranes for pallet handling, gantry cranes for container handing, cargo winches, windlass, and constant-tension mooring winches
- Oil pumps and water pumps
- Compressor pumps in refrigerators and air conditioners
- Air ventilation fans
- Dynamic positioning of ships or floating platforms
- Propulsion motors (synchronous, induction, or other types)
- Compressor fans in the antiheeling system

The antiheeling system damps the rolling motion of a ship in rough seas. It uses two-wing ballast tanks on two sides of the ship that are connected to each other. When the ship begins to heel on one side, the antiheeling fans turn on, moving a large volume of air into the ballast tank, effectively pressurizing it. This causes the ballast water to flow through the connection into the other wing tank, providing a weight offset that damps the severe rolling. The fans are powered by VFDs that vary the fan speed according to the rolling rate and the heeling angle change.

## 4.4   VFD FOR MEDIUM-SIZE MOTOR

The widely used VFD configuration for most medium-size motors—other than large propulsion motors—is the dc-link frequency converter shown in Figure 4.7. It can use the current source inverter (CSI) or the voltage source inverter (VSI) design depending on the type of motor used for the overall system-level optimization. The VFD takes 60-Hz or 50-Hz power from the generator, rectifies it into dc, and then inverts into a lower-frequency, lower-voltage output, which is applied

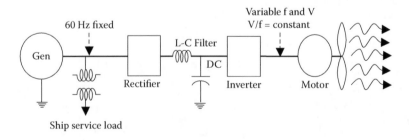

**FIGURE 4.7**   DC-link VFD schematic for ac motors.

to the induction or synchronous motor. In general, the high-frequency harmonics produced by the switch-mode rectifier and inverter are filtered by an L-C filter. However, the induction motor VFD using a CSI needs a large inductor, whereas the synchronous motor VFD using a VSI needs a large capacitor. In high-power propulsion motor VFDs, the design drivers are the torque and speed characteristics as they relate to the propeller efficiency, in addition to the size, weight, total system cost, and availability of the high-power devices. Beginning in late 1980s, high-power VFDs used high-voltage (5000-V) gate turnoff (GTO) thyristors until the advent of the insulated gate bipolar transistor (IGBT) along with the gate-commutated thyristors (GCTs), both of which are now gradually replacing the GTO thyristors.

Figure 4.8 is an interior view of a 300-kW induction motor drive, whereas Figure 4.9 shows an exterior view of VFDs in various kilowatt ratings. The VFD designs at present use various configurations shown in Figure 4.10. Their alternate names and key features are as follows:

*Synchroconverter or current source inverter (CSI):* This converter (Figure 4.10a) is typically used with a synchronous motor in ships other than icebreaker and dynamic positioning ships. As compared to the cycloconverter, its advantages are (a) much fewer components, hence it is simple and reliable; (b) lower volume and weight; (c) better power factor, around 0.85–0.90; (d) lower harmonics; and (e) improved overall system efficiency.

*PWM (pulse width modulated) converter or voltage source inverter (VSI):* It is often used with induction and permanent magnet motors and sometimes

**FIGURE 4.8**    Interior view of 300-kW IGBT VFD for induction motor.

**FIGURE 4.9**  Exterior views of VFDs in various kilowatt ratings. (With permission from Converteam, Inc.)

with a synchronous motor. The voltage source drive (Figure 4.10b) can be designed with IGBTs, GTO thyristors, or IGCTs. The PWM drive can be designed to operate at a power factor near unity over the entire range of the motor speed, compared to an 0.85 power factor for the CSI and 0.75 for the cycloconverter. The control voltage signal frequency $V_{control}$ is compared to the triangular waveform, as shown in Figure 3.10. It reflects low harmonics on the ac source side, and lower-order harmonics are absent on the load side. It is typically used for an azimuthing stern drive (ASD) with a substantial weight saving compared to the conventional design.

*Cycloconverter:* This direct ac-ac frequency converter (Figure 4.10c) is used mainly with the synchronous motor having a greater air gap as opposed to the induction motor. The cycloconverter typically uses many more power electronics devices, making it more expensive and less efficient. Also, it adds lag in the motor current due to the phase control modulation. Its typical power factor is around 0.75, which is at the low end of the 0.7–0.9 range that many ship classification societies allow. The cycloconverter is an old

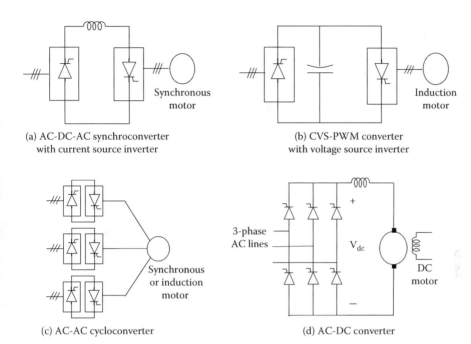

**FIGURE 4.10**   Alternative variable-speed drive topologies.

converter technology whose time had passed for a while but is now getting new attention for certain benefits. It works well in icebreakers and dynamic positioning ships requiring very high torque at low speed. However, its use is limited for output frequency up to 30–40% of the supply frequency.

*DC motor drive:* The dc motor speed is controlled simply by changing the motor terminal voltage while maintaining the constant field current to maintain constant air gap flux. Unlike the ac motor drive, the dc motor drive (Figure 4.10d) has a simple circuit configuration. It is discussed in further detail in Section 4.11.

## 4.5   CONSTANT *V/F* RATIO OPERATION

All VFDs maintain the output voltage-to-frequency (*V/f*) ratio constant at all speeds for the reason that follows. The motor phase voltage $V$, frequency $f$, and magnetic flux $\varphi$ are related by Equation (4.4):

$$V = 4.444fN\varphi_m \qquad \text{or} \qquad V/f = 4.444N\varphi_m \qquad (4.4)$$

where $N$ = number of turns per phase. If we apply the same voltage at reduced frequency, the magnetic flux would increase and saturate the magnetic core, significantly distorting the motor performance. The magnetic saturation can be avoided only by keeping $\varphi_m$ constant. Moreover, the motor torque is the product of the stator flux and the rotor current. The rated torque can be maintained at all speeds only by maintaining the flux constant at its rated value, which is done by keeping the $V/f$ ratio constant. That requires lowering the motor voltage in the same proportion as the frequency to avoid magnetic saturation due to high flux or lower-than-rated torque due to low flux.

Maintaining the $V/f$ ratio constant at all speeds requires controlling the rectifier firing delay angle in proportion to the desired speed. At very low speed, when the frequency and voltage are also low, the inductive reactance is negligible, and the voltage setting the air gap flux is $V - I_{stator}R_{stator}$. Therefore, for constant air gap flux, the $(V - I_{stator}R_{stator})/f$ ratio must be maintained constant. This requires a voltage boost of $I_{stator}R_{stator}$ to compensate for the voltage drop in the motor armature resistance. At low speeds, therefore, the motor terminal voltage must be boosted by $I \times R$ drop in the motor armature as shown in Figure 4.11.

The exact VFD design with a constant $V/f$ ratio is somewhat different for the induction motor than for the synchronous motor drive to optimize the system performance with the somewhat different nature of the two motors. In either case, controlling the motor terminal voltage requires controlling the rectifier firing delay angle in proportion to the desired speed as discussed next.

With a three-phase, six-pulse rectifier with variable firing angle to vary $V_{dc}$ and hence the ac output voltage of the VFD, Equation (3.16) gave (with all ac voltages fundamental rms line-to-line values),

$$V_{dc} = 1.35\ V_{LLline}\cos\alpha \qquad (4.5)$$

With a three-phase, six-pulse VSI, Equation (3.35) gave the inverter output voltage at the motor terminal:

$$V_{LLmotor} = 0.78\ V_{dc} = 0.78 \times 1.35\ V_{LLline}\cos\alpha = 1.05\ V_{LLline}\cos\alpha \qquad (4.6)$$

With a three-phase, six-pulse CSI, Equation (3.39) gave the inverter output voltage at the motor terminal:

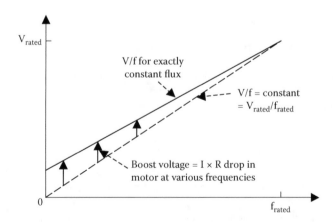

**FIGURE 4.11**   Voltage boost requirement in VFD at low speeds.

$$V_{LLmotor} = \frac{\sqrt{3}\,V_{dc}}{\sqrt{6}\cos\theta} = \frac{1}{\sqrt{2}\cos\theta}1.35V_{LLline}\cos\alpha = \frac{0.955}{\cos\theta}V_{LLline}\cos\alpha \qquad (4.7)$$

where $\cos\theta$ = motor power factor and $\alpha$ = rectifier firing delay angle. With a typical value of induction motor power factor of 0.9, Equations (4.6) and (4.7) give almost identical voltage at the motor terminals. It can further be shown that

$$\text{Incoming line pf with } CSI \;=\; 0.955\cos\alpha \;=\; 0.955\left(\frac{\text{Actual Speed}}{\text{Rated speed}}\right) \qquad (4.8)$$

And, Equation (4.6) gives the actual motor speed as follows:

$$\frac{\text{Actual Speed}}{\text{Rated Speed}} = \frac{f_{motor}}{f_{rated}} = \frac{V_{LLmotor}}{V_{LLline}} = 1.05\cos\alpha \qquad (4.9)$$

Thus, the motor speed is linearly related with $\cos\alpha$. When using the CSI with an induction motor, Equation (4.8) indicates that the incoming line power factor is good near the rated speed but degrades significantly at lower speeds due to the excessive delay in firing angle $\alpha$ that is necessary to reduce the voltage to a low value. In practice, this disadvantage can be overcome by controlling $V_{dc}$ not by $\alpha$, but by a step-down dc-dc buck converter.

## Example 4.4

A 100-hp, three-phase, 60-Hz, four-pole, 1750-rpm, 0.85-pf induction motor operates with a VFD connected to 460-V lines. The VFD design has CSI fed from a dc link. If the motor needs to run at 980 rpm, determine (a) the rectifier firing delay angle and (b) the VFD output voltage at the motor terminals.

### SOLUTION

Using Equation (4.9), we have

$$\frac{980}{1750} = 1.05\cos\alpha$$

from which we have $\cos\alpha = 0.533$ and $\alpha = 57.77°$.
    Then, using Equation (4.7),

$$V_{LLmotor} = \frac{0.995}{0.85} \times 460 \times 0.533 = 287.2 \ V$$

Since the *V/f* ratio of the motor input is kept constant at reduced speed, the air gap flux remains constant, and the torque is linearly proportional to the armature current, which varies with the rotor slip rpm. For blade-type loads (fan, pump, and propeller) driven by the induction motor with constant flux, we have, with different values of constant $k$ in different equations,

$$\text{Torque} = k \times \text{Current} = k \times \text{Slip rpm} \tag{4.10}$$

For blade-type loads, we know the following:

$$\text{Torque} = k \times \text{Speed}^2 = k \times \text{Frequency}^2 \tag{4.11}$$

At stable operation, Equations (4.10) and (4.11) give slip rpm = $k \times$ Frequencsy$^2$, from which

$$\text{Perunit slip} = \frac{\text{slip rpm}}{\text{synchronous speed}} = \frac{k \, x \, freqeuncy^2}{frequency} = k \, x \, frequency \tag{4.12}$$

Therefore, we can write the following:

For blade-type loads,

$$\frac{\text{pu slip at new frequency}}{\text{pu slip at old frequency}} = \frac{\text{new frequency}}{\text{old frequency}} \tag{4.13}$$

For constant-torque loads—slow-moving loads such as via conveyor belt, crane, and winches that must overcome a constant friction—this is not true. With constant flux at constant $V/f$ ratio, the rotor current must remain constant, hence the slip. Therefore, for constant-torque load below the rated frequency with constant flux, we have

$$\text{Slip rpm} = \text{Constant} \quad \text{and} \quad \frac{\text{hp at reduced frequency}}{\text{hp at rated frequency}} = \frac{\text{reduced frequency}}{\text{rated frequency}} \tag{4.14}$$

For a constant-torque load above the rated frequency, the motor is operated at the rated line voltage that produces lower flux, which in turn produces lower torque. The high speed and proportionately lower torque at high frequency results in constant horsepower. Therefore, we have

Above rated frequency,

$$\text{HP} = \text{Constant}, \quad \frac{\text{torque at higher frequency}}{\text{torque at rated frequency}} = \frac{\text{rated frequency}}{\text{higher frequency}} \tag{4.15a}$$

and

$$\frac{\text{slip rpm at higher frequency}}{\text{slip rpm at rated frequency}} = \frac{\text{higher frequency}}{\text{rated frequency}} \tag{4.15b}$$

Figure 4.12 depicts the motor operation from 0% to 200% of the synchronous speed as the frequency varies from 0 to 2 times the rated value. The motor operates at constant torque at subsynchronous speeds and at constant horsepower at supersynchronous speeds.

## Example 4.5

A 500-hp, 460-V, three-phase, 60-Hz, four-pole, 1728-rpm induction motor is driving a large ventilating fan. When it is running from VFD output of 230 V, 30 Hz with constant *V/f* ratio, determine (a) approximate speed and (b) exact speed.

### SOLUTION

At 60-Hz, a four-pole ac motor has synchronous speed = 120 × 60/4 = 1800 rpm. The motor has the rated speed of 1728 rpm, at which the rotor slip = 1800 − 1728 = 72 rpm, which is 4% of the synchronous speed. The induction motor speed depends on the frequency, as does the running speed.

#### APPROXIMATE ESTIMATE

At 30-Hz operation of a 60-Hz motor,

Approximate speed = (30/60) × Rated speed = ½ × 1728 = 864 rpm

The synchronous speed at 30 Hz is 900 rpm, so the slip = 900 − 864 = 36 rpm, which is 4% of the new synchronous speed, the same percentage slip as that at 60 Hz.

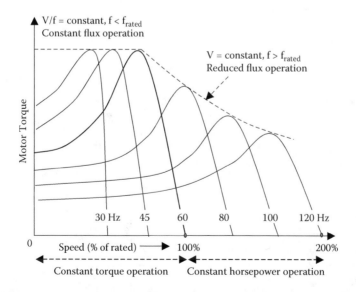

**FIGURE 4.12**   Induction motor torques at different frequencies below and above 60 Hz.

### EXACT SOLUTION

The fan load torque varies with speed$^2$. With speed approximately half, the load torque will be close to a quarter of the rated torque.

Since the $V/f$ ratio of the motor input is kept constant at reduced speed, its air gap flux remains constant. To produce a fourth of the torque at the same flux, the rotor current must be a fourth of the rated value. At constant flux, the rotor current linearly varies with the rotor slip rpm, which will become $\frac{1}{4} \times 72 = 18$ rpm. This is 2% of the new synchronous speed, as we also would have derived from Equation (4.13).

$\therefore$ At 30 Hz, exact rotor speed = 900 − 18 = 882 rpm

The exact speed (882 rpm) is about 2% higher than the approximate speed (864 rpm). This difference may matter in some applications and may not matter in others, such as in ventilating fans for which only the approximate airflow rate within a few percent matters for comfort.

## Example 4.6

A VFD drives a three-phase, 60-Hz, 100-hp, 1750-rpm induction motor. Determine (a) power, torque, and full-load speed when operating at 20 Hz; and (b) power, torque, and full-load speed when operating at 80 Hz.

### SOLUTION

(a) We use the following relation that is valid for any machine delivering mechanical power (motor, diesel engine, turbine, etc.):

$$Horsepower = \frac{Torque_{lb.ft} \times Speed\ in\ rpm}{5252}$$

Below the rated frequency, the motor operates at constant $V/f$ ratio, that is, at constant flux; hence, it will produce constant torque at the rated current. So,

Torque at 20Hz = Torque at 60 Hz = 100 hp × 5252 ÷ 1750 rpm = 300 lb-ft.

The 60-Hz, 1750-rpm motor must be a four-pole motor with a synchronous speed of 1800 rpm and a rotor slip speed = (1800 − 1750) = 50 rpm. At 20 Hz, the motor will run at the same 50-rpm slip to produce the rated current to produce a rated torque at the rated flux. So, the speed at 20 Hz is (600 rpm synchronous speed − 50 rpm slip) = 550 rpm.

$\therefore$ 20-Hz hp = 550 × 300 ÷ 5252 = 31.42 hp

This reduced horsepower is approximately in the frequency ratio, that is, (20/60) × 100 = 33.33 hp.

(b) Above the rated frequency, the motor operates at constant horsepower since the flux is lower at the rated voltage and the current is kept at the rated value to limit heating. At 80 Hz, the slip rpm can increase to (80/60) × 50 = 66.67 rpm. With the new synchronous speed of 2400 rpm, the rotor speed = 2400 − 66.67 = 2333.3 rpm, and torque = 100 × 5252/2333.3 = 225 lb-ft, which is the same as the 300 × 60/80 = 225 lb-ft that also could have been derived from Equation (4.15a).

## 4.6 COMMUTATION AND CONTROL METHODS

*Forced commutated VFD:* With a lagging power factor induction motor load, the inverter thyristors need forced commutation with capacitors and diodes. The leakage inductance of the motor plays a significant role in the commutating circuit design for the inverters. Therefore, such a VFD is designed for use with only a specific motor. On the positive side, it can be used for regenerative braking without additional circuits.

*Load-commutated VFD:* The load-commutated thyristors in a CSI inverter are widely used with large multimegawatt synchronous motors for pumps, compressors, rolling mills, and ship propulsion. Figure 4.13 is such a VFD with a six-pulse rectifier and a six-pulse inverter. For high-power drives, such as for electric propulsion of ships, 12-pulse converters are used to lower the harmonic content. The VFD is load commutated by the leading power factor of the synchronous motor that can be achieved by overexciting the rotor field to obtain higher back emf induced in the motor armature. The frequency and the phases of the three phase currents are synchronized to the rotor position, and the current commutation in the inverter thyristor is accomplished using the motor back emf. Slow-speed thyristors can be used on both sides. This makes the inverter simple, cost effective, energy efficient, and reliable compared to the CSI with an induction motor.

At low speeds, however, the back voltage is not sufficient to achieve the force commutation, and the commutation is provided using the thyristors in the rectifier operating at large $\alpha$ to make $I_{dc} = 0$, thus turning off the inverter thyristors. Also, since large synchronous motors have a few percent higher efficiency than the comparable size induction motors, the load-commutated

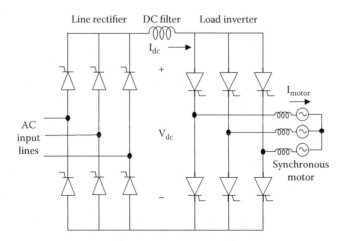

**FIGURE 4.13** Load-commutated inverter drive for synchronous motor.

VFD system with a synchronous motor offers a few percent efficiency advantage, which can make a significant positive difference in energy consumption in multimegawatt systems running almost continuously over the year.

*VFD feeding power back to the source:* When the VFD is used to feed power back to fixed-frequency grid lines, such as in wind or photovoltaic energy farms, or while regenerative braking of a metro train, certain grid interface issues must be addressed. At the grid interface, the power flow direction and the magnitude depend on the voltage magnitude and the phase relation of the inverter voltage with respect to the grid voltage. The grid voltage being fixed, the inverter voltage must be controlled both in magnitude and in phase to feed power to the grid when available and to draw from the grid when needed. If the inverter is already included in the system for frequency conversion, the magnitude and phase control of the inverter output voltage are achieved with the same inverter with no additional hardware cost. Such controls are accomplished as follows:

*Voltage control:* Since the magnitude of ac voltage output from the inverter is proportional to the dc voltage input from the rectifier, the motor voltage can be controlled by operating the inverter with a variable dc link voltage from a phase-controlled rectifier. However, there are two problems associated with this scheme. At reduced output voltage, this method gives a poor distortion power factor and high harmonic content that require heavy filtering before feeding to the inverter. Also, in circuits deriving the load-commutating current from the commutating capacitor voltage from the dc link, the commutating capability decreases when the output voltage is reduced. This could lead to an operational difficulty when the dc link voltage varies over a wide range, such as in a motor drive controlling the speed in a ratio exceeding 4 to 1. In variable-speed wind power generation, such commutation difficulty is unlikely as the speed varies over a narrow range.

*Frequency control:* The output frequency of the inverter solely depends on the rate at which the switching thyristors or transistors are triggered into conduction. The triggering rate is determined by the reference oscillator producing a continuous train of timing pulses, which are directed by logic circuits to the thyristor gating circuits. The timing pulse train is also used to control the commutating circuits. The frequency stability and accuracy requirements of the inverter dictate the selection of the reference oscillator. A simple temperature compensated R-C relaxation oscillator gives frequency stability within 0.02%. When better stability is needed, a crystal-controlled oscillator and digital counters may be used, which can provide stability of 0.001% or better. The frequency control in a stand-alone power system is an open-loop system. The steady-state or transient load changes do not affect the frequency. This is a major advantage of the power electronics inverter over the old electromechanical means of frequency conversion using a motor-generator set.

## 4.7 OPEN-LOOP CONTROL SYSTEM

A simple open-loop volt/hertz control system for the induction motor drive using the PWM VSI is shown in Figure 4.14. It has no feedback and hence no instability concern in operation. Therefore, it is widely used in the industry. However, it has somewhat inferior performance. Its front end is a rectifier with a filter to obtain constant $V_{dc}$. The speed or frequency is the command signal, from which $V_s^*$ is derived for the constant air gap flux. At low frequency, the boost voltage $V_{boost}$ is added to compensate for a large voltage drop in the stator armature resistance.

The three-phase sinusoidal voltages are derived from the voltage magnitude and angle command signals from the PWM inverter. The motor speed can be increased or decreased by slowly ramping up or down the speed command signal. While decelerating, the motor works as a generator, and the power is dissipated in the rotor resistance. When the command speed exceeds the base (full) speed, $V_s$ saturates, the frequency proportionality is lost, the drive enters the field-weakening region, and the developed torque is decreased due to the flux reduction. The motor delivers constant horsepower in such operation.

## 4.8 VECTOR CONTROL DRIVES

On ships, the vector control is often used for mooring winches and windlass. Winches are designed to hold mooring lines to the pier in a constant-tension mode. The operator can set the desired tension in tens of thousand pounds by adjusting a potentiometer on the control panel. A windlass is a device that uses a rope or cable

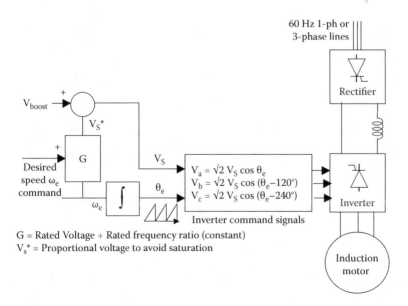

**FIGURE 4.14** Open-loop V/f control schematic of induction motor with dc-link frequency converter.

wound around a revolving drum to pull or lift things and to raise and lower the anchor. Typically, a 50- to 100-hp induction motor is fed from a vector drive to operate the windlass. Some windlass failures with the vector drive at low-speed operation at low load have been reported in the past.

The induction motor draws lagging current having two components: (a) in phase with voltage that produces the real power and (b) lagging the voltage by 90° that produces the magnetic flux. The basic idea of the vector control is to control these two components separately to control the torque and the motor speed independently. For that reason, the induction motor with vector control behaves essentially like the dc motor. A true vector control requires information on the exact position of the rotor and the motor electrical parameters for a feedback control system shown in Figure 4.15. The rotor position sensor is basically a permanent magnet on the rotor and a stationary sensor coil, the output of which is fed back in the speed and torque control loop. The three feedback loops are the output current, motor speed, and rotor position. The current loop prevents operation of the motor above the rated current for a set time period. Above the base speed, the flux component is reduced for a constant-horsepower operation. Another key point is that the frequency of the voltage applied to the motor is calculated from the slip frequency and the motor speed. To simplify the design, some manufacturers have developed open-loop control software that, once tuned with the motor, can operate the motor in the open loop.

The vector-controlled inverter separates the motor current into the flux- and the torque-producing components and maintains a 90° relationship between them, in what is also known as the field-oriented control. The decoupled control of the flux and the torque offers the advantages of higher efficiency and full torque control. To separate the current into the flux and the torque components, the drive must know the position of the motor-magnetizing flux. An encoder attached to the motor shaft relays this information to the converter. The use of vector control produces the maximum torque from the base speed down to zero speed. The other algorithm that could be used is scalar control, which would provide a constant volt/hertz output.

The vector control gives precise speed and position control over a wider operating range. If a VFD without vector control can control the motor speed in the 40:1 range, an open-loop vector control can control in the 120:1 speed range with speed regulation of 0.1%. A true vector control can provide the rated torque at zero speed,

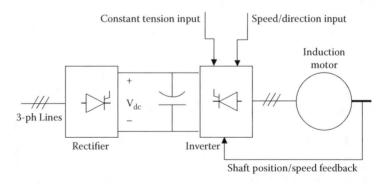

**FIGURE 4.15**  Vector control of induction motor for constant-torque applications.

just like the dc motor, while the open-loop vector control can provide full torque up to about 15 rpm in a four-pole motor. Vector control also offers faster response to the torque and speed change command.

## 4.9 PROPULSION WITH A TWELVE-PULSE VFD

Compared to the rarely used three-phase, three-pulse converter, the three-phase, six-pulse converter produces low harmonic distortion. It is like a three-pulse converter using a three-phase center-tap transformer, where each of the six voltages are out of phase by $360/6 = 60°$.

The harmonic content in the output ac can be further reduced by pulsing the inverters 12 times instead of 6 times. Both the 6-pulse and 12-pulse converters (often incorrectly called 6-phase and 12-phase converters) use 3-phase power mains. The 12-pulse inverter uses a three-winding transformer with one primary and two secondaries of equal ratings, one connected in $Y$ and the other in $\Delta$. The outputs of the two secondaries are 30° out of phase, pulsing the one bloc of the inverter six times and the other block also six times, with total pulses in the combined output 12 times. The main advantage of the 12-pulse design is elimination of the 5th, 7th, 17th, 19th, ... harmonics, thus improving the quality of the output power. Among the inverters commonly used for high-power applications, 12-pulse, line-commutated, full-wave topology prevails.

The VFDs for propulsion and thruster motors are often separate sets as in Figure 4.16, which shows one pair of 12-pulse VFDs for large propulsion motors and one pair of 6-pulse VFDs for small thruster motors. In each VFD pair, one is for the port side and the other for the starboard side. Thus, in total, there are four separate VFDs. The 12-pulse VFDs for the propulsion motors are obtained by using the three-winding $\Delta$-$\Delta Y$ transformers from the high-voltage switchboard as explained. The 12-pulse design is obtained by using the 60° phase difference in the output voltage of the $\Delta$ and $Y$ secondaries, instead of 120° in the conventional 6-pulse converters for the thruster motors.

For further harmonic reduction, 18-pulse converters can be designed using a four-winding transformer (one primary and three secondaries), and 24-pulse converters can be designed using two sets of three-winding transformers in parallel with 15° phase shift from each other on the primary side.

It is possible to design a transformer-less electrical propulsion system using generators with multiphase isolated outputs to propulsion power converters having isolated inputs to provide variable-frequency, variable-voltage power to the ship propulsion motors and separate converters for the ship service power at 60 Hz, 50 Hz, or dc. With this arrangement, power converter transformers are eliminated, thereby reducing the size and weight of the distribution system.

## 4.10 SPECIAL VFD CABLES

Since the VFD output and input power contain high-frequency harmonics, the radiated electromagnetic interference (EMI) from the VFD cables can be of concern in some ships with harmonic-sensitive equipment, especially in the confined spaces of submarines and some high-power combat ships. Moreover, additional harmonic

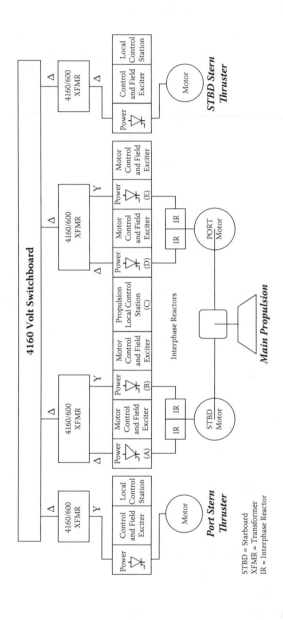

**FIGURE 4.16**  Electric propulsion power system with 12-pulse VFDs for main propulsion and 6-pulse VFDs for thruster motors. STBD = Starboard. XFMR = Transformer, IR = Interphase Reactor. (From Dana Walker, U.S. Merchant Marine Academy.)

heating and high *dv/dt* and *di/dt* stresses degrade the cable insulation at a higher rate than in the conventional sinusoidal equipment. The remedy is to use cables with higher-grade insulation and an enhanced EMI shield. Figure 4.17 is one such cable suitable for marine applications in high-voltage class up to 15 kV. It is available in size AWG 14 to 777 kcmil in ampacity up to 472 A in temperature class up to 110°C. The resistance, inductance, and weight per 1000 feet of 2-kV class VFD cable are listed in Table 4.1, those of 8-kV class cable in Table 4.2, and those of 15-kV class cable in Table 4.3.

37-105VFD

## Type MMV-VFD
## Power Cable

Three Conductor: 8 kV–15 kV ● 133% Insulation Level ● Rated 90°C          **Oil & Gas Cables**

**Conductors (3)**
Soft annealed flexible stranded tinned copper per IEEE 1580 Table 11.

**Insulation**
Extruded thermosetting 90°C Ethylene Propylene Rubber (EPR), meeting UL 1309 (Type E), IEEE 1580 (Type E) and UL 1072.

**EMI Shield**
Overall tinned copper braid plus aluminum/polyester tape providing 100% coverage

**Insulation Shield**
Composite shield consisting of 0.0126″ tinned copper braided with nylon providing 60% copper shielded coverage meeting UL 1309, IEEE Std. 1580 and UL 1072. The nylon is colored for easy phase identification (three conductor = black, blue, red) without the need to remove the shield to find an underlying colored tape.

**Conductor Shield**
A combination of semi-conducting tape and exruded thermosetting semi-conducting material meeting UL 1309, IEEE 1580, and IL 1072.

**Insulation Shield**
Semi-conducting layer meeting UL 1309, IEEE 1580 and UL 1072.

**Symmetrical Insulated Grounding Conductors**
Three soft annealed flexible stranded tinned copper conductors per IEEE 1580 Table 11. Gexol insulation sized per Table 23.2 of UL 1072. Color: Green

**Jacket**
A black, arctic grade, flame retardant, oil, abrasion, chemical and sunlight resistant thermosetting compound meeting UL 1309/CSA 245, IEEE 1580, and UL 1072. Colored jackets for signifying different voltage levels are available on special request (orange = 8 kV and red = 15 kV).

**Armor (optional)**
Basket weave wire armor per IEEE 1580 and UL 1309/CSA 245, Bronze standard. Tinned copper available by request.

**Sheath (optional)**
A black, arctic grade, flame retardant, oil, abrasion, chemical and sunlight resistant thermosetting compound meeting UL 1309/C SA 245, IEEE 1580, and UL 1072, Colored jackets for signifying different voltage levels are also available on special request (orange = 8 kV and red = 15 kV).

**Low smoke halogen-free jacket avilable on request.**

### Termination Kits
AmerCable Systems offers per-sized and pre-formed termination kit packages specifically for VFD cable constructions

**Ratings & Approvals**
■ UL Listed as Marine Shipboard Cable (E111461)
■ American Bureau of Shipping (ABS)
■ Det Norske Veritas (DNV) Pending
■ Lloyd's Register of Shipping (LRS) Pending
■ 90°C Temperature Rating
■ Voltage Rating – 8 kV to 15 kV (25 kV available on request)

**Applications**
A flexible, braid and foil shielded, power cable specifically engineered for use in medium voltage variable frequency AC drive (VFD) applications.

**Features**
■ Flexible stranded conductors, braided shields and a braided armor (when armored). Suitable for applications involving repeated flexing and high vibration.

■ Small minimum bending radius (6 × OD for unarmored cables and 8 × OD for armored cables) for easy installation.

■ Insulation resists the repetitive 3 × voltage spikes from VFDs and reduces drive over current trip problems due to cable charging current.

■ Overall braid and foil shield provides 100% coverage containing VFD EMI emissions.

■ Symmetrical insulated ground conductors reduce induced voltage imbalances and carry common mode noise back to the drive.

■ High strand count conductors and braid shield design is much more flexible, easier to install and more resistant to vibration than type MC cable.

■ Severe cold durability: exceeds CSA cold bend/cold impact (−40°C/−35°C).

■ Flame retardant: IEC 332-3 Category A and IEEE 1202.

■ Suitable for use in Class I, Division 1 and Zone 1 environments (armored and sheathed).

**FIGURE 4.17** Three-conductor 8- to 15-kV class shielded VFD cable with heavy insulation for propulsion motors. (With permission from Gexol-insulated marine shipboard cable, a product of AmerCable, Inc.)

**TABLE 4.1**

**Three-Phase VFD Power Cable 2 kV, 110°C Class**

37-102VFD
**Type VFD Power Cable**
**Gexol® Insulated**
Three Conductor • 2kV • Rated 110°C

| Size AWG/ kcmil | Unarmored | | | Armored | | | Armored & Sheathed (BS) | | | Green insulated Grounding Conductor (x3) Size (AWG) | Ampacity | | | | |
|---|---|---|---|---|---|---|---|---|---|---|---|---|---|---|---|
| | Part No. 37-102 | Nominal Diameter inches* | Weight Per 1000 Ft. | Part No. 37-102 | Nominal Diameter inches* | Weight Per 1000 Ft. | Part No. 37-102 | Nominal Diameter inches* | Weight Per 1000 Ft. | | 110°C | 100°C | 95°C | 90°C | 75°C |
| 14 | -508VFD | 0.540 | 194 | -508BVFD | 0.590 | 281 | -508BSVFD | 0.725 | 356 | 18 | 27 | 25 | 22 | - | 18 |
| 12 | -516VFD | 0.590 | 224 | -516BVFD | 0.646 | 321 | -516BSVFD | 0.772 | 401 | 18 | 33 | 31 | 27 | - | 24 |
| 10 | -506VFD | 0.633 | 308 | -509BVFD | 0.694 | 412 | -506BSVFD | 0.820 | 497 | 14 | 44 | 41 | 36 | - | 33 |
| 8 | -309VFD | 0.764 | 441 | -309BVFD | 0.920 | 565 | -309BSVFD | 0.938 | 702 | 14 | 56 | 52 | 48 | - | 43 |
| 6 | -310VFD | 0.865 | 570 | -310BVFD | 0.925 | 708 | -310BSVFD | 1.090 | 865 | 12 | 75 | 70 | 64 | 98 | 58 |
| 4 | -312VFD | 1.072 | 986 | -312BVFD | 1.125 | 1061 | -312BSVFD | 1.296 | 1243 | 12 | 99 | 92 | 85 | 122 | 79 |
| 2 | -314VFD | 1.215 | 1421 | -314BVFD | 1.271 | 1618 | -314BSVFD | 1.440 | 1822 | 10 | 131 | 122 | 113 | 159 | 105 |
| 1 | -315VFD | 1.340 | 1517 | -315BVFD | 1.395 | 1743 | -315BSVFD | 1.560 | 1966 | 10 | 153 | 143 | 131 | 184 | 121 |
| 1/0 | -316VFD | 1.443 | 1808 | -316BVFD | 1.493 | 2027 | -316BSVFD | 1.666 | 2327 | 10 | 176 | 164 | 152 | 211 | 145 |
| 2/0 | -317VFD | 1.572 | 2153 | -317BVFD | 1.622 | 2399 | -317BSVFD | 1.854 | 2840 | 10 | 201 | 188 | 175 | 243 | 166 |
| 4/0 | -319VFD | 2.053 | 3463 | -319BVFD | 2.103 | 3785 | -319BSVFD | 2.395 | 4347 | 8 | 270 | 252 | 235 | 321 | 223 |
| 262 | -320VFD | 2.193 | 4175 | -320BVFD | 2.243 | 4522 | -320BSVFD | 2.475 | 5120 | 6 | 315 | 294 | 267 | 365 | 254 |
| 313 | -321VFD | 2.370 | 4727 | -321BVFD | 2.420 | 5104 | -321BSVFD | 2.652 | 5747 | 6 | 344 | 321 | 299 | 408 | 287 |
| 373 | -322VFD | 2.501 | 5415 | -322BVFD | 2.551 | 5809 | -322BSVFD | 2.846 | 6674 | 6 | 387 | 361 | 334 | 451 | 315 |
| 444 | -323VFD | 2.670 | 6707 | -323BVFD | 2.721 | 7141 | -323BSVFD | 3.014 | 8059 | 6 | 440 | 411 | 372 | 499 | 350 |
| 535 | -324VFD | 2.972 | 7480 | -324BVFD | 3.022 | 7966 | -324BSVFD | 3.316 | 8961 | 6 | 498 | 443 | 418 | - | 390 |
| 646 | -326VFD | 3.164 | 9916 | -326BVFD | 3.214 | 9428 | -326BSVFD | 3.508 | 10504 | 4 | 558 | 516 | 470 | - | 431 |
| 777 | -327VFD | 3.388 | 10385 | -327BVFD | 3.438 | 10940 | -327BSVFD | 3.732 | 12088 | 4 | 602 | 562 | 529 | | 473 |

*Cable diameters are subject to a +/- 5% manufacturing tolerance

*Source:* With permission from Gexol-insulated marine shipboard cable, a product of AmerCable, Inc.

## 4.11    VARIABLE-VOLTAGE DC MOTOR DRIVE

The term *variable-frequency drive* (VFD) applies to the ac motor drive that requires a variable-frequency power source, whereas the term *variable-speed drive* applies to the dc motor drive that requires only a variable-voltage power source. The dc motor speed is linearly related to the applied voltage and inversely with the flux (field current). The VSD is, therefore, basically a dc-dc full-wave converter with the firing delay angle control to vary the output voltage that is applied to the motor. The electric propulsion system using a VSD may be found on old ships with a dc motor driving fixed-pitch propellers, where the thrust is controlled by the motor speed.

In the VSD, a full-wave thyristor rectifier with variable-voltage output feeds the dc motor. The motor voltage is varied by the thyristor firing delay angle $\alpha$ between $0°$ and $180°$. The field winding is excited with a constant regulated field current. Therefore, the speed of the motor is proportional to the dc armature voltage. Such drive is generally limited to 2- to 3-MW power ratings at 600 V ac or 750 V dc. Its disadvantages for high-power applications are (a) high maintenance cost of the dc

## TABLE 4.2
### Three-Phase VFD Power Cable 8 kV, 90°C, 133% Insulation Class

#### Three Conductor Type MMV-VFD Marine Medium Voltage – 8kV • 133% Insulation Level

| Size AWG/ kcmil | mm2 | Unarmored | | | Armored & Sheathed (BS) | | | Ampacity | | DC Resistance at 25°C (ohms/1000 ft.) | AC Resistance at 90°C, 60Hz (ohms/1000 ft.) | Inductive Reactance (ohms/1000 ft.) | Voltage Drop (Volts per amp per 1000 ft.) | Green Insulated Grounding Conductor (3x) Size (AWG) |
|---|---|---|---|---|---|---|---|---|---|---|---|---|---|---|
| | | Part No. 37-185 | Nominal Diameter (inches) | Weight (Lbs./ 1000 ft.) | Part No. 37-185 | Nominal Diameter (inches) | Weight (Lbs./ 1000 ft.) | In Free Air (amps) | Single Banked in Trays (amps) | | | | | |
| 6 | 12.5 | -332VFD | 1.541 | 1349 | -332B8VFD | 1.879 | 2048 | 88 | 75 | 0.445 | 0.556 | 0.048 | 0.820 | 10 |
| 4 | 21 | -333VFD | 1.728 | 1770 | -333B8VFD | 2.069 | 2548 | 116 | 99 | 0.300 | 0.376 | 0.043 | 0.564 | 10 |
| 2 | 34 | -334VFD | 1.939 | 2335 | -334B8VFD | 2.283 | 3201 | 152 | 129 | 0.184 | 0.230 | 0.040 | 0.359 | 10 |
| 1 | 43 | -335VFD | 2.031 | 2664 | -335B8VFD | 2.378 | 3570 | 175 | 149 | 0.147 | 0.184 | 0.038 | 0.294 | 8 |
| 1/0 | 54 | -336VFD | 2.133 | 3065 | -336B8VFD | 2.481 | 4014 | 201 | 171 | 0.117 | 0.147 | 0.037 | 0.242 | 8 |
| 2/0 | 70 | -337VFD | 2.269 | 3593 | -337B8VFD | 2.621 | 4680 | 232 | 197 | 0.093 | 0.117 | 0.036 | 0.199 | 8 |
| 3/0 | 86 | -338VFD | 2.370 | 4064 | -338B8VFD | 2.723 | 5113 | 266 | 226 | 0.074 | 0.094 | 0.035 | 0.166 | 6 |
| 4/0 | 109 | -339VFD | 2.511 | 4770 | -339B8VFD | 2.866 | 5878 | 306 | 260 | 0.058 | 0.075 | 0.033 | 0.139 | 6 |
| 262 | 132 | -340VFD | 2.691 | 5544 | -340B8VFD | 3.119 | 6936 | 348 | 296 | 0.046 | 0.063 | 0.032 | 0.121 | 6 |
| 313 | 159 | -341VFD | 2.841 | 6340 | -341B8VFD | 3.273 | 7805 | 386 | 328 | 0.040 | 0.053 | 0.032 | 0.106 | 6 |
| 373 | 189 | -342VFD | 3.058 | 7435 | -342B8VFD | 3.494 | 9097 | 429 | 365 | 0.034 | 0.045 | 0.031 | 0.094 | 4 |
| 444 | 227 | -343VFD | 3.218 | 8596 | -343B8VFD | 3.650 | 10250 | 455 | 387 | 0.028 | 0.039 | 0.030 | 0.085 | 4 |
| 535 | 273 | -344VFD | 3.403 | 9900 | -344B8VFD | 3.836 | 1164 | 528 | 449 | 0.024 | 0.033 | 0.030 | 0.076 | 4 |

*Source:* With permission from Gexol-insulated marine shipboard cable, a product of AmerCable, Inc.

## TABLE 4.3
### Three-Phase VFD Power Cable 15 kV, 90°C, 133% Insulation Class

#### Three Conductor Type MMV-VFD Marine Medium Voltage – 15kV • 133% Insulation Level

| Size AWG/ kcmil | mm2 | Unarmored | | | Armored & Sheathed (BS) | | | Ampacity | | DC Resistance at 25°C (ohms/1000 ft.) | AC Resistance at 90°C, 60Hz (ohms/1000 ft.) | Inductive Reactance (ohms/1000 ft.) | Voltage Drop (Volts per amp per 1000 ft.) | Green Insulated Grounding Conductor (3x) Size (AWG) |
|---|---|---|---|---|---|---|---|---|---|---|---|---|---|---|
| | | Part No. 37-185 | Nominal Diameter (inches) | Weight (Lbs./ 1000 ft.) | Part No. 37-185 | Nominal Diameter (inches) | Weight (Lbs./ 1000 ft.) | In Free Air (amps) | Single Banked in Trays (amps) | | | | | |
| 2 | 34 | -357VFD | 2.474 | 3375 | -357B8VFD | 2.829 | 4468 | 156 | 133 | 0.184 | 0.230 | 0.0440 | 0.364 | 10 |
| 1 | 43 | -358VFD | 2.561 | 3748 | -358B8VFD | 2.917 | 4876 | 178 | 151 | 0.147 | 0.184 | .0430 | 0.299 | 8 |
| 1/0 | 54 | -359VFD | 2.663 | 4184 | -359B8VFD | 3.090 | 5562 | 205 | 174 | 0.117 | 0.147 | .041 | 0.246 | 8 |
| 2/0 | 70 | -360VFD | 2.795 | 4764 | -360B8VFD | 3.226 | 6208 | 234 | 199 | 0.093 | 0.117 | 0.0390 | 0.203 | 8 |
| 3/0 | 86 | -361VFD | 3.013 | 5622 | -361B8VFD | 3.447 | 7171 | 269 | 229 | 0.074 | 0.094 | .038 | 0.170 | 6 |
| 4/0 | 109 | -362VFD | 3.155 | 6404 | -362B8VFD | 3.591 | 8024 | 309 | 263 | 0.058 | 0.075 | 0.037 | 0.142 | 6 |
| 262 | 132 | -363VFD | 3.168 | 6925 | -363B8VFD | 3.607 | 8457 | 352 | 299 | 0.046 | 0.063 | 0.035 | 0.124 | 6 |
| 313 | 159 | -364VFD | 3.276 | 7549 | -364B8VFD | 3.666 | 9211 | 389 | 331 | 0.040 | 0.053 | 0.034 | 0.109 | 6 |
| 373 | 189 | -365VFD | 3.396 | 8529 | -365B8VFD | 3.832 | 10271 | 432 | 367 | 0.034 | 0.045 | 0.034 | 0.097 | 4 |
| 444 | 227 | -366VFD | 3.548 | 9589 | -366B8VFD | 3.985 | 11405 | 456 | 388 | 0.028 | 0.039 | 0.033 | 0.08 | 4 |

*Source:* With permission from Gexol-insulated marine shipboard cable, a product of AmerCable, Inc.

motor commutator surface and brushes, and (b) the presence of high EMI radiated from the motor commutator can interfere with other equipment (navigation equipment, computers, etc.).

Many speed control strategies are used in the industry. For speed below the rated speed, the dc link current that determines the motor current is varied with torque load at a given speed, while simultaneously controlling the field current to keep the air gap flux and the torque constant at the rated values. The horsepower output therefore decreases with decreasing speed. Above the rated speed, the motor torque capability degrades due to the armature's current limitation due to heating, but the motor can supply constant-rated horsepower, which is the product of the torque and the speed, as shown in Figure 4.18.

At present, most drives in the industry are ac, whereas dc drives are used selectively for applications for which the system offers special benefits. The dc motor speed can be changed over a wide range by combinations of voltage or field control. As stated, the drive delivers constant hp above the rated speed and constant torque below the rated speed all the way down to zero speed. Thus, it is possible for the dc machine to deliver the full rated torque at zero speed, that is, produce a holding torque for dynamic positioning. Such operation is possible with an induction motor only when driven by vector control, which is complex and costly in design and in operation.

## 4.12   VARIABLE-SPEED DRIVE IN METRO TRAINS

Modern railway traction incorporates the advances made in motor and power electronics technologies. In large-city metro lines, the power is typically dc, which is fed by a trolley wire or a third rail typically at 600 or 750 V dc. Some new installations have used 1500 V dc and even higher voltages up to 3000 V dc. In metro trains with frequent stops and starts, the drive must have four-quadrant operating capability for regenerative braking in both forward and reverse rotations. The train dc drives in the past have used thyristor-based designs, but the newer metro train dc drives use GTO thyristors, and most recent designs use IGBTs for lower cost and higher efficiency. For 750-V dc drives, 1200-V series-connected IGBTs or 1700-V IGBTs have been found adequate in a simple two-level inverter design. For 1500-V dc drives, 3.3-kV IGBTs have been available since the early 2000s. Even for higher 3000-Vdc traction systems, 6.6-kV IGBTs have been available since 2000 for two-level inverter designs.

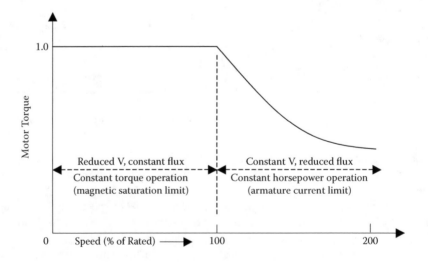

**FIGURE 4.18**   DC motor speed control regions with VSD.

The bullet trains in Japan are successful high-speed trains implemented in the mid-1960s. In the early version, the trains used dc drives from single-phase ac lines through tap-changing transformers and diode rectifiers. These were later replaced by thyristor phase-controlled rectifier dc drives. Since the 1990s, the trains have used GTO thyristor-based PWM rectifiers with two-level inverters with induction motor drives. In some parts of Kolkata, India, and Santiago, Chile, rheostat-controlled dc motors have been used in some sections where the station distances are greater than a mile or so but are not used anymore due to their poor efficiency.

In both dc and ac cases, dc motors have been used in the past, but they are being replaced with induction motors or ac series motors. For example, many long-distance trains in India use 440-V ac series motors specially design for the high torque needed in railway traction applications. In some new long-distance trains, 25- to 35-kV main frequency ac power is used.

## 4.13 VFD AS LARGE-MOTOR STARTER

Small induction motors can be started directly from the lines. However, the starting in-rush current in a large motor with high inertia can be high and for a prolonged time, causing the motor to overheat. Moreover, lights may flicker under excessive voltage dip in the lines. The starting current can be reduced by starting the motor with reduced voltage using a $Y$-$\Delta$ starter, autotransformer, or a phase-control voltage step-down converter. The last, however, is the least-preferred method due to its unacceptably high harmonic current. All these methods result in low starting torque proportional to the applied voltage squared and are not really suitable for starting a large motor with high initial torque load. The wound rotor induction motor can give a high starting torque with an external rotor resistance temporarily introduced through the slip rings. However, the cost of a wound rotor can be high and reliability low due to the slip rings and brushes. The squirrel cage induction motor—the most widely used ac motor—can be most effectively started with full rated torque using a variable-frequency starter. Such a scheme incurs additional cost but has the following advantages:

- It gives high starting torque compared to other methods.
- Line current can be made sinusoidal with displacement power factor unity with a PWM rectifier.
- The same converter can be used in a time-sharing mode for other motors.
- The converter can be bypassed after starting the motor to eliminate the converter losses.
- At light load running, the voltage can be adjusted to improve the running efficiency.

The variable-frequency starters—very similar to VFDs in design—are gradually becoming widely used due to decreasing power electronics prices and stringent power quality standards on the line harmonics imposed on the variable-frequency motor drives.

*Shared motor starter*: The motor takes less than a minute to start from zero speed and come up to full speed. Therefore, many motors can share a variable-frequency starter by staggering the starting operations. The rating of the power electronics motor starter in such shared applications must correspond to the largest motor in the group. Each motor can be brought to full speed in an open-loop control mode with a constant $V/f$ ratio. The motor can be induction or wound rotor synchronous or permanent magnet synchronous.

During soft starting of a motor, the frequency and voltage start from a small value and are ramped up to full rated value in a linear ramp. However, near zero speed and frequency at starting, the inductive reactance is negligible, and the voltage setting the air gap flux is $V - I_{stator}R_{stator}$. For constant air gap flux, then, the $(V - I_{stator}R_{stator})/f$ must be maintained constant. As discussed earlier, this requires a voltage boost of $I_{stator}R_{stator}$ during the start. The required boost decays as the motor speeds up under increasing frequency. This results in a $V/f$ ratio linearly ramping up as shown in Figure 4.11. The rate at which $V$ and $f$ are ramped up depends on the load inertia. The ramping is done over a longer time for high-inertia loads than for low-inertia loads.

*Synchronous motor with starter*: The variable-frequency power electronics starter for a synchronous motor is possible in theory but is hardly used in the synchronous motor mode. Instead, the synchronous motor is normally started as the induction motor using the damper bars on the pole faces. A small permanent magnet synchronous motor can be started as the induction motor directly online using the cage bars, but a large motor may be started softly using a variable-frequency power electronics starter. A large synchronous motor can also be started using a pony motor on the shaft. Either way, the synchronous motor is first started and brought near the synchronous speed, often at no load or light load to keep the rating of the starting device low. The field is then excited, which brings the motor to the full synchronous speed, and then the load is applied after bypassing the starting device and connecting the motor to the main supply lines.

## 4.14 CONVERTER TOPOLOGIES COMPARED

The synchronous motor is generally more efficient than the induction motor. It is often used in very large ratings that run for more hours per year, so that the energy savings are significant at the end of the year.

As for the drives, the synchronous motor generally uses the load commutated inverter (LCI), whereas the large induction motor generally uses CSI. In construction, a typical CSI inverter uses six thyristors, six diodes, and six capacitors. Since the induction motor is a lagging power factor load, the load commutation from one phase of the thyristor to another phase requires a forced commutation circuit made of diodes and capacitors, tuned to the specific motor leakage inductance. This makes it difficult to design a general-purpose drive for multiple motor ratings. Therefore, the CSI is used only with large induction motors and is designed specifically for a given motor.

The LCI typically uses only six thyristors. The load commutation is done using the back emf of the motor. However, at speeds below 10%, the back emf is not enough for the current commutation in the load converter. The commutation is therefore provided by the line converter by going into the inverter mode and forcing $I_{dc}$ to become zero to commutate the load inverter. The LCI control scheme requires sensing the rotor position for commutating the load current, making the control more complex.

The LCI with a synchronous motor and the CSI with an induction motor compare as follows:

- LCI has low cost and high reliability since it requires fewer devices.
- High efficiencies of both the inverter and synchronous motor make the LCI a much more efficient drive system.
- The LCI is not a motor-specific design, whereas the CSI must match with the leakage inductance of the induction motor.
- The LCI requires a complex control scheme.
- Both drives are capable of regenerative braking if desired.

The CSI with a forced commutated converter has the following advantages:

- Independent control of active and reactive powers
- Absence of commutation failure
- Lower harmonic distortion factor
- Lower adverse impact of large wind energy farms on weaker grid
- Ability of startup from zero load (i.e., from cold network)

The cycloconverter, CSI, and LCI converters have seen recent decline in use in ship propulsion power systems, for which most of the drives installed at present tend to be the VSI-PWM type. Advance high-voltage motors optimized with the power electronics drives are being developed that do not require a propulsion transformer for voltage matching. Transformer-less drives at 6.6 and 11 kV may be available soon.

Regarding the LCI versus VSI versus any other power electronics converter, the debate will perhaps continue on the most optimum design, more so when the total system with the motor is considered. If the propulsion plant is separate from the ship service power plant, then designing the transformer-less system is easy using a 12-phase or 15-phase generator and designing all the converters accordingly. The integrated power plant is another matter, depending on the proportion of the propulsion power and the ship service power.

## 4.15  NOTES ON VFDS

*High electrical stresses on motor insulation induced by power electronics:* In many applications using fast-switching IGBTs in PWM inverters, the motor winding insulation experiences high *dv/dt* stress due to the steep wave front of the PWM wave. Capacitive ground leakage current flowing in the winding insulation under high-frequency harmonics causes additional heating.

The phase-to-phase, phase-to-ground, and turn-to-turn insulations all see high dielectric stress at high temperature, causing accelerated aging and premature insulation failure. With long cables between the inverter and the motor, overvoltages can also result from the reflected waves. The common-mode stray current in the bearings is another deteriorating effect. A soft-switched inverter or a low-pass filter can minimize such problems.

*Torsional stress in the motor shaft in a variable-speed motor drive*: The variable-speed motor drives were rapidly adopted in oil refineries due to significant energy-saving potentials in their large and continuous oil-pumping requirements. The early drives had high harmonic content due to lack of an industry standard limiting the harmonics. This resulted in a high pulsating torque superimposed on the steady load torque on the motor shaft. The torsional vibrations in the shaft due to such torque pulsations shortened the fatigue life, leading to premature mechanical failures of the shaft. The failures were often seen particularly at the shaft key slots, where the mechanical stress concentration was high due to sharp corners. The problem gradually became less severe as the harmonic content was reduced in newer designs and the shafts were redesigned for a longer fatigue life by reducing the stress concentration.

*Induction motor under single phasing*: The power electronics device may fail open or short, but it often fails by short circuit and then opening up the converter phase by fuse clearing. This results in a single-phase supply to the three-phase motor between two lines. With a three-phase induction motor drive, the motor will continue to run as a single-phase motor but will introduce a high degree of pulsating torque. At full frequency of 60 or 50 Hz, the torque pulsation has a double-frequency (120 or 100 Hz, respectively) component superimposed on the average steady torque. The pulsation frequency is high enough to filter out the speed pulsations due to the mechanical inertia. However, at low frequency (i.e., at low motor speed), the pulsations may be excessive and may be felt in the load.

*VSD in air-conditioning*: In conventional air-conditioning, the single-phase induction motor is used at a fixed supply frequency. The motor runs essentially at constant speed, and the temperature is controlled by a thermostat. The efficiency of a constant-speed air conditioner is poor. The load-proportional VSD can be more efficient. In countries like Japan, where electrical power cost is high, over 90% of the home air conditioners use VSDs with a permanent magnet synchronous motor, which gives even better efficiency compared to the induction motor drive. Refrigerators and washing machines can also use such drives for improving energy efficiency.

*Conducting band around permanent magnet rotor*: In the permanent magnet synchronous motor, a metal band is often used around the permanent magnets to brace the rotor parts against the centrifugal force. Such a band made of a conducting metal works like a cage winding and may provide a starting torque and the damping torque during transient oscillations of the rotor following a step load change. It has no effect on synchronous operation since no current is induced in it at synchronous speed. However, in a motor with VFD feeding the harmonic voltages or harmonic currents, the band carries harmonic frequency currents and generates high power losses and heating on the rotor. Nonconducting bands, such as a fiber-epoxy composite band, in such cases are preferred.

*Bearing currents and heating*: The currents induced in the motor bearings are due to $dv/dt$ across the stray capacitance of the motor air gap. The air gap capacitance varies inversely with length of the air gap. Therefore, the bearing currents can be of concern in the induction motor, where the air gap is typically very small compared to the synchronous motor (permanent magnet or wound rotor).

## PROBLEMS

*Problem 4.1*: A water pump driven by an 1800-hp induction motor pumps 200,000 m³/day at the rated speed. If the motor were run at a slower speed using a VFD to reduce the flow rate to 60% over a longer time to pump the same quantity of water per day, determine the percentage savings in kilowatt-hours energy consumed per day.

*Problem 4.2*: A ship was designed to travel at 15 knots with a dedicated 4.6-kV, 30-MW$_e$ propulsion power plant consisting of the generators, transformers, VFDs, and motors rated accordingly. A new line of ship is being designed with essentially the same hull, except it has a higher speed of 20 knots. Determine the rating of the new propulsion power plant and the recommended voltage level.

*Problem 4.3*: A ship travels 3000 nautical miles at 20 knots speed using 400,000 gallons of fuel oil. If the speed were reduced by 15%, determine the percentage change in total gallons of fuel consumption.

*Problem 4.4*: A 200-hp, three-phase, 60-Hz, six-pole, 1150-rpm, 0.90-pf induction motor operates with a VFD connected to 480-V lines. The VFD design has CSI fed from a dc link. If the motor needs to runs at half the rated speed, determine (a) the rectifier firing delay angle and (b) the VFD output voltage at the motor terminals.

*Problem 4.5*: A 300-hp, 480-V, three-phase, 50-Hz, four-pole, 1446-rpm induction motor is driving a large ventilating fan. When it is running from a VFD output of 25 Hz with a constant *V/f* ratio, determine the (a) approximate speed and (b) exact speed.

*Problem 4.6*: A VFD drives a three-phase, 60-Hz, 100-hp, 1760-rpm induction motor. Determine power, torque, and full-load speed when operating at (a) 40 Hz and (b) 80 Hz.

*Problem 4.7*: A small training ship *Kings Pointer* designed to travel at 10 knots needs 2000 kW$_e$ propulsion power and 500 kW$_e$ for ship service loads. If it is desired to redesign the power plant for a higher speed of 15 knots, determine the total (a) kilovolt-ampere rating of the new electrical generators, (b) horsepower rating of the new diesel engines, and (c) diesel fuel consumption in U.S. gallons for a round trip of 1000 nautical miles at 15 knots.

Assume (1) the ship service load remains the same, (2) the new generators have a 90% power factor and 92% efficiency, and (3) the thermodynamic cycle efficiency of the engines is 50%.

## QUESTIONS

*Question 4.1*: In an unthrottled fluid pipeline, explain why you should expect a higher percentage reduction in the motor input power than the percentage reduction in the flow output rate.

*Question 4.2*: List key types of inverters that are used in VFDs.

*Question 4.3*: What is the primary reason for maintaining a constant *V/f* ratio at the motor terminal at all speeds?

*Question 4.4*: Why is a voltage boost needed in the VFD output voltage when starting and at low motor speeds?

*Question 4.5*: With a VFD, explain how the motor torque varies below and above the rated speed.

*Question 4.6*: With a VFD, explain how the motor horsepower varies below and above the rated speed.

*Question 4.7*: How does the VSD for a dc motor differ for the VFD for an ac motor in construction and in operation?

*Question 4.8:* Explain how the VFD can also be used for soft starting a large motor and the benefits of doing so.

*Question 4.9*: Why the metal bands on the permanent magnet synchronous motor rotor pose an overheating risk? What is the remedy?

*Question 4.10*: Explain the construction and working of the special transformer used to build a 12-pulse VFD.

## FURTHER READING

Barnes, M. 2003. *Practical Variable Speed Drives and Power Electronics.* Burlington, MA: Elsevier.

Bose, B. 2006. *Power Electronics and Motor Drives.* Burlington, MA: Academy Press.

Mohan, N., T.M. Undeland, and W. Robbins. 2003. *Power Electronics Converters, Applications, and Design.* New York: John Wiley & Sons.

Murphy, M. 2001. Variable speed drives for marine electric propulsion. *Transactions of the Institute of Marine Engineers*, 108(Part 2), 97–107.

Wildi, T. 2002. *Electrical Machines, Drives, and Power Systems.* Upper Saddle River, NJ: Prentice Hall.

Wu, Bin. 2006. *High-Power Converters and AC Drives.* New York: John Wiley & Sons.

# 5 Quality of Power

Although the quality of power has recently become a part of the industry standards for electrical power that the utility companies and other power providers are expected to deliver at the user's terminals, the term *quality of power* has no single universally accepted definition and measurement at present. However, organizations such as the Institute of Electrical and Electronics Engineers (IEEE), International Electrotechnical Commission (IEC), and North American Reliability Council have developed working definitions that include the following key performance measures: Transient voltage deviations during system faults and disturbances
- Periodic dips in voltage that may cause light flickers
- Transient change in voltage due to large step loading or motor starting
- Steady-state voltage regulation (voltage rise at light load)
- Harmonic distortion (primarily generated by power electronics converters)

The discussion of power quality actually addresses the *compatibility* between the power delivered by the bus and the ability of the connected load equipment to perform as specified. Any gap in the compatibility between the two has two alternative solutions: (a) to clean up the bus power or (b) to make the equipment tougher to deliver full performance even with a poor-quality power input.

## 5.1 POWER QUALITY TERMINOLOGY

The ac bus voltage should ideally be sinusoidal with amplitude and frequency set by the national standards for the utility mains, or by the system specifications for a stand-alone bus that is independent of the mains, such as on ships or on an island. The bus voltage may deviate from normal in many ways. The commonly used terms in describing the bus voltage deviations are as follows:

*Sag (American term) or dip (British term):* The rms voltage falling below the rated value by 10–90% for a duration lasting for 0.5 cycle to 1 minute.

*Swell:* The rms voltage rising above the rated value by 10–90% for a time duration lasting for 0.5 cycle to 1 minute.

*Undervoltage:* The voltage dropping by more than 10% of the rated value for more than 1 minute (also known as brownout). This term is commonly used to describe a reduction in system voltage that requires the utility or system operator to decrease the load demand or to increase the system operating margin.

*Overvoltage:* The voltage rising by more than 10% of the rated value for more than 1 minute. Some lights may burn out, electromagnetic equipment may increase hum due to magnetic saturation, and many equipment may overheat due to increased power losses.

*Transient voltage deviation:* The voltage may deviate within the acceptable band defined by voltage-versus-time (v-t) limits on both sides of the nominal system voltage. For example, the American National Standards Institute (ANSI) requires that the system voltage be maintained within the v-t envelope shown in Figure 5.1. The narrow right-hand side of the band is set primarily from the steady-state performance considerations of the motor and transformer-like loads. The middle tapering portion of the band comes from visible light flicker annoyance considerations, and the broad left-hand side of the band comes from electronics load susceptibility considerations. In general, the deviation that can be tolerated by user equipment depends on its magnitude and the time duration. Smaller deviation can be tolerated for a longer time than a larger deviation. ANSI requires the steady-state voltage of the utility source to be within 5% and the short-time frequency deviations to be less than 0.1 Hz.

Computer and business equipment using microelectronics circuits are more susceptible to voltage transients than rugged power equipment like motors and transformers. They require a narrower band than that shown in Figure 5.1, which allows deviations only for microseconds based on the power supply's volt-second or flux limitation in the magnetic core to avoid saturation (since $v \times dt = N \times d\varphi$ by Faraday's law).

*Flicker:* The random or repetitive variations in the rms voltage between 90% and 110% of nominal value can cause flickers in lights, especially in fluorescent lights. Light flickers cause unsteadiness of visual sensation on the human eye and can be a source of annoyance. Flickers are often caused by frequent switching on and off large loads. In small systems, such as on ships, switching a relatively large load may result in a large voltage flicker that may have an impact on other neighboring equipment and distract people working in the area. To minimize flickers in the neighboring

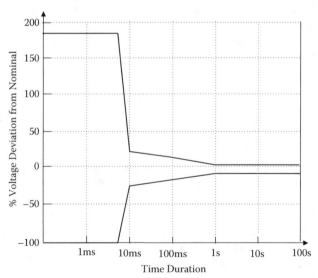

**FIGURE 5.1**    Allowable voltage deviation versus time duration in utility power (ANSI).

equipment, the grid company may restrict large loads from switching on and off to no more than three to four times per hour. Flicker limits are specified in IEEE Standards 141 and 519, which have served the industry well for many years. Cooperative efforts among the IEC, EPRI (Electric Power Research Institute), and IEEE have resulted in updated standards as documented in IEC Standard 61000-3.

*Spike, impulse, or surge:* Abrupt but brief (in microseconds) increase in voltage, generally caused by a large inductive load being turned off within the system or by a lightning surge coming from outside via cables. The surge suppressor at the power outlet used by many small users to protect computer-type sensitive equipment does not eliminate spikes or solve the power quality problem. It merely diverts large voltage spikes to ground that would otherwise enter the equipment, possibly causing damage.

*Frequency deviation:* The frequency deviation from the rated value is caused mainly by the prime mover speed regulation or other reasons.

*Harmonics:* The deviation from a pure sinusoidal wave shape that contains high-frequency components superimposed on the fundamental frequency voltage or current.

*Bus, source, or line resistance:* The Thevenin equivalent series resistance behind the bus terminals. It causes the steady-state voltage to drop when the load current increases.

*Bus, source, or line inductance:* The Thevenin equivalent series inductance behind the bus terminals. It causes a transient dip or spike in the voltage when the load current suddenly changes in one step.

Each of these power quality problems has a different cause and solution. Some problems result from using a shared generator, transformer, or line feeder by multiple users. A problem on one user equipment may cause the bus voltage transient to affect other users on the same bus. Other problems, such as harmonics, arise within the user's own installation and may or may not propagate onto the bus to affect other users. Harmonic distortion can be minimized by a combination of good practice in the harmonic reduction at the source and the harmonic filter design.

## 5.2 ELECTRICAL BUS MODEL

Some power quality issues at the bus terminals can be better understood in terms of the Thevenin model of the bus as a single generator, although multiple generators could be working in parallel behind the bus in the power grid or on the ship. The Thevenin equivalent source model consists of a single source voltage $V_s$ with single internal impedance $Z_s$ in series as shown in Figure 5.2a, where

Thevenin source voltage $V_s$ = Open circuit voltage at load terminals 1 and 2    (5.1a)

Thevenin source impedance $Z_s$ = Voltage drop per ampere of load current    (5.1b)

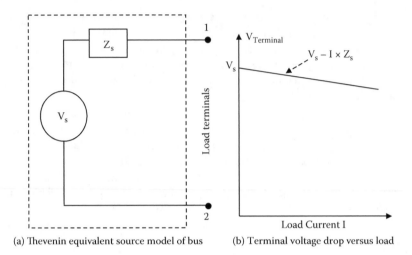

(a) Thevenin equivalent source model of bus     (b) Terminal voltage drop versus load

**FIGURE 5.2**   Thevenin equivalent source model and bus terminal voltage droop versus load current.

If $Z_s$ is derived under a steady-state power frequency tests, then $Z_s$ is called the steady-state or static bus impedance. Under steady load current $I$, the internal bus impedance $Z_s$ causes the bus voltage to drop by $I \times Z_s$ as shown in Figure 5.2b. In absence of any feedback control to regulate the bus voltage, this drop will remain steady. A source with low $Z_s$ will have a low internal voltage drop but will result in a high short circuit current, making the circuit breaker rating unmanageable. The design engineer must strike a balance between these two conflicting requirements.

The bus voltage drops under load current due to the internal voltage drop in the Thevenin source impedance $Z_s$ in series with the voltage source. If $Z_s = R + jX$ Ω/ phase, then the voltage drop under current $I$ at power factor $pf = \cos \theta$ is given by the following formula, which is fairly accurate for all practical power factors:

$$V_{drop} = I(R \; \cos\theta + X \; \sin\theta) = I\{R \times pf + X \times \sqrt{1 - pf^2}\} \text{ volts/phase} \qquad (5.2)$$

Under a sudden step load increase of $\Delta I$ shown in Figure 5.3, the voltage momentarily drops by $\Delta V_d = \Delta I \times Z_d$ and settles to $\Delta I \times Z_s$ after several transient oscillations of the R-L-C circuit. Here, $Z_d = \Delta V/\Delta I$ is called the *dynamic bus impedance*, which is quite different from the static bus impedance $Z_s$. If a synchronous generator is the dominant contributor to the bus impedance $Z_d$, then the *d*-axis subtransient reactance $X_d''$ of the generator contributes in $Z_d$, and the *d*-axis synchronous reactance $X_d$ contributes in $Z_s$. The voltage dip $V_d$ falling below a certain value at the lower end may cause a light flicker. In a system with feedback voltage control, the bus voltage slowly moves toward the rated value within the controller's regulation limit. Under a step load decrease, the opposite happens; the voltage rises, oscillates, and then moves toward the rated value. Such oscillations are often called *ringing*.

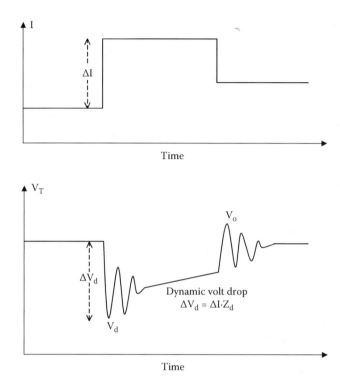

**FIGURE 5.3**    Voltage deviation $\Delta V$ after step change $\Delta I$ in load current.

## Example 5.1

A three-phase, 1-MW bus has the Thevenin source voltage of 520 V and the Thevenin source impedance of 5 + $j$35 m$\Omega$/phase. Determine the bus voltage when delivering the rated current at 0.85 $pf$ lagging.

### SOLUTION

Assuming a Y-connected source (generator),

Rated current in each line (phase) = 1,000,000 ÷ ($\sqrt{3}$ × 480) = 1203 A

Using Equation (5.2), voltage drop in an impedance is given fairly accurately at all practical power factors by $R\cos\theta + X\sin\theta$ volts per phase per ampere.
For this machine,

$V_{drop}$ = 0.005 × 0.85 + 0.035 × $\sqrt{1-0.85^2}$ = 0.0227 V/phase per ampere

∴ Voltage drop = 1203 × 0.0227 = 27.3 V/phase or $\sqrt{3}$ × 27.3 = 47.3 $V_{LL}$

Terminal line voltage = 520 − 47.3 = 472.7 V

## 5.3  HARMONICS

The utility power source is primarily sinusoidal with one dominant frequency, called the fundamental frequency, which is 60 or 50 Hz. *Harmonics* is the term used to describe higher-frequency sinusoids superimposed on the fundamental wave. The power electronics converter is the most common source of harmonics in the electrical power system. The magnetic saturation in power equipment also generates harmonics. The generator and transformer cores are normally designed to carry flux near the magnetic saturation limit where the magnetizing current is slightly nonlinear and nonsinusoidal, peaking more than the normal peak value of $\sqrt{2}$ × rms value.

As discussed in Chapter 3, Section 3.1.3, any nonsinusoidal alternating current (or voltage) can be decomposed into a Fourier series of sinusoidal current $I_1$ of the fundamental frequency and sinusoidal harmonic currents $I_h$ of higher frequencies, where $h = 3, 5, 7, 9, \ldots$ . All even harmonics are absent in practical power systems. The frequency of the $h$th-harmonic current = $h$ × fundamental frequency. For example, the frequency of the ninth harmonic in a 60-Hz power system = 9 × 60 = 540 Hz.

If a three-phase load is fed by a $\Delta$-connected transformer, all triplen (multiples of three) harmonics are absent in the line current, that is, $I_h = 0$ for $h = 3, 9, 15$, and so on. The six-pulse, full-wave inverter circuit contains harmonics of the order $h = 6k \pm 1$, where $k = 1, 2, 3, \ldots$ . Therefore, major harmonics present in a six-pulse inverter are 5, 7, 11, 13, 17, 19, 23, and 25. The 12-pulse, full-wave inverter circuit contains harmonics of the order $h = 12k \pm 1$, where $k = 1, 2, 3, \ldots$ . Therefore, major harmonics present in a 12-pulse inverter are 11, 13, 23, and 25, which are fewer than in a six-pulse inverter. In both cases, the harmonic current magnitude is inversely proportional to the harmonic order $h$, that is,

$$I_h = \frac{I_1}{h} \tag{5.3a}$$

where $I_1$ is the fundamental sinusoidal current. This formula gives approximate harmonic contents in 6- and 12-pulse inverters, as given in the second and third columns of Table 5.1. It clearly shows the benefits of using the 12-pulse converter.

The harmonics can be determined either by the circuit calculations leading to the converter output wave and then going through the Fourier series analysis or by measurements using a harmonic spectrum analyzer or a power quality analyzer—equipment that displays and prints percentage harmonics of the voltage or current fed to it. IEEE Standard 519 gives the actual measured current harmonics in typical three-pulse and six-pulse converters as listed in the right-most column of Table 5.1, which are lower than those given by Equation (5.2) and listed in the second column. The harmonic current induces harmonic voltage on the bus, which is given by

$$V_h = I_h \cdot Z_h \tag{5.3b}$$

where $Z_h$ = harmonic impedance of the bus for the $h$th-harmonic frequency. It is similar to the dynamic impedance $Z_d$ and can be determined from tests at the harmonic frequency of interest. The skin effect makes $R_h$ higher than $R_1$. The generator

**TABLE 5.1**

**Harmonic Currents in Six-Pulse and Twelve-Pulse Converters (% of the Fundamental)**

| | Theoretical Value | | Actual Value |
|---|---|---|---|
| Harmonic Order $h$ | 6-Pulse Converter Eq. (5.2) | 12-Pulse Converter Eq. (5.2) | 3-Pulse and 6-Pulse Converters (IEEE Std. 519) |
| 5 | 20 | — | 18.5 |
| 7 | 14.5 | — | 11.1 |
| 11 | 9.1 | 9.1 | 4.5 |
| 13 | 7.7 | 7.7 | 2.9 |
| 17 | 5.9 | 5.9 | 1.5 |
| 19 | 5.3 | 5.3 | 1.0 |

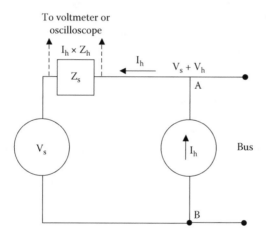

**FIGURE 5.4** Harmonic impedance determination by injecting harmonic current source.

$d$-axis synchronous reactance $X_d$ contributes to $X_1$ and the subtransient reactance $X_d''$ contributes to $X_h$ for high-frequency harmonics. The direct way of measuring $Z_h$ is by injecting $I_h$ ampere of harmonic current in the bus from an outside source as shown in Figure 5.4 and measuring the harmonic voltage rise $V_h$ at the bus that is given by $I_h \times Z_h$. Then, $Z_h = V_h/I_h$. The presence of harmonics in a waveform is often indicated by the crest factor (CF), defined as

$$\text{Crest factor} = \frac{\text{Peak of waveform}}{\text{rms value of waveform}} \tag{5.4}$$

For an ideal sine wave, CF $= \sqrt{2} = 1.414$. The CF deviating away from this value would indicate the presence of harmonics. For an ideal square wave, CF $= 1.0$.

### 5.3.1 Harmonic Power

The performance of a system driven by a source voltage with multiple harmonics is determined by superimposing the system performance under each harmonic separately. For such an analysis, the system is represented by a series of equivalent circuits, each for the corresponding harmonic frequency. For example, when a motor is driven by a voltage containing multiple harmonics as shown in Figure 5.5a, we first determine the motor performance from the equivalent circuit model for each source frequency separately as shown in Figure 5.5b, one at a time, and then superimpose all of them to get the total performance, that is

$$\text{Total performance} = \text{Performance}_1 + \Sigma \, \text{Performance}_h \qquad (5.5)$$

where $\text{Performance}_1 = $ fundamental frequency performance, and $\text{Performance}_h = $ performance under the $h$th-harmonic voltage, $h = 5, 7, 11, 13, \ldots$. Most motors are $\Delta$-connected, in which the triplen harmonics are zero, as explained in Section 5.5.2.

To analyze the harmonic power in any equipment, we write the bus voltage and current as the sum of their respective fundamentals plus the harmonics, that is,

$$V(t) = V_1 + \Sigma_{all.h} V_h \quad \text{and} \quad I(t) = I_1 + \Sigma I_h \qquad (5.6)$$

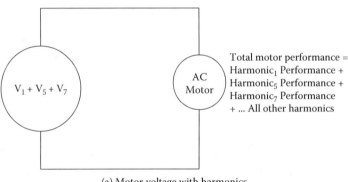

(a) Motor voltage with harmonics

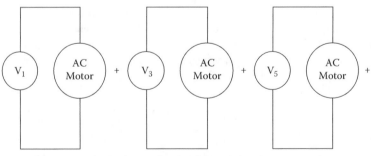

(b) Motor under fundamental and each harmonic voltage separately

**FIGURE 5.5**    Superposition model of motor powered by voltage with multiple harmonics.

The power is the product of voltage and current, that is, $p(t) = v(t) \times i(t)$. Using the rms values in Equation (5.6), we write

$$P(t) = V_1 \times I_1 + \Sigma_{all.h} V_h \times I_h + (V_1 \times I_h + V_h \times I_1) \qquad (5.7)$$

On the right-hand side of Equation (5.7), the first term $V_1 \times I_1$ is the real active fundamental power. The third and fourth terms in parentheses each amount to zero average power since $V$ and $I$ are of different frequency (known as the orthogonal functions in Fourier series). Their multiplication has nonzero instantaneous value but zero average value. It works like reactive power, adding into VAR that is unproductive and undesirable. The second term $V_h I_h$ is the harmonic power due to each harmonic with average value $\neq 0$, which adds undesirable high-frequency pulsations in equipment performance. For example, it would produce high-frequency pulsations in the motor torque and resulting torsional vibrations in the shaft that may lead to early fatigue failure.

All harmonic currents produce additional $I^2R$ power loss in the conductor. The harmonic power loss can be significantly high due to skin and proximity effects since the high-frequency current concentrates near the conductor surface. The effective skin depth $\delta_h$ of the conductor for the $h^{th}$-harmonic current is given by

$$\delta_h = \frac{\delta_1}{\sqrt{f_h}} \qquad (5.8)$$

For copper conductor, $\delta_1$ for a 60-Hz fundamental frequency current is about 9 mm, $\delta_7$ for the seventh-harmonic (420-Hz) current is 3.4 mm, and $\delta_{25}$ for the 1500-Hz current is 1.8 mm. The thin skin effectively reduces the current-carrying area of the conductor, making the high-frequency resistance significantly higher than the 60-Hz value. Therefore, the harmonic power loss can be much higher than the main frequency power loss in thick conductors, particularly in heavy bus bars made with solid conductors.

Thus, harmonics produce pulsating power, reactive power, kVARs, and additional power loss in the conductor, all of which are undesirable. Therefore, poor-quality power also results in a system with poor efficiency.

## 5.3.2  TOTAL HARMONIC DISTORTION AND POWER FACTOR

To compare the harmonic content in various power buses, we recall the results from Chapter 3, Section 3.1.4. The total harmonic distortion (THD) factor in current is defined as

$$\text{THD}_i = \frac{\text{Total harmonic current rms}}{\text{Fundamental current } I_1 \text{ rms}} = \frac{I_{Hrms}}{I_{1rms}} \qquad (5.9)$$

Also we recall that the total rms value is obtained by the *root sum square* (rss) of the component rms values, that is,

$$\text{Total rms current } I_{Total} = \sqrt{I_1^2 + I_3^2 + I_5^2 + ....} = \sqrt{I_1^2 + I_H^2} \qquad (5.10)$$

where the total harmonic current

$$I_H = \sqrt{I_3^2 + I_5^2 + I_7^2 + ....}$$
(5.11)

The THD factor in the current waveform, using all rms values, is given by

$$THD_i = \frac{I_H}{I_1} = \sqrt{\frac{I_H^2}{I_1^2}} = \sqrt{\frac{I_3^2}{I_1^2} + \frac{I_5^2}{I_1^2} + \frac{I_7^2}{I_1^2} ..} = \sqrt{\left(\frac{I_3}{I_1}\right)^2 + \left(\frac{I_5}{I_1}\right)^2 + \left(\frac{I_7}{I_1}\right)^2 ..}$$
(5.12)

The $TDH_i$ can be easily derived from the harmonic current ratios often given to the power engineer by the converter manufacturer. Similar calculations apply to the THD factor $THD_v$ in the voltage waveform. The THD is useful in comparing the quality of ac power at various locations. In a pure sine wave ac source, THD = 0. A higher value of THD indicates a more distorted sine wave, resulting in a system with lower efficiency.

## Example 5.2

The actual values of the harmonic currents in a practical six-pulse converter are given in the $I_h/I_1$ ratios below, which are less than the theoretical values of $I_h/I_h$.

| Fundamental | 5 | 7 | 11 | 13 | 17 | 19 | 23 | 25 |
|---|---|---|---|---|---|---|---|---|
| $I_h/I_1$ ratio | 0.173 | 0.108 | 0.041 | 0.032 | 0.021 | 0.015 | 0.011 | 0.009 |

Determine the $THD_i$ with the actual and theoretical harmonic currents.

### SOLUTION

Recall that the total effective rms value is the rss value the (fundamental + harmonics), that is,

$$I_T = \sqrt{I_1^2 + I_3^2 + I_5^2 + I_7^2 + ....25^{th}}$$

The table in this example gives the actual harmonic current ratios, from which we obtain, using Equation (5.12),

$$THD_i = \sqrt{0.173^2 + 0.108^2 + 0.041^2 + 0.032^2 + 0.021^2 + 0.015^2 + 0.011^2 + 0.009^2}$$

$$= 0.212 \text{ pu or } 21.2\%$$

Similar calculations with theoretical values of $I_h = I_1/h$ would give $THD_i = 28.7\%$, which is significantly higher than the actual $THD_i$ found in practical converters.

As seen, the harmonic current $I_h$ drawn by any nonlinear load on the bus causes the harmonic distortion in the bus voltage that is given by $V_h = I_h \times Z_h$. The $V_h$ in turn causes the harmonic current to flow even in a purely linear resistive load connected to the same bus, called the *victim load*. If the power plant is relatively small, as in many ships, a nonlinear power electronics load may cause significant distortion on the bus voltage, which then supplies distorted current to the linear resistance-type loads. IEEE Standard 519 limits the $THD_v$ for the utility-grade voltage to less than 5%. $THD_v$ above 5% is considered unacceptable, and above 10% it needs major correction.

As mentioned, a rough measure of $THD_v$ is the crest factor—the ratio of the peak to rms voltage—measured by the true rms voltmeter. In a pure sine wave, this ratio is $\sqrt{2} = 1.414$. Most acceptable bus voltages will have this ratio in the 1.35 to 1.45 range, which can be used as a quick approximate check on the $THD_v$ at any location in the system. A true (or total) rms meter is another way of identifying the power quality. The true rms value of a voltage with harmonic contents is the total rms value given by $V_{Trms} = (V_1^2 + V_3^2 + V_5^2 + \ldots + V_n^2)^{\frac{1}{2}}$ where $V_1, V_3, V_5, \ldots, V_n$ are the rms values of the fundamental and all harmonic voltages present. In terms of the $THD_v$, we can write

$$V_{Trms} = V_{1rms}\sqrt{1 + THD_v^2} \tag{5.13}$$

High harmonic content distorts (degrades) the load power factor also since only the fundamental component of the total is doing the useful work. *True (total) power factor = Fundamental power factor × Distortion power factor*, where the distortion power factor is defined as

$$\text{Distortion pf} = \frac{V_{1rms}}{V_{Trms}} = \frac{1}{\sqrt{1 + THD_v^2}} \tag{5.14}$$

$$\therefore \text{True pf} = \frac{\text{Fundamental pf}}{\sqrt{1 + THD^2}} \tag{5.15}$$

The Canadian Standards Association and IEC defined the harmonic distortion factor based on the total $I_{rms}$, that is, $THD_{True} = I_{Hrms}/I_{Trms}$. The IEEE and other U.S. standards, on the other hand, define the $THD$ based on the fundamental current, that is, $THD = I_{Hrms}/I_{1rms}$, which is somewhat higher than the $THD_{True}$. The two are related as follows:

$$THD_{True} = \frac{THD}{\sqrt{1 + THD^2}} \quad \text{where } THD \text{ is as defined by the U.S. standards} \tag{5.16}$$

or,

$$THD_{international} = \frac{THD_{usa}}{\sqrt{1 + THD_{usa}^2}} \tag{5.17}$$

Although the technical definition of $THD$ is the same in the international and U.S. standards, the bases are different. This may create a legal tangle when the $THD$ is specified in the contract without a clear base. The contract should have a clear definition of THD, specifically defining the base that can be the fundamental or the total (true) rms value.

## Example 5.3

A three-phase, *Y*-connected load draws current containing the following Fourier harmonics (all rms amperes) in each line:

| Order h | 1 | 3 | 5 | 7 | 9 | 11 | 13 | 15 |
|---------|----|----|----|----|----|----|----|----|
| $I_h$ rms A | 50 | 30 | 20 | 15 | 11 | 9 | 7 | 6 |

Determine (a) the total (true) rms value of the current, (b) *THD* based on the fundamental current, (c) true $THD_T$ as defined in some international standards, and (d) current in the neutral conductor.

### SOLUTION

(a) Total (true) rms $I_{Trms} = \sqrt{[50^2 + 30^2 + ... + 6^2]} = 65.67$ A

(b) The fundamental current $I_{1rms} = 50$ A and

Total harmonic current $I_{Hrms} = \sqrt{[30^2 + ... + 6^2]} = 42.57$ A

∴ Based on the fundamental current, THD = 42.57/50 = 0.8514 or 85.14%

Based on the total rms current, $THD_T = 42.57/65.67 = 0.648 = 64.8\%$, which is much lower than a *THD* of 85.14%. The $THD_T$ could have been derived from Equation (5.17) also. The difference between *THD* and $THD_T$ must be well understood and clarified, particularly in international contracts, to avoid any litigation.

All harmonics other than triplen have 120° phase difference between line currents; hence, they cancel out, requiring no return conductor. The triplen harmonics (i.e., 3, 9, and 15 in this case), on the other hand, are all in phase in three lines and require the neutral conductor to carry the sum that is equal to three times the rss value of all triple harmonics in each phase, that is,

$$I_{neutral} = 3 \times \sqrt{[30^2 + 11^2 + 6^2]} = 97.53 \text{ A rms}$$

This neutral current is much higher than the fundamental current of 50 A rms in each phase line.

So, the neutral conductor for this load has to be much heavier than the line conductor. This is not uncommon in systems containing harmonic currents.

## 5.3.3  K-RATED TRANSFORMER

As seen in Section 5.3.1, harmonics of different orders—being orthogonal functions in the Fourier series—do not contribute in delivering real average power but produce additional $I^2R$ heating due to skin and proximity effects. Such additional heating in generators, motors, and transformers causes higher temperature rise due to their limited cooling areas, as opposed to long running cables. For this reason, the National Electrical Code (NEC®) requires all distribution transformers to state the K rating on a permanent nameplate. This is useful in buying a proper transformer for harmonic-rich power electronics loads. The K-rated transformer does not eliminate line harmonics. The K rating merely represents the ability of the transformer to tolerate harmonics within the design temperature limit of the conductor.

Since the magnetic power loss in the core and the eddy current loss in the conductor vary with frequency squared, the harmonic losses also increase with frequency squared. Therefore, the higher-order harmonic currents produce additional power losses in the transformer, leading to a higher temperature rise. To limit such temperature rise, the transformer catering to high harmonic currents is designed with thicker conductors and thinner core laminations, both of which cost more. The transformer designed this way is designated by its K rating, which is the indication of the ability of the transformer to deliver power to a power electronics-type harmonic-rich load without exceeding its rated operating temperature. The K rating is defined as

$$K = \frac{\sum h^2 \times I_h^2}{I_{rms}^2} = \sum_{all\, h} h^2 \left( \frac{I_h}{I_{rms}} \right)^2 \text{ for }$$

$$h = 1, 3, 5, 7 \text{ (includes fundamental)} \tag{5.18}$$

Underwriters Laboratories (UL) has designated the K factor as a means of rating the ability of a transformer to handle loads that generate harmonic currents. The UL recognizes K-factor values of 4, 9, 13, 20, 30, 40, and 50, which are based on information contained in ANSI/IEEE C57.110-1986, *Recommended Practice for Establishing Capability When Supplying Nonsinusoidal Load Currents*. The K-factor number tells us how much a transformer must be derated to handle a definite nonlinear load or, conversely, how much it must be oversized to handle the same load.

The transformer designed to power a sinusoidal load from a sinusoidal source requires a K rating of 1.0. Typical transformer K ratings required for a power electronics load are K-4, K-9, K-13, K-20, up to K-50. A K-9-rated transformer, for example, means that it can withstand nine times a 60-Hz loss without exceeding the design temperature limit. The K-rated transformer is used anywhere nonlinear loads are present. Typically, a K-13-rated transformer is sufficient in most power electronics systems. The following rules are generally acceptable in selecting the K rating of the transformer:

- Use a standard transformer where the harmonic-producing equipment kilovolt-ampere (KVA) rating is less than 15% of the source rating.
- A K-4-rated transformer is needed where the power electronics equipment represent 30–40% of the total load.
- Use a K-13-rated transformer where the power electronics equipment represent 70–80% of the total load.
- A K-20-rated transformer is needed where supplying a 100% power electronics load, such as for the variable-frequency motor drive.
- Higher K-factor ratings are generally reserved for specific pieces of equipment where the harmonic spectrum of the load is known to be high.

If it is desired to use an old existing transformer without a K rating (i.e., with a K-1 rating), the transformer must be derated so it does not exceed the design temperature. The advantage of using a K-rated transformer is that it is usually more economical than using a derated, oversized transformer.

## Example 5.4

A power electronics converter draws the fundamental and harmonic currents from a transformer as follows:

| Fundamental | 5th $h$ | 7th $h$ | 11th $h$ | 13th $h$ | 17th $h$ | 19th $h$ |
|---|---|---|---|---|---|---|
| 20 A | 5 A | 3 A | 2 A | 1.5 A | 1 A | 1 A |

Determine (a) the total rms current in the transformer output and (b) the K rating of the transformer needed to power this converter.

### SOLUTION

The total rms value of the transformer output current is given by the rss value of the fundamental plus all harmonic currents, that is,

$$I_{T.rms} = \sqrt{20^2 + 5^2 + 3^2 + 2^2 + 1.5^2 + 1^2 + 1^2} = 21.03 \text{ A}$$

The K rating of the transformer derived from Equation (5.18) is

$$K = 1^2\left(\frac{20}{21.03}\right)^2 + 5^2\left(\frac{5}{21.03}\right)^2 + 7^2\left(\frac{3}{21.03}\right)^2 + 11^2\left(\frac{2}{21.03}\right)^2$$

$$+ 13^2\left(\frac{1.5}{21.03}\right)^2 + 17^2\left(\frac{1}{21.03}\right)^2 + 19^2\left(\frac{1}{21.03}\right)^2$$

Therefore, $K = 6.72$; hence, the transformer must have a K rating of 7 or higher.

## 5.3.4 MOTOR TORQUE PULSATIONS

In the induction motor driven by a variable-frequency drive (VFD), the third-harmonic currents in three phases have $3 \times 120° = 360°$ or $0°$ phase difference from each other, that is, they are in phase, and so are the sixth, ninth, ... (all triplen) harmonics in phase. The in-phase triplen harmonic currents cannot flow in Δ-connected motor lines since there is no return conductor to and from the motor terminals. Therefore, the largest current harmonics going in the motor are the fifth and seventh. However, the actual torque harmonics on the motor shaft are at six times the fundamental frequency, as explained next.

The fundamental three-phase currents are, say, $I_{a1} = 100 \cos \omega t$, $I_{b1} = 100 \cos (\omega t - 120)$, and $I_{c1} = 100 \cos (\omega t - 240)$, where all angles are in degrees. Such currents produce a magnetic flux rotating at the synchronous speed, driving the motor in that direction also, say, the forward direction.

The fifth-harmonic three-phase currents are $I_{a5} = 100 \cos (5\omega t)$, $I_{b5} = 100(5\omega t -600) = 100 \cos (5\omega t + 120)$ and $I_{c5} = 100 \cos (5\omega t -1200) = 100 \cos (5\omega t + 240)$. These currents produce a flux *rotating backward* at five times the synchronous speed.

The seventh-harmonic three-phase currents are $I_{a7} = 100 \cos (7\omega t)$, $I_{b7} = 100 (7\omega t - 840) = 100 \cos (7\omega t - 120)$, and $I_{c7} = 100 \cos (7\omega t - 1680) = 100 \cos (7\omega t - 240)$. These currents produce a flux *rotating forward* at seven times the synchronous speed. The speed and rotation of the stator flux created by various harmonic currents are given in Table 5.2.

The three-phase induction motor runs near the synchronous speed in the forward direction. The fifth-harmonic currents set up a magnetic field rotating backward at five times the synchronous speed with respect to the stator and at about six times the synchronous speed with respect to the rotor. Therefore, the rotor would have sixth-harmonic currents, producing the sixth-harmonic torque. The seventh-harmonic currents set up a magnetic field rotating forward at seven times the synch speed with respect to the stator and at about six times the synchronous speed with respect to the rotor. Therefore, the rotor currents would have sixth-harmonic currents, producing the sixth-harmonic torque.

So, both the fifth- and seventh-harmonicons currents produce the sixth-harmonic (6 × 60 = 300 Hz) torque pulsations that add up to cause significant shaft jitters (Figure 5.6) and other performance degradations. The total torque has the 300-Hz pulsations, giving the speed pulsations ripple amplitude as follows:

$$Speed\ ripple\ amplitude = Constant \times \frac{Torque\ ripple\ amplitude}{Ripple\ frequency \times Mass\ inertia} \quad (5.19)$$

The impedance of the rotor circuit to each harmonic is different due to much different slips, such as approximate slip = 6 pu for the 5th and 7th harmonic, slip = 12 pu for the 11th and 13th harmonics, and so on. This is because the synchronous

## TABLE 5.2
## Speed and Direction of Rotation of Magnetic Flux Created by Various Harmonic Currents

| Harmonic Order | Frequency in 60-Hz System (Hz) | Synchronous Speed in 4-Pole, 60-Hz Motor (rpm) | Direction of Rotation |
|---|---|---|---|
| 1 (fundamental) | 60 Hz | 1,800 rpm | Forward |
| 5 | 300 | 5 × 1800 | Backward |
| 7 | 420 | 7 × 1800 | Forward |
| 11 | 660 | 11 × 1800 | Backward |
| 13 | 780 | 13 × 1800 | Foreword |
| 17 | 1020 | 17 × 1800 | Backward |
| 19 | 1140 | 19 × 1800 | Forward |
| 23 | 1380 | 23 × 1800 | Backward |
| 25 | 1500 | 25 × 1800 (45,000 rpm) | Forward |

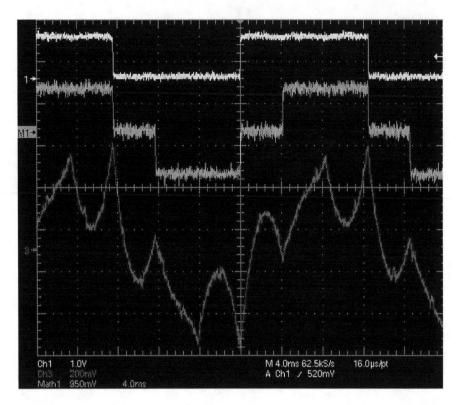

**FIGURE 5.6** The 6th harmonic motor torque pulsations produced by fifth and seventh harmonic currents. (From Bill Veit, U.S. Merchant Marine Academy.)

speed of the harmonic flux is much greater than the actual rotor speed. Therefore, the harmonic impedance is primarily due to the leakage reactance at the harmonic frequency, with the ratio (harmonic resistance ÷ harmonic slip) approaching zero. The harmonic current, however, produces high $I^2R$ loss due to the skin effect at high frequency. And, the harmonic flux produces high loss in the magnetic core. Thus, the total increase in power loss in the motor driven by VFD is significant and must be considered in selecting the motor type and the rating. When operating at lower than rated speed, the low-frequency operation of the inverter produces relatively higher harmonics. Up to a certain speed, the constant $V/f$ ratio maintains the rated flux and rated torque. Below that speed, the torque loading on a given motor should be proportionately reduced as shown in Figure 5.7 to limit the harmonic-related heating and to allow for the reduced self-cooling at lower speed. At a speed and frequency above the rated values, the torque is inversely proportional to the speed, and the motor delivers constant horsepower.

For fan and pump motors, which generally operate over a limited speed range of 2:1 or 3:1, the horsepower output varies as the speed cubed. So, the motor load naturally falls off considerably at lower speed, and derating from the rated horsepower does not require any action from the user.

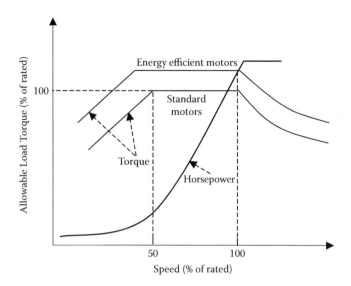

**FIGURE 5.7** Motor load torque derating versus speed with VFD harmonics.

### 5.3.5 HARMONIC-SENSITIVE LOADS ON SHIPS

Both the navy and merchant ships have many harmonic-sensitive loads in ac as well dc systems. The dc is increasingly seen as free from harmonic-related problems, but ripples superimposed on the average dc have the same adverse effects as the harmonics in ac.

On ships with electric propulsion, the power electronics motor drives shown as nonlinear load-2 in Figure 5.8 draw a substantial percentage of the total power system capacity. This makes the harmonic problems much more severe on ships than typically seen in large land-based power systems. The harmonic distortion in the main bus voltage, even with clean generator voltage, could be pronounced and could be inflicted on clean linear load-1. Such a situation is also true in land-based systems in the neighborhood of a large nonlinear load, such as steel mills and other industrial plants using power electronics converters for process controls. Referring to Figure 5.8, if the clean load-1 draws $I_{L1}$ and the power electronics load-2 draws $I_{L2} + I_h$, the total current drawn for the generator is $I_{L1} + I_{L2} + I_h$. The voltage at the generator bus is the clean generated voltage minus the voltage drop in the internal source impedance, that is,

$$V_{AB} = V_{gen.clean} - (I_{L1} + I_{L2} + I_h) \times Z_s = \{V_{gen.clean} - (I_{L1} + I_{L2}) \times Z_s\} - I_h \times Z_s \quad (5.20)$$

The term on the right-hand side in curly brackets is clean voltage, whereas $I_h \times Z_s$ is harmonic voltage at the bus output that will cause harmonic currents in the clean load-1 as well, making it the victim load.

In the marine industry, a special harmonic-related problem may occur in fishing ships. The fishery electronics on such ships work on the hydroacoustics principle by sending out surveying signals of high-frequency power in the 20- to 800-kHz frequency range. Each fish species echoes differently, giving information on the type of fish present in the survey area. The very high order harmonics in the power source can distort the signals going out, and the echo readings on the type of fish present may also be distorted.

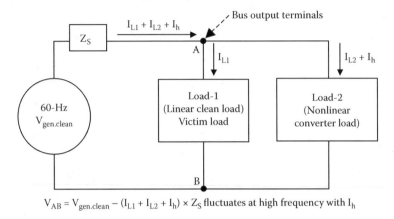

$$V_{AB} = V_{gen.clean} - (I_{L1} + I_{L2} + I_h) \times Z_S \text{ fluctuates at high frequency with } I_h$$

**FIGURE 5.8**  Bus voltage contamination by power electronics load current harmonics.

## 5.4  POWER QUALITY STUDIES

The power engineers can use the following means to identify poor power quality:

- The true rms meter can identify the THD factor when compared with the fundamental rms value. A ratio of true to fundamental rms values close to 1.0 indicates good power quality.
- The ratio of the peak to rms value can be a quick indication of power quality. The peak/rms ratio of $\sqrt{2} = 1.414$ indicates the perfect sinusoidal power quality. A generally acceptable quality would have the ratio in the 1.35–1.45 range, and poor unacceptable quality would have the ratio outside this range.
- The current sensor in the neutral wire showing a significant value indicates the presence of triplen harmonic currents that are in phase and return through the neutral wire.
- The oscilloscope display of the voltage or current wave shape that has many high-frequency ripples would surely indicate poor power quality.
- A harmonic analyzer, often available in a handheld version, gives a variety of useful information, such as real, reactive, and apparent power; crest factor; true power factor; THD as percentage of the fundamental or true rms; and the entire harmonic spectrum. The Fluke F41 model (Figure 5.9) has the capability of downloading information to a serial port of a personal computer.
- Power quality analyzers not only offer the harmonic analysis described but also monitor and record transient events for days at a time. They are available in a three-phase version in larger bench-type or handheld units, such as one shown in Figure 5.10, which is the PowerPad-3945 model of AEMC Instruments.

Shipboard electrical power quality evaluation service is provided by many companies and by the Naval Sea Systems Command, Washington, D.C.

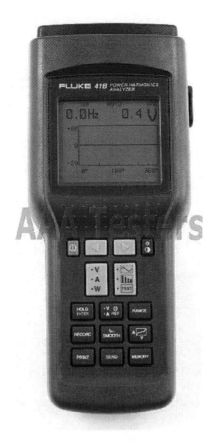

**FIGURE 5.9**   Harmonic analyzer Fluke F41 handheld model.

## 5.5   HARMONIC REDUCTION

Harmonic currents can propagate from one piece of equipment to another by solid conductors or by electromagnetic coupling. Equipment not connected by conductors cannot inflict other equipment by conduction. However, high-frequency harmonic currents can inflict other proximal equipment by electromagnetic interference (EMI) through magnetic coupling via leakage flux in air. Therefore, in addition to filtering out major harmonic currents, we need to protect harmonic-sensitive loads from both the conducted harmonics by not connecting through wire and the radiated EMI by minimizing the leakage flux. Some of the design and operational methods used to minimize the harmonic problems in shipboard systems are described next.

### 5.5.1   HARMONIC FILTER

The harmonics can be filtered from the system by a series L-C circuit that is called the harmonic filter. A single-phase version of the filter is shown by the single-line diagram in Figure 5.11. The harmonic current is then supplied or absorbed by the

### AEMC 3945B PowerPad
### 3-Phase Power Quality Analyzer

- True RMS single, dual and three phase measurements at 256 samples/cycle plus DC
- Real time color display of waveforms including transients
- Easy on-screen graphics for setup
- True RMS voltage and current
- Display voltage, current and power harmonics to the 50th order, including direction, in real time
- Phasor diagram display
- Nominal frequency from 40 to 70 Hz
- VA, VAR and W per phase and total
- KVAh, VARh and kWh
- Neutral current
- Crest factors for current & voltage
- K-factor for transformer
- Power factor, displacement PF and tangent
- Captures up to 50 transients
- Short-term flicker and voltage
- Phase unbalance
- Harmonic distortion (Total and individual)
- Includes DataView configuration and analysis software
- 3945 has a CAT III safety rating
- 3945-B has a CAT IV safety rating

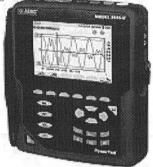

**AEMC 3945B Shown**

| Product Information | |
|---|---|
| Datasheet | 📄 **PDF** <br> 1.4 MB |
| Interactive Demo | Flash |

AEMC 3945 PowerPad is a hand-held three phase power and power quality meter with a large easy-to-read graphical color display. Measurements are displayed numerically and graphically with colored waveforms. The meter is menu driven with pop-up functions that are activated at the push of a button. All necessary measurements are available for a comprehensive power system check or analysis to 830Vrms; 6500 Arms.

**Screenshots**

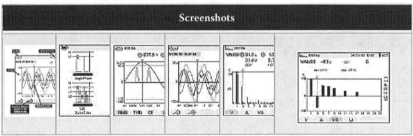

**FIGURE 5.10** Three-phase power quality analyzer. (With permission from Chauvin Arnoux, Inc./AEMC Instruments, Dover, NH.)

filter rather than by the bus, thus improving the quality of power at the bus. The $L$ and $C$ values are tuned to the harmonic frequency desired to be filtered out. The series resistance $R$ (often internal to $L$) is kept small by design. If we ignore $R$ for simplicity here, then the values of $L$ and $C$ should be such that they are in series resonance at the harmonic frequency $f_h$, such that

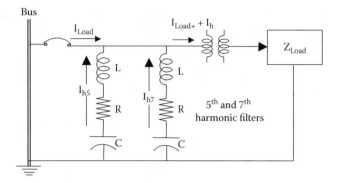

**FIGURE 5.11**    Load harmonic filter keeps bus free from harmonic currents.

$$f_h = \frac{1}{2\pi\sqrt{L \, x \, C}} \qquad \text{Hz} \qquad\qquad (5.21)$$

Then, the harmonic impedance of the filter $Z_h = 0$, which behaves like a shorted wire. All the harmonic current $I_h$ will pass through the filter and bypass the bus. The bus that was delivering $I_{Load} + I_h$ current before to the load will now see only the pure sine wave $I_{Load}$. Thus, the harmonic filter does not eliminate the harmonic current; it merely bypasses it so it does not reach to the bus, hence keeping the bus voltage clean. If we wish to filter out multiple harmonics, there must be one such filter tuned to each harmonic frequency. Or, for cost considerations, two dominant harmonics of adjacent orders can be combined into one filter design at the midrange frequency.

If $R \neq 0$, then a small $I_h$ will be drawn from the bus, which is determined by the current-sharing rule between the load impedance $Z_{Load}$ and the harmonic filter resistance $R$. In designing a harmonic filter, the values of $L$ and $C$ can be any as long as they meet Equation (5.21) and can carry the harmonic current without overheating.

The harmonic filter can be placed near the bus on the transformer primary side or near the load on the transformer secondary side. Both have certain advantages. Placing the filter near or on the high voltage bus is preferred when many small power electronics loads without individual harmonic filters are connected to a common bus where one central filter can be cost effective. On the other hand, placing the filter next to the load prevents harmonics going into the load side transformer and cables as well, eliminating overheating in both.

In designing a harmonic filter, the series resonance frequency of the filter is kept about 5% lower than the dominant harmonic frequency. For a typical six-pulse converter with dominant fifth-harmonic (300-Hz) current, the filter series resonance frequency is kept around 285 Hz. Since the filter now supplies the harmonic current (instead of coming from the bus), its $L$ and $C$ components must be designed to carry the harmonic current, which can be calculated by the current divider rule between the source impedance in parallel with the filter impedance looking from the harmonic load point. If the kilovolt and kVAR rating of the bus frequency capacitor are known, then we derive its capacitance value as follows (the formula remains the same with single-phase or three-phase line ratings):

$$C = \frac{kVAR_{cap}}{1000 \, x \, kV_{cap}^2 \, x \, 2\pi f_{bus}} \text{ farads/phase} \tag{5.22}$$

Then, Equation (5.21) gives the required inductance value in the harmonic filter,

$$L_h = \frac{1000 \, kV_{cap}^2 \, f_{bus}}{kVAR_{cap} \, 2\pi f_h^2} \text{ henrys/phase} \tag{5.23}$$

When a large power electronics load draws power from the utility grid, there are certain interface issues that must be addressed to maintain the quality of power for the rest of the users drawing power from the same bus or the same substation of the grid. On ships, the propulsion motor power electronics drive drawing power from the source (generators) can contaminate the quality of power for all other users by introducing a high harmonic content. For this reason, the bus itself may require a central harmonic filter to benefit all loads connected to the bus. Figure 5.12 shows a harmonic filter placed on the bus.

## Example 5.5

A 480-V bus in an industrial plant supplying power to several feeders has the Thevenin source impedance behind the bus equal to $0.003 + j0.015$ $\Omega$/phase. A power factor correcting capacitor rated 600 $kVAR_{3ph}$ is connected to the bus as shown in Figure E5.5. Determine the parallel resonance frequency and identify the precaution needed for connecting a power electronics load to this bus.

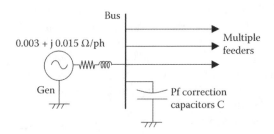

### SOLUTION

Using Equation (5.22), we obtain the capacitance of 600 $kVAR_{3ph}$ capacitor bank,

$$C = \frac{600}{1000 \text{x} 0.480^2 \text{x} 2\pi \text{x} 60} = 6.91 \text{x} 10^{-3} \text{ F/phase}$$

The inductance behind the bus,

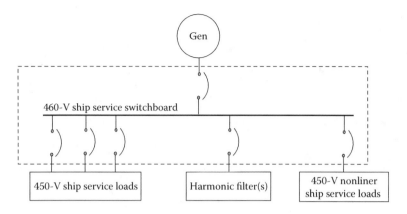

**FIGURE 5.12** Ship service power distribution with harmonic filter on main bus.

$$L = \frac{0.015}{2\pi x 60} = 39.5 \times 10^{-6} \; H/phase$$

From Equation (5.21), the undamped parallel resonance frequency (ignoring resistance),

$$f_o = \frac{1}{2\pi\sqrt{39.5 \times 10^{-6} \times 6.91 \times 10^{-3}}} = 305 \; Hz$$

which is 5.08 × 60 Hz.

PRECAUTION

Since the parallel resonance frequency of 305 Hz is essentially equal to the fifth-harmonic frequency in a 60-Hz system, any power electronics load that draws significant fifth-harmonic current from the bus would resonate and severely distort the bus voltage. The next example determines such distortion.

## Example 5.6

The bus in Example 5.5 supplies a 200-kVA, 480-V power electronics load that has the harmonic spectrum shown in Table E5.1, which also shows the harmonic impedances behind the bus. Determine (a) the rss and THD of the source current without the 600-kVAR capacitor bank and (b) the rss and THD of the bus voltage with a 600-kVAR capacitor connected to the bus.

**TABLE E5.1**

**Bus Impedances and Voltage Drops at Bus Due to Harmonic Currents**

| Harmonic Order | Frequency (Hz) | Line Current $I_h$ (A) | $R_h$ (Ω/ph) | $X_h$ (Ω/ph) | $Z_h$ (Ω/ph) | Volt Drop/ph ($V_h = I_h Z_h$) |
|---|---|---|---|---|---|---|
| 5 | 300 | 50 | 0.003 | 0.075 | 0.0751 | 3.76 |
| 7 | 420 | 30 | 0.003 | 0.105 | 0.105 | 3.15 |

(Continued)

## TABLE E5.1 (CONTINUED)
### Bus Impedances and Voltage Drops at Bus Due to Harmonic Currents

| Harmonic Order | Frequency (Hz) | Line Current $I_h$ (A) | $R_h$ (Ω/ph) | $X_h$ (Ω/ph) | $Z_h$ (Ω/ph) | Volt Drop/ph ($V_h = I_h Z_h$) |
|---|---|---|---|---|---|---|
| 11 | 660 | 15 | 0.003 | 0.165 | 0.165 | 2.48 |
| 13 | 780 | 7 | 0.003 | 0.195 | 0.195 | 1.37 |

### SOLUTION

The fundamental impedance behind the bus, as given in Example 5.5, is $Z_1 = 0.003 + j0.015$ Ω/phase. In the table, the harmonic resistance is constant (ignoring the skin effect), and $X_h = h \times X_1$, which rises with the harmonic frequency.

Fundamental load current = $200{,}000 \div (\sqrt{3} \times 480) = 240.6$ A

Total rms current is given by the rss of all harmonic rms values, and the THD of the source current and the bus voltage are

$$I_{rss} = \sqrt{240.6^2 + 50^2 + 30^2 + 15^2 + 7^2} = 248 \ A$$

$$\therefore \ THD_i = \frac{\sqrt{50^2 + 30^2 + 15^2 + 7^2}}{240.6} = 0.252 \ or \ 25.2\%$$

as defined by the U.S. standards.

Since the harmonic voltages are volts per phase, $THD_v$ calculations must use the phase voltage of the bus, which is $480/\sqrt{3} = 277$ V.

$$\therefore \ THD_v = \frac{\sqrt{3.76^2 + 3.15^2 + 2.48^2 + 1.37^2}}{277} = 0.0204 \ or \ 2.04\%$$

With a 600-kVAR capacitor bank connected to the bus, the Thevenin source impedance behind the bus is in parallel with the capacitor impedance on the bus. The total equivalent impedance of the parallel combination is then calculated for each harmonic, which gives the following results:

| Frequency (Hz) | $Z_{Total.h}$ (Ω/ph) | $I_h$ (A) | $V_h$ (V/ph) |
|---|---|---|---|
| 300 | 1.58 | 50 | 79 |
| 420 | 0.116 | 30 | 3.48 |
| 660 | 0.045 | 15 | 0.675 |
| 780 | 0.035 | 7 | 0.245 |

The rss value and THD of the bus voltage are

$$V_{bus.rss} = \sqrt{277^2 + 79^2 + 3.48^2 + 0.675^2 + 0.245^2} = 288\ V$$

$$THD_v = \frac{\sqrt{79^2 + 3.48^2 + 0.675^2 + 0.245^2}}{277} = 0.285\ or\ 28.5\%$$

Thus, the harmonic distortion with the *pf* correction capacitor is much higher—28.5% versus 2.04% without the capacitor—due to the parallel resonance between the bus impedance and the capacitor at one of the power electronics load harmonics, which is the fifth harmonic in this example. The amplification of the fifth-harmonic voltage as seen in the table has resulted in such unacceptable harmonic distortion on the bus voltage. The system study should identify such a possibility and avoid it to maintain good-quality power.

### Example 5.7

Determine the value of filter inductance required to filter out the 11th harmonic from a 60-Hz bus voltage that supplies a 12-pulse converter with a 100-kVAR, 4160-V bus capacitor.

#### SOLUTION

As generally practiced by power engineers, we tune the harmonic filter at 5% lower than the harmonic frequency, that is, $f_h = 0.95 \times 11 \times 60 = 627$ Hz for the 11th harmonic. Then, using Equation (5.23), we get

$$L_{11} = \frac{1000\ x\ 4.160^2 x 60}{100\ x\ 2\pi x 627^2} = 0.0042\ H\ or\ 4.2\ mH/phase$$

### 5.5.2 Very Clean Power Bus

In ships with electric propulsion, high-power electronics converters switching on and off high currents produce high harmonic distortion on the bus that can have an adverse impact on the performance of the harmonic-sensitive loads, such as the navigation instruments, radar, science instruments, computers, and so on. In such ships, very clean, harmonic-free, electrically isolated power is derived by a dedicated motor-generator set shown in Figure 5.13. Since there is no electrical or magnetic connection between the generator and the motor, the motor-generator set output is very clean and free of all harmonics.

### 5.5.3 Δ-Connected Transformer

If the load draws fundamental plus triplen and other harmonic currents, the triplen harmonics can be effectively eliminated from the bus lines by using a Δ-Y transformer

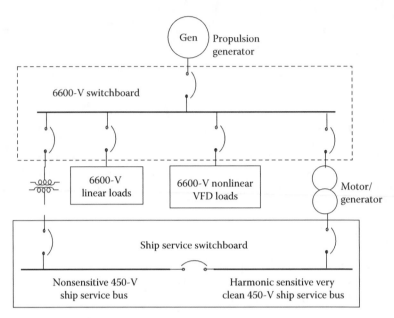

**FIGURE 5.13**   A 6,600-V electric propulsion bus with 460-V motor-generator set for very clean power for harmonic-sensitive loads.

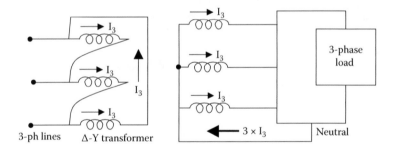

**FIGURE 5.14**   Triplen harmonics elimination by using Δ-*Y* transformer.

with neutral as shown in Figure 5.14. It works as follows, where all angles are in degrees:

The balanced 3-phase fundamental currents are 120° out of phase from each other, that is,

$$I_{a1} = 100 \cos \omega t \quad I_{b1} = 100 \cos (\omega t - 120) \quad I_{c1} = 100(\omega t - 240) \quad (5.24)$$

The neutral current is the phasor sum of three phase currents, that is, $I_{N1} = I_{a1} + I_{b1} + I_{c1} = 0$, that is, no fundamental neutral current flows in the neutral wire. The triplen harmonic currents are a different matter. If we have 40% third-harmonic current in each phase, then the third-harmonic current in the three phases are

$$I_{a3} = 40 \cos \omega t$$

$$I_{b3} = 40 \cos 3(\omega t - 120) = 40 \cos (3\omega t - 360) = 40 \cos 3\omega t$$

$$I_{c3} = 40 \cos 3(\omega t - 240) = 40 \cos (3\omega t - 720) = 40 \cos 3\omega t \qquad (5.25)$$

Equation set (5.25) shows that the third-harmonic currents in three lines are in phase, and their sum in the return wire is $I_{N3} = 120 \cos 3\omega t$, which has an amplitude of 120 A. So, in the four-wire $Y$-connected system, the neutral wire will carry 120 A, 20% higher than the 100-A fundamental current in the main lines for this typical example.

The triplen harmonic currents in the $\Delta$-connected primary coils come from the $Y$-connected secondary coils by the transformer turn ratio. However, these currents are absorbed in the $\Delta$-windings and do not propagate further to the main lines on the source side since there is no fourth wire for the return current. However, the triplen harmonic currents in the windings increase the internal power loss and the operating temperature. They reduce the load capability of the transformer and hence are important for selecting the K rating of the transformer.

If the neutral wire were not provided in the $Y$-connected system, then also $I_{N3} = 0$. Thus, in a three-wire $Y$- or $\Delta$-connected system, all triplen harmonics are zero (filtered out from the lines), although they circulate within the phase coils of $\Delta$. With all triplen harmonic filtered, the remaining harmonics are $h = 6k \pm 1$ where $k = 1$, 2, 3, ... , that is, $h = 5, 7, 11, 13, 17, 19, 23,$ and 25 in practical systems. For special 12-pulse connections of rectifier and inverter using a three-winding transformer with $\Delta/\Delta Y$ connections, the harmonics orders are $h = 12k \pm 1$, where $k = 1, 2, 3, ...$, that is, $h = 11, 13, 23,$ and 25. The harmonic current magnitudes $I_h = I_1/h$, so the higher-order harmonics have inherently lower magnitudes. In most applications, the 23rd- and higher-order harmonics can be ignored without significant loss of accuracy.

### 5.5.4 Cable Shielding and Twisting

The EMI—sometimes called the high-frequency or radio-frequency electrical noise—is generally caused by the leakage flux coming from one electrical equipment (culprit) linking to another equipment (victim) in the same proximity as shown in Figure 5.15a. The leakage flux—although small but of high frequency—induces voltage in the victim equipment per Faraday's law. This can distort the rated performance of the victim equipment, which may pose a severe problem if there is much equipment in a small confined space, such as in an engine room with high power electronics loads (motor drives). Delicate low-power equipment in close proximity to high-current cables may be affected more severely.

The leakage flux from the power cables carrying harmonic currents can also inflict serious ill effects on nearby control signal wires carrying telephone communications or data. Higher-order harmonics having higher frequencies pose a greater problem due to higher interference voltage induced in the signal wires. For this reason, many systems require the power cables to have a metallic shield that will prevent the leakage flux from escaping the Faraday cage formed by the shield. In addition,

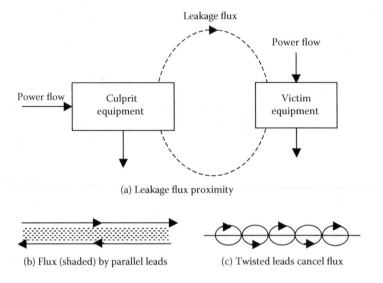

(a) Leakage flux proximity

(b) Flux (shaded) by parallel leads          (c) Twisted leads cancel flux

**FIGURE 5.15**   EMI from culprit equipment inflicting victim equipment.

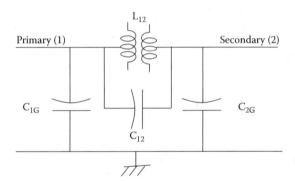

**FIGURE 5.16**   Transformer model for EMI studies.

the data and signal cables are required to be in separate metal trays placed with sufficient physical distance between them.

The amount of leakage flux radiated from or received by wire loops depends on the wire loop area, which is large for lead and return wires placed far apart. It can be minimized by placing the lead and return wires as close as possible and using twisted wires in both culprit and victim equipment as shown in Figure 5.15c.

### 5.5.5 ISOLATION TRANSFORMER

An isolation transformer can inhibit the propagation of conducted EMI. Having no direct conduction of current by metallic wire, it provides electrical isolation, hence the name. However, some energy transfer from the input to the output can still take place either by the inductive or capacitive coupling between the coils. The circuit mode of the transformer for this purpose is shown in Figure 5.16, where $L_{12}$ = leakage inductance, $C_{12}$ = capacitance between the primary and secondary windings, $C_{1G}$ = capacitance of the primary coil to the ground, and $C_{2G}$ = capacitance of the secondary coil to the ground. The ground is usually the static shield foil placed between the primary and secondary coils.

At high frequency, the high value of $X_L = \omega L$ blocks the high-frequency current from going further, and the low value of $X_C = 1/\omega C$ diverts the high-frequency current to the ground. Thus, both $X_L$ and $X_C$ jointly block the high frequency current and voltage spikes going from one side to the other side of the transformer. The isolation transformer has three purposes: (a) safety of the personnel working on the low-voltage side, (b) safety of the equipment from switching or accidental overvoltages in wires on the high-voltage side, and (c) filtering of the high-frequency magnetic and electrical noise.

As per the often-used definition, the isolation transformer has two separate coils with a 1:1 turn ratio. Its sole purpose is to isolate two sides without changing the voltage. However, many power engineers use the term for any two-winding transformer with any turn ratio. Such transformer does isolate, but it also steps up or down the voltage on the other side. The autotransformer with a metallic connection between the primary and secondary coils is not an isolation transformer.

## 5.6 IEEE STANDARD 519

The earlier version of IEEE Standard 519 specified the THD limits on the voltage provided by the utility company but said nothing about the customer's load current causing the voltage distortion. The new version recognizes that both the utility and the customer share the responsibility for maintaining power quality in the area served. We recall that the customer may draw a sinusoidal current as in a transformer

**TABLE 5.3**
**IEEE Standard 519 Limits on THD$_i$ in Individual Customer Load Current**

| SCR | $h < 11$ | $h = 11-15$ | $h = 17-21$ | % THD |
|-----|------|---------|---------|-------|
| <20 | 4.0 | 2.0 | 1.5 | 5.0 |
| 20–50 | 7.0 | 3.5 | 2.5 | 8.0 |
| 50–100 | 10.0 | 4.5 | 4.0 | 12.0 |
| 100–1000 | 12.0 | 5.5 | 5.0 | 15.0 |
| >1000 | 15.0 | 7.0 | 6.0 | 20.0 |

or a chopping current as in a power electronics converter, which would cause the bus voltage harmonics for other area customers. On that basis, the IEEE standard now defines the harmonics current limits listed in Table 5.3 for individual customers. Because a large customer can cause more distortion on the system voltage than a small customer, the standard allows a lower harmonic to large customers than to the small one. The customer size is measured by the *short-circuit ratio* (SCR), which is defined as

$$\text{Short circuit ratio} = \frac{\text{Short circuit kVA at service point}}{\text{Customer's maximum kVA demad}} \qquad (5.26)$$

where

$$\text{Short circuit kVA} = \text{Line voltage} \times \text{Short circuit current at service point} \div 1000$$

$$= (\text{Line voltage})^2 \div (\text{Source impedance at service point} \times 1000) \qquad (5.27)$$

The source impedance is also known as the Thevenin impedance or the internal bus impedance. The maximum demand KVA is known from the monthly utility bill of the medium and large power user. Large customers have a low SCR since they demand relatively higher KVA compared to the system capacity. They can cause

**TABLE 5.4**

**Acceptable Line Voltage Variation at Medium-Voltage Distribution Points**

| Country | Acceptable Range[a] |
|---|---|
| United States | ±5% |
| France | ±5% |
| United Kingdom | ±6% |
| Spain | ±7% |

[a] Low-voltage consumers may see wider variations.

**TABLE 5.5**

**Allowable Step Change in Voltages a Customer Can Cause by Step Loading or Unloading**

| Country | Allowable Range |
|---|---|
| France | ±5% |
| United Kingdom | ±3% |
| Germany | ±2% |
| Spain | ±2% for grid-connected renewable systems |
|  | ±5% for stand-alone power systems |

**TABLE 5.6**

**Voltage Dips, Interrupts, and Variations on European Power Distribution Networks**

| Voltage Drop | Number of Events per Year for Various Duration of Disturbances | | | |
| | 10–100 msec | 100 msec to 0.5 sec | 0.5–1 sec | 1–3 sec |
|---|---|---|---|---|
| 10–30% | 61 | 66 | 12 | 6 |
| 30–60% | 8 | 36 | 4 | 1 |
| 60 to <100% | 2 | 17 | 3 | 2 |
| 100% | 0 | 12 | 24 | 5 |

*Source:* Adapted from Lutz, M., and Nicholas, W. 2004. *Conformity Magazine,* November, p. 12.

more distortion to other area customers and hence have a smaller THD limit in their load current. A small customer can hardly distort the system voltage even by continuously drawing high harmonics in small load current; hence, they do not draw much attention.

## 5.7  INTERNATIONAL STANDARDS

The generally acceptable line voltage variations at a medium-voltage distribution point in four countries are listed in Table 5.4. The permissible variations are higher in developing countries like China, India, and many Eastern European and South American countries. The allowable step changes in voltage a customer can cause on loading or unloading are listed in Table 5.5. The number of voltage disturbances of various durations in a typical European land-based power distribution system (Table 5.6) shows what is expected; small voltage deviations are more frequent, and vice versa.

## PROBLEMS

*Problem 5.1:* A 50-Hz, 800-kW bus has the Thevenin source voltage of 480 V and the Thevenin source impedance of $3 + j25$ m$\Omega$/phase. Determine the bus voltage when delivering rated current at 0.90 pf lagging.

*Problem 5.2:* The values of the harmonic currents measured in a six-pulse converter are given in the $I_h/I_1$ ratios that follow, which are less than the theoretical values of $I_1/I_h$.

| Harmonic | 5 | 7 | 11 | 13 | 17 | 19 | 23 | 25 |
|---|---|---|---|---|---|---|---|---|
| $I_h/I_1$ ratio | 0.18 | 0.12 | 0.05 | 0.04 | 0.02 | 0.015 | 0.01 | 0.01 |

Determine the $THD_i$ with the actual and theoretical harmonic currents.

*Problem 5.3:* A three-phase, *Y*-connected load draws current containing the following Fourier harmonics (all rms amperes) in each line:

| Order $h$ | 1 | 3 | 5 | 7 | 9 | 11 | 13 | 15 | 17 |
|---|---|---|---|---|---|---|---|---|---|
| $I_h$ rms A | 80 | 48 | 32 | 25 | 17 | 15 | 10 | 9 | 8 |

Determine (a) the total (true) rms value of the current, (b) THD based on the fundamental current, (c) true $THD_T$ as used in some international standards, and (d) current in the neutral conductor.

*Problem 5.4:* A power electronics converter draws the fundamental and harmonic currents from a transformer as follows:

| Fundamental | 5th h | 7th h | 11th h | 13th h | 17th h | 19th h | 23rd h | 25th h |
|---|---|---|---|---|---|---|---|---|
| 30 A | 7.5 A | 4.5 A | 3 A | 2 A | 1.5 A | 1.3 A | 1.1 | 1.0 |

Determine (a) the total rms current in the transformer output and (b) the K rating of the transformer needed to power this converter.

*Problem 5.5:* A 460-V bus in a factory supplying power to several feeders has the Thevenin source impedance behind the bus equal to $0.002 + j0.01$ Ω/phase. A power factor-correcting capacitor rated 500 $kVAR_{3ph}$ is connected to the bus as shown in Figure P5.5. Determine the parallel resonance frequency and identify the precaution needed for connecting a power electronics load to this bus.

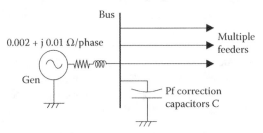

*Problem 5.6:* It is desired to filter out the fifth harmonic from the 60-Hz supply lines using six-pulse converters in an industrial power distribution system. Design the harmonic filter capacitor using a 100-mh inductor coil that is readily and economically available for use.

*Problem 5.7:* For a 60-Hz, 4160-V bus that supplies power to 12-pulse converters, determine two separate harmonic filter inductance values (a) to filter out the 11th harmonic with a 100-kVAR capacitor and (b) to filter out the 13th harmonic with a 120-kVAR capacitor.

*Problem 5.8:* The product of the fundamental voltage and the fundamental current results in a net average power delivered to the loads. On the other hand, the product of the fundamental voltage and a harmonic current results in zero net average power over one cycle. Using Excel, plot the product of 100-$V_{rms}$ 60-Hz fundamental voltage and 80-$A_{rms}$ third-harmonic current in 15° increments over one cycle and verify that their product averages out to be zero over one cycle.

*Problem 5.9*: A three-phase, 460-V transformer is to be procured to power a motor drive that will draw line current containing the following harmonics as per the VFD supplier:

| Harmonic order h | 1 | 5 | 7 | 11 | 13 | 17 | 19 | 23 | 25 |
|---|---|---|---|---|---|---|---|---|---|
| Harmonic amps rms | 25 | 5 | 4 | 2.5 | 2 | 1.5 | 1.5 | 1 | 1 |

Determine the three-phase KVA rating and the K factor of the transformer that must be specified in the procurement contract.

*Problem 5.10:* A 120-V power source with no feedback voltage regulator has static source impedance of $0.2 + j0.6\ \Omega$. Determine its steady-state voltage drop at its terminals under an increase of 8 A in the load current.

*Problem 5.11:* The utility line voltage displayed on a harmonic spectrum analyzer shows the harmonic content of 10%, 7%, 5%, 3%, and 1% of the fundamental—percentages based on the fundamental magnitude—for the 5th, 7th, 11th, 13th, and 17th harmonics, respectively. Determine the $THD_v$ factor for this line voltage as per the U.S. standards and as per some international standards.

## QUESTIONS

*Question 5.1:* In a power distribution system, the line voltage often drops by 50% for 10 ms. Does it meet the ANSI standard on quality of power?

*Question 5.2:* What is the major cause of harmonic currents in the electrical power system?

*Question 5.3:* Explain the difference between static source impedance and dynamic source impedance.

*Question 5.4:* Explain the term *ringing* in the bus voltage and indicate when it occurs.

*Question 5.5:* Where and how do we use the superposition theorem when the source voltage has many harmonics?

*Question 5.6:* What is the transformer K rating, and why does the higher K-rated transformer cost more?

*Question 5.7*: Identify equipment around your work area that may severely suffer in performance due to poor-quality power.

*Question 5.8:* Identify major ill effects of poor power quality on the performance of a large induction motor.

*Question 5.9:* How is poor power quality identified anywhere (home, office, industry, ships, etc.)?

*Question 5.10:* What are the triplen harmonic currents? Identify where they can and cannot flow.

*Question 5.11:* Indicate why the triplen harmonic currents cannot flow in the Δ-connected system.

*Question 5.12:* Explain how very clean harmonic-free power is obtained on navy ships.

*Question 5.13:* Explain how cable shielding and twisting significantly reduce EMI.

## FURTHER READING

Baggini, A. 2010. *Handbook of Power Quality*. New York: John Wiley & Sons.

Fuchs, E., and M. A. Masoum. 2008. *Power Quality in Power Systems and Electrical Machines*. Burlington, NY: Elsevier Academic Press.

Santoso, S. 2010. *Fundamentals of Electric Power Quality*. Austin: University of Texas.

Vedan, R.S., and M.S. Sarma. 2009. *Power Quality*. Boca Raton, FL: CRC Press/Taylor & Francis.

# 6 Power Converter Cooling

The power rating of a piece of equipment, electrical or mechanical, is the maximum continuous output power it can deliver while keeping the steady-state operating temperature below the allowable limit in normal use. The cooling is therefore an integral part of the equipment design for limiting the temperature rise under a rated load. The heat is generated inside the equipment by internal power loss, which must be carried outside and eventually dissipated in the ambient air. We recall that an operating temperature higher than rated significantly reduces equipment life. The 10°C rule for half-life for power equipment holds approximately true for power electronics devices also. Other adverse effects of high temperature on power electronics components are summarized in Table 6.1.

Conventional power equipment, such as motors, generators, and dry-type transformers, are cooled primarily by air circulated inside either by natural convection or by forced air from cooling fans. In power electronics devices, the heat is generated inside a small semiconducting device having a small wafer-thin volume around the junction area. This results in a high thermal gradient from the heat-generating junction to the ambient air, where the heat is finally dissipated. Therefore, limiting the semiconductor junction temperature is much more challenging. It generally requires a heat sink (finned metal) with large surface area around the semiconducting device to keep the junction temperature below the allowable limit. The heat sink, commonly made of aluminum, removes heat from the inside junction to the outside surface by conduction and then to the ambient air by convection and radiation. Figure 6.1 shows a small power electronics subassembly mounted on a heat sink.

## 6.1 HEAT TRANSFER BY CONDUCTION

The heat transfer rate by conduction from one surface to another (Btu/hour or joules/second [watts]) is given by

$$Watts = \frac{\sigma \cdot A \cdot \Delta T}{d} = \frac{\Delta T}{d / \sigma \cdot A} = \frac{\Delta T}{R_{Th}} \quad \text{where} \quad R_{Th} = \frac{d}{\sigma \cdot A} \qquad (6.1)$$

where

$\sigma$ = thermal conductivity of medium between two surfaces
$A$ = area available for heat transfer between two surfaces
$\Delta T$ = temperature difference (gradient) between two surfaces
$d$ = distance between two heat transfer surfaces
$R_{Th}$ = thermal resistance between two surfaces (°C rise/watt transferred or *thermal ohms*)

## TABLE 6.1
## Adverse Effects of High Temperature in Power Electronics Equipment Performance

| Semiconductor Devices | Capacitors | Magnetic Components |
|---|---|---|
| • Unequal power sharing in parallel or series devices. | • Electrolyte evaporation rate increases significantly with temperature increases and thus shortens lifetime. | • Losses increase above 100°C even at constant power input |
| • Reduction in breakdown voltage in some devices. | | • Winding insulation (lacquer or varnish) degrades above 100°C |
| • Increase in leakage currents. | | |
| • Increase in switching times. | | |

**FIGURE 6.1**     Heat sink of thin aluminum fins with power electronics subassembly.

Recalling the electrical Ohm's law: $I = V/R$, Equation (6.1) can be viewed as Ohm's law in heat transfer by conduction. The analogy between the electrical and thermal Ohm's laws is given in Table 6.2. Once this analogy is understood, the heat conduction in a complex system can be broken down in various series-parallel paths, which can be reduced to one equivalent thermal resistance by using the same formulas as for the electrical resistances in series-parallel combinations.

**TABLE 6.2**
**Analogy between Electrical and Thermal Circuits**

|  | Electrical | Thermal |
|---|---|---|
| Ohm's law | $I = V/R$ | $P = \Delta T/R_{Th}$ |
| Flow | Current $I$ (amperes or coulombs/sec) | Heat $P$ (watts or joules/sec) |
| Driver | Potential difference voltage (volts) | Temperature gradient $\Delta T$ (°C) |
| Resistance | Electrical resistance $R = \dfrac{length}{\sigma A}$ (ohms) | Thermal resistance $R_{Th} = \dfrac{distance}{\sigma A}$ (thermal ohms) |

## 6.2 MULTIPLE CONDUCTION PATHS

Once the heat sink conducts the heat from the inside junction of a power electronics device to the outside surface, the heat is then dissipated by natural radiation and convection in the ambient air. Over the operating temperature range of power electronics equipment, the radiation and convections can be combined into one expression similar to Equation (6.1), that is,

$$Watts = \frac{\theta}{R_{\theta sa}} \qquad (6.2)$$

where $\theta$ = temperature gradient $\Delta T$ (we use $\theta$ for $\Delta T$ for simplicity in writing equations from this point), and $R_{\theta sa}$ = heat sink surface-to-air thermal resistance.

From the junction to outside air, there are three thermal resistances in series as shown in Figure 6.2, where

$R_{\theta jc}$ = Thermal resistance (in conduction mode) from device junction to case

$R_{\theta cs}$ = Thermal resistance (in conduction mode) from device case to heat sink

$R_{\theta sa}$ = Thermal resistance (in convection and radiation modes) from heat sink to ambient air

The $R_{\theta jc}$ value is given by the power electronics device vendor, and the $R_{\theta cs}$ value depends on the materials and methods used by the design engineer to mount the device on the heat sink. The value of $R_{\theta sa}$ comes from the technical product data sheet of the heat sink manufacturer. Representative data for selected heat sinks from one manufacturer are given in Table 6.3.

The three temperature rises (gradients) in Figure 6.2 are determined by the power loss multiplied by the respective thermal resistance. The actual operating temperatures are then given by

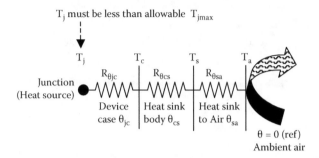

**FIGURE 6.2**  Temperature gradients from power electronics device junction to ambient air.

---

**TABLE 6.3**

**Thermal Resistance of Selected Heat Sinks**

| Heat Sink No. | Thermal Resistance $R_{\theta sa}$ °C/watt | Heat Sink Volume (cm³) |
|---|---|---|
| 1 | 3.2 | 75 |
| 3 | 2.2 | 180 |
| 6 | 1.7 | 300 |
| 9 | 1.25 | 600 |
| 12 | 0.65 | 1300 |

---

Operating temperature of the heat sink surface $T_s = T_a + \theta_{sa}$

Operating temperature of the device case $T_c = T_s + \theta_{cs}$

Operating temperature of the device junction $T_j = T_c + \theta_{jc}$ (6.3)

The total thermal resistance from junction to air is the sum of all three resistances in series, that is,

Total resistance from junction to air    $R_{\theta ja} = R_{\theta jc} + R_{\theta cs} + R_{\theta sa}$ (6.4)

The junction temperature rise above the ambient air is then

$$\theta_{ja} = \frac{Power \ loss \ in \ watts}{R_{\theta ja}}$$ (6.5)

Knowing the maximum power loss under permissible overload and the maximum possible ambient air temperature, the heat sink must be selected to keep

$$R_{\theta ja} < \frac{T_{jmax} - T_{amax}}{Maximum \ watt \ loss}$$ (6.6)

The analysis of steady-state heat transfer by conduction through solid metal can show that the optimum heat sink shape is a cone with the heat-generating devices at the base as shown in Figure 6.3. Therefore, many heat sinks that are solid (i.e., not made of sheet metal) are approximately triangular.

## Example 6.1

A power electronics device has an on-state power loss of 40 W and a switching loss in watts equal to 1.1 times the switching frequency in kilohertz. The junction-to-case thermal resistance is 1.85°C/W, and the maximum allowable junction temperature $T_{jmax}$ is 150°C. If the device case is mounted on a heat sink to limit the case temperature to 50°C, determine the maximum allowable switching frequency for this device on this heat sink.

### SOLUTION

With switching frequency $f_{sw}$ in kilohertz, the total power loss = $40 + 1.1\, f_{sw}$. The maximum allowable thermal gradient from the device case to the junction is $150 - 50 = 100°C$.

$\therefore$ Maximum power loss this device can withstand = $100° C \div 1.85°C/W = 54.05$ W, which should be equal to $40 + 1.1 \times f_{sw}$.

Equating the two, we get $54.05 = 40 + 1.1\, f_{sw}$ or $f_{sw} = 12.78$ kHz, which is the maximum allowable switching frequency for this device.

## Example 6.2

A transistor junction has total power loss of 25 W, and the junction-to-case thermal resistance is 0.9°C/W. The case is mounted on heat sink number 9 with 50-μm

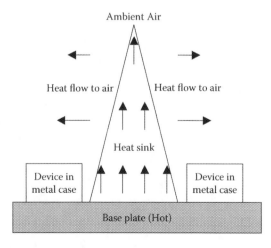

**FIGURE 6.3** Optimum shape of solid metal heat sink is approximately conical.

thick mica insulation with thermal grease, resulting in a case-to-heat sink thermal resistance of 0.5°C/W. The air temperature inside the converter cabinet can be as high as 55°C. Determine the maximum junction temperature in this device

## SOLUTION

We read from Table 6.3 that the thermal resistance from the heat sink surface to air for heat sink number 9 is 1.25°C/W.

∴ Total thermal resistance of three elements in series = 0.9 + 0.5 + 1.25 = 2.65°C/W

The junction temperature rise above cabinet air = 25 × 2.65 = 66.25° C

and

Maximum total junction temperature = 66.25 + 55 = 121.15°C

## 6.3 CONVECTION AND RADIATION

Conduction is the only mode of heat transfer from the interior of the semiconducting device junction to the heat sink surface. From there, the heat is dissipated by convection and radiation to the ambient air inside the cabinet, which also rises in temperature above that of the ambient air in the room. The total internal power loss must eventually be dissipated from the outer surface of the equipment to the ambient air. The heat is transferred from a heated body to the surrounding air by three modes: (a) conduction to solids in contact that can be ignored since most cabinets stand on thin legs with negligible contact with the ground or other solid surfaces, (b) convection in air, and (c) radiation in space. If $\theta$ = temperature difference (rise) between the heated body and the cooler surrounding air, then the total heat dissipation in the typical operating temperature range can be expressed with some approximation as

$$\text{Watts dissipated} = K_1\theta + K_2\theta^\alpha + K_3\theta^\beta \tag{6.7}$$

where $K_1$, $K_2$, $K_3$ are the conduction, convection, and radiation constants, respectively, which depend on the areas and thermal conductivities of the materials involved in the cooling path, and $\alpha$ and $\beta$ are the convection and radiation exponents, respectively.

The temperature rise is determined by the surface area available for dissipating the internal power loss that heats the equipment body. Air-cooled equipment dissipate the internal heat mostly by convection, some by radiation, and negligibly by conduction. A heated surface with the same exposed area for convection and radiation in the normal operating range dissipates approximately equal heat by convection and radiation, and its temperature rise in °C above the ambient air is given by an empirical relation:

$$\theta_{rise°C} = \frac{540 \times H_{cm}^{0.1} \times \left(\text{watts}\right)^{0.8}}{cm^2} \tag{6.8}$$

where watts = power loss to be dissipated, cm² = surface area (both horizontal and vertical) exposed to ambient air, and $H_{cm}$ = height in centimeters.

Greater height results in greater $\Delta T$ because of reduced convention cooling at the upper height, where the heated air is close to the surface temperature, thereby reducing the convective heat transfer rate. Having determined the temperature rise from Equation (6.8), the enclosure operating temperature of the heat-dissipating body is then

$$T_{operating} = \theta_{rise} + T_{room} \qquad (6.9)$$

The equipment cabinet may be sealed if it is in an area with dust or combustible vapor. In that case, the cooling calculations are done in two stages, one from the equipment to the cabinet inside air and then from the cabinet surface to the room air. Industry standards require that the enclosure surface temperature be less than the specified limit considered safe for human touch in a given application.

## Example 6.3

The power train devices in a power electronics converter are mounted on a metal baseplate 10 cm high by 20 cm wide. The total switching power loss in the devices is 40 W. Determine the temperature rise of the plate due to convection and radiation cooling from both sides of the plate. The ambient air circulating inside the converter cabinet is at 50°C. Also, determine the effective thermal resistance of the plate in °C/W units.

### SOLUTION

Plate surface area on both sides = 2 × 10 × 30 = 600 cm²

Using Equation (6.8), the plate temperature rise

$$\theta_{rise} = \frac{540 \times 10^{0.1} \times 40^{0.8}}{600} = 21.7 \,^{\circ}C$$

∴ Effective thermal resistance = 21.7/40 = 0.54°C/W

and

Surface temperature of the plate = 21.7 + 50 = 71.7°C

The junction temperature of the devices will be higher than 71.7°C by the conductive thermal gradients from the plate surface to the device case and then from the device case to the junction.

## 6.4  THERMAL TRANSIENT

On turning on the equipment, the transient temperature rise $\theta$ above the ambient air at any time can be determined from the differential equation

$$\frac{d\theta}{dt} = \frac{(P - Sk\theta)}{GC_p} = \frac{\dfrac{P}{Sk} - \theta}{\dfrac{GC_p}{Sk}} = \frac{(\theta_m - \theta)}{\tau} \qquad (6.10)$$

where $P$ = power loss in equipment at rated load, $S$ = dissipating surface area, $\lambda$ = dissipation rate perunit area per degree centigrade, $G$ = equipment mass, $C_p$ = specific heat of mass (average), $\tau = GC_p/S\lambda$ = time constant, and $\theta_m = P/(Sk)$ = final temperature rise the equipment will reach.

For equipment starting from initial room temperature, the particular solution to Equation (6.10) is

$$\theta = \theta_m \left[ 1 - e^{-\frac{t}{\tau}} \right] \qquad (6.11)$$

which is in the same form as the capacitor voltage or inductor current rising during the charging period (engineering fields have many such similarities). In the reverse, when the equipment is turned off, we can show that the body temperature above the air decays as

$$\theta = \theta_m e^{-\frac{t}{\tau}} \qquad (6.12)$$

Equation (6.12) gives an exponential decay rate $d\theta/dt = -1/\tau\theta_m e^{-\frac{t}{\tau}}$ at any time $t$, and the initial rate of decay at $t = 0$ is equal to $-\theta_m/\tau$. If the body continues to cool at this initial rate of temperature decay, it will reach room temperature (i.e., $\theta = 0$) in one time constant $\tau$. This leads to an alternative useful way of defining the thermal time constant as follows:

Thermal time constant $\tau$ = Time taken by body under cooling to reach room temperature if the initial cooling rate continues until end                     (6.13)

From electrical circuit theory, we know that the capacitor and inductor charging and discharging take five time constants ($5 \times \tau$) for the transient to reach the final value (99.67%). That applies to thermal heating and cooling as well. The following is a simple and practical use of this definition:

Say we wish to turn off a piece of equipment and start an urgent fix as soon as it reaches room temperature. To estimate how long we must wait, we can measure the initial surface temperature of the equipment body at $t = 0$ (say it is 75°C) and then measure it again at $t = 1$ minute after it is turned off (say it is 73°C). This gives the decay rate of 2°C per minute. If the room temperature is 25°C, the surface temperature has to fall a total of $75 - 25 = 50$°C, which gives the time constant of $\tau = (50$°C total decay $\div 2$°C per minute initial decay rate$) = 25$ minutes. The final room temperature will then be reached in $5 \times 25 = 125$ minutes, when we can start working on the repairs.

In the reverse, after turning on the equipment, we can predict the time to reach the final temperature of the equipment by measuring the temperature rise in the first minute of turning on at full load. If the rated temperature rise is given on the equipment nameplate (it usually is), then dividing it by the first minute rise gives the time constant $\tau$ of the equipment enclosure, and the time to reach the final operating temperature will then be $5 \times \tau$.

## 6.5 WATER COOLING

Modern power electronics equipment handle high power and yet have a compact volume. Although the efficiency is high, a 10-MW, 95% efficient converter would generate $10,000 \times \{1/0.95 - 1\} = 526.3$ kW heat internally in a small volume around the device junctions. Since cooling such high-power equipment by natural air convection can be a design challenge, they are often cooled by internally circulating water through copper tubes. The water is a much more effective coolant than air because of its high density and specific heat compared to air. Large water-cooled generators have stator coils wound with hollow copper conductors that carry both the electrical current and the cooling water. Figure 6.4 shows the cross section of a water-cooled stator conductor

**FIGURE 6.4** Cross section of hollow water-cooled stator conductor of three-phase, 1200-MVA, 30-kV generator ($6 \times 8$ cm overall dimensions and 23,100 $A_{rms}$ rated current).

of a large three-phase, 1200-MVA (1100-MW), 30-kV synchronous generator for a large power plant. It has 6 × 8 cm overall dimensions and 23,100 $A_{rms}$ rated current.

The heat transfer from the wall of a thermally conductive metal tube carrying circulating water depends on many fluid dynamics variables (e.g., the Nusselt, Reynolds, and Prandtl numbers; whether the flow is streamline or turbulent, etc.). However, we take a simplified view of the fluid flow and heat transfer to the circulating water in conductor tubes running through the heat-generating equipment (power electronics converter or electrical machine). The basic equation that determines the water temperature rise is

Heat transferred to water = Water mass × specific heat × Temperature rise

which leads to

$$Water\ temperature\ Rise = \frac{Watts\ transferred\ to\ water}{Water\ flow\ rate\ kg\ per\ second \times Specific\ heat} \quad (6.14)$$

Measuring the temperature rise of water between inlet and outlet in degrees centigrade, Equation (6.14) reduces to a simplified relation between the power loss and the water quantity flow rate in two systems of units as follows:

$$Liters/minute\ \frac{14.3 \times kW\ power\ loss}{Water\ temperature\ rise\ in\ {}^{\circ}C}$$

$$U.S.\ gallons\ per\ minute = \frac{3.75 \times kW\ power\ loss}{Water\ temperature\ rise\ in\ {}^{\circ}C} \quad (6.15)$$

Equation (6.15) gives the water flow rate for a given piece of equipment assuming that all water is in contact with the cooling tube. In a large-diameter tube with laminar flow, only a fraction of the total water flow is in contact with the tube wall. Therefore, even with the required quantity of flow rate passing through the tubes, the *heat may not be removed* unless the tube is narrow enough to have high-velocity turbulent flow with all water in contact with the tube. On the other hand, narrow tubes result in a high drop in pressure. The design engineer may take several iterations before arriving at a satisfactory design that optimizes the heat transfer and the pressure drop.

### 6.5.1 Cooling Tube Design

The cooling water progressively heats up while passing through copper tubes in the heat-generating equipment. The water enters one end at room temperature (10–25°C) and leaves the other end about 30–40°C hotter. Therefore, the temperature range of the cooling water in the equipment could be 10–65°C. The properties of water (density, specific heat, etc.) would change with temperature but by a negligible amount over this temperature range. Therefore, we assume the water density and specific heat constant to simplify the temperature rise estimate. Over the practical temperature range, we use a constant value for water density of 1000 kg/m³ and specific heat of 4180 J/(kg °C). With these values, a fluid flow and heat transfer analysis will lead to the following relation for the cooling water path parameters shown in Figure 6.5:

$$\text{Watts}/(L\theta) = 148.2\,(VD)^{0.8} \tag{6.16}$$

where

Watts = Equipment power loss transferred to water
$L$ = Total length of cooling tube in one series path (m)
$\theta$ = Mean temperature difference, $\theta_{tube(constant)} - \theta_{water(average)}$ (°C)
$V$ = Velocity of water in tube (m/s)
$D$ = Inside diameter of cooling tube (cm)

In terms of the water quantity flow rate $Q$ in liters/minute,

$$Q = \frac{\pi(D/2)^2\,100V}{1000} \times 60 \text{ or } V = \frac{Q}{1.5\pi D^2} \tag{6.17}$$

Using Equation (6.17) in (6.16), we obtain

$$\frac{\text{Watts}}{L\times\theta} = 148.2\left(\frac{Q}{1.5\times\pi\times D}\right)^{0.8} = 42.9\left(\frac{Q}{D}\right)^{0.8}$$

$$\therefore Q = \frac{Liters}{minute} = D\left(\frac{Watts}{42.9\,L\times\theta}\right)^{1.25} \tag{6.18}$$

$$\text{Water velocity } V_{m/s} = 0.212\frac{Liters/minutes}{D^2_{cm}} \tag{6.19}$$

The left-hand side of Equation (6.18) suggests that for the same power loss, tube length, and temperature difference, the $Q/D$ ratio must be maintained constant. If we wish to have fewer liters of water circulated, then the diameter of the tube must be reduced. However, it would increase both the water temperature and the tube

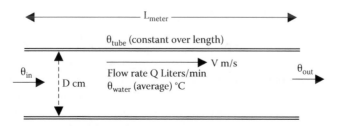

**FIGURE 6.5**   Water cooling tube parameters for thermal analysis.

wall temperature. It would also increase the pressure drop through the tube running through the equipment, which would require a larger pump and higher pumping power. These are the design trades the engineer makes to meet the overall system requirements.

As the cooling water circulates through the tube, its temperature rises by $\Delta\theta_{water} = \theta_{in} - \theta_{out}$, where $\theta_{in}$ = inlet temperature of water, and $\theta_{out}$ = outlet temperature of water. If $\theta_{tube}$ = tube temperature (assumed constant), then the mean temperature difference between tube and water

$$\theta = \frac{1}{2}\{(\theta_{tube} - \theta_{in}) + (\theta_{tube} - \theta_{out})\} = \theta_{tube} - \frac{1}{2}(\theta_{in} + \theta_{out}) \tag{6.20}$$

These equations are accurate within 5% for $\theta_{in}/\theta_{out} < 2$.

### 6.5.2 Pressure Drop

Determining the water pump rating requires calculating the pressure difference (drop) between the water inlet and outlet. Various sources of pressure drop in a fluid flow through a series of cooling tubes are frictions at (a) the tube wall along its length, (b) tube entrance and exit, and (c) each bend in the tube routing path. The water is generally drawn from a large header (or reservoir) and is discharged into a large header (or reservoir). The entry and exit velocities are therefore extremely low and can be ignored. The pressure drop is then approximately given by the following:

$$\Delta P \text{ in pascals} = 251\frac{5.55\,L \times V^{1.8}}{D_{cm}^{1.2}} + (750 + 50 \times \text{Number of } 90° \text{ bends})\,V^2 \tag{6.21}$$

In Equation (6.21), the multiplier 50 with the number of 90° bends assumes that the bend radius to the tube diameter ratio is greater than three (soft bends). The multiplier will be 75 for less-soft bends with the ratio of two and 100 for the ratio of one (sharp bends). For 180° bends (U turns), the number of 90° bends to use in Equation (6.21) = 2 × 180° bends.

In SI units, $\Delta P$ is measured in pascals (newtons/m²). In practice, it is often expressed in the water head in meters, which is

$$\Delta H \text{ in meters of water head} = \frac{\Delta P \text{ in pascals}}{9.81 \times 1000} \tag{6.22}$$

*Parallel cooling paths:* Multiple parallel paths may be required for limiting the pressure drop in a long tube. If $N$ = number of identical parallel paths, then the previous formulas apply to each path with its own kilowatt power loss, length, and water quantity per path. However, the pressure drop $\Delta P$ is the same for all parallel paths. For nonidentical parallel paths, the kilowatt power loss, water flow rate $Q$, and water velocity will be different in each path, but the $\Delta P$ would be the same.

*Noncircular tubes:* The cooling water tubes are often round copper or aluminum tubes in most applications. When they are other shapes (rectangular, square, oblong, or triangular tubes or even two parallel plates with a narrow gap), their equivalent diameter is used in the formulas. It is derived to give the same water flow rate for a given $\Delta P$.

$$Equivalent\ tube\ diameter = \frac{4 \times Cross\ section\ of\ the\ water\ flow}{Wetted\ perimeter\ of\ the\ water\ flow\ path} \quad (6.23)$$

The power electronics converter subassemblies are often mounted on a solid metal plate—called a baseplate or the cold plate—with a labyrinth of water tubes as shown in Figure 6.6. Parallel plate heat exchangers are also used at many places on ships, such as in large propulsion power converters and lubricating oil cooling systems. Figure 6.7 is one such parallel plate heat exchanger for cooling the main lubricating oil.

## Example 6.4

A power converter assembly with 800-W power loss is mounted on a 20-cm square baseplate cooled by water. The water tubes are 2-mm internal diameter placed on a 15-cm square with four 90° bends as shown in Figure E6.4. If the inlet and outlet water temperature difference is to be limited to 25°C, determine the water flow rate in liters per minute.

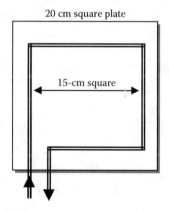

20 cm square plate

15-cm square

### SOLUTION

For the cooling tube length $L = 4 \times 15 \div 100 = 0.6$ m, $D = 0.2$ cm, and $\theta = 25°C$, Equation (6.15) gives the cooling water flow rate

Liters/minute $= 14.3 \times 0.80 \div 25 = 0.458$

From Equation (6.19),

**Metal Base Plate Cooled by Water**

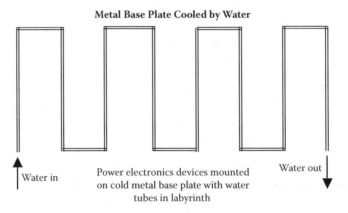

Water in

Power electronics devices mounted
on cold metal base plate with water
tubes in labyrinth

Water out

**FIGURE 6.6**   Labyrinth metal tubes for water-cooled power electronics converter assembly.

**FIGURE 6.7**   Parallel plate heat exchanger for main lubricating oil cooler. (From Raul
P. Osigian, U.S. Merchant Marine Academy.)

Water velocity in tubes = 0.212 × 0.458 ÷ 0.2² = 2.43 m/s

From Equations (6.21), and (6.22)

$$\text{pressure drop} = 251\frac{5.55\times0.6\times2.43^{1.8}}{0.2^{1.2}}+(750+50\times4)\times2.43^{2}$$

$$= 28,614 + 5,610 = 34,224 \text{ pascals}$$
$$= 34,224 \div (9.81\times1000) = 3.49 \text{ m water head}$$

## Example 6.5

A piece of electrical equipment is cooled by a labyrinth (Figure 6.6) of 1.0-cm inside diameter water tubes made of 15 straight sections, each 0.6-m long joined at ends by 90° connectors and 0.1-m straight bridging pieces. The cooling water enters the labyrinth from a large header and discharges to a large header such that the velocity at both ends is negligible. If the water velocity in the tubes is 0.8 m/s, determine the pressure drop and heat removal capacity of the labyrinth with a 30°C water temperature rise.

### SOLUTION

Total tube length = (15 × 0.6 ) + (14 × 0.1) = 10.4 m

Number of bends = 14 × 2 = 28

Equation (6.21) gives the pressure drop

$$\Delta P = 251\frac{5.55\times10.4\times0.8^{1.8}}{1.0^{1.2}}+(750+50\times28)\times0.8^{2} = 9695+1376$$

$$= 11,071 \text{ pascals}$$

Pressure drop in meters of water head = 11,071 ÷ (9.81×1000) = 1.13 m = 1.6 psi (small)

From Equation (6.19), we get

Liter/minute = 0.8 × 1.0²/0.212 = 3.774

Then, from Equation (6.15), we obtain the heat removal capacity:
kW = 3.774 × 30/14.3 = 7.917

## Example 6.6

A vertical heat exchanger dissipates 1 MJ/hour per pair of cooling plates that are 25 cm wide and 50 cm high and separated by a 1.5-mm water gap. Determine the water flow rate and pressure drop to limit the water temperature rise to 20°C.

## SOLUTION

Heat exchange rate = $1 \times 10^6 \div 3600 = 278$ J/s = 0.278 kW

From Equation (6.15),

Water flow rate = $14.3 \times 0.278/20 = 0.2$ liter/minute

Actual water velocity between plates = $0.2 \times 1000/(25 \times 0.15) = 53.3$ cm/s = 0.533 m/s

From Equation (6.23),

Equivalent diameter of flow cross Section $= \dfrac{4 \times (25 \times 0.15)}{2 \times (25 + 0.15)} = 0.298$ cm

Using Equation (6.19),

Effective velocity = $0.212 \times 0.2 \div 0.298^2 = 0.480$ m/s (close match with the actual 0.533 m/s)

Length of the water flow = 50 cm = 0.5 m

Then, Equation (6.21) gives

Pressure drops $\Delta P = 251 \dfrac{5.55 \times 0.50 \times 0.48^{1.8}}{0.298^{1.2}} + 0$ for no bends $= 793$ pascal

$$= \frac{793}{1000 \times 9.81} = 0.08 \text{ m of water head}$$

The pressure drop here is negligible, as expected in a plate heat exchanger. The pump rating is primarily determined by the pressure drops in other plumbing that brings the water to the plates.

### 6.5.3 COOLING WATER QUALITY

The electrically live hollow stator conductors (Figure 6.4) in large utility-scale synchronous generators are cooled with deionized distilled water, which gives an equipment life of 25 to 30 years. For a reasonably long life of the cooling tubes that are not electrically live, the circulating water inside must be at least clean freshwater. Shipboard equipment dissipating the heat ultimately in the ocean may require an intermediate freshwater or deionized water heat exchanger. For example, the U.S. Coast Guard *Healy* icebreaker cycloconverter is water cooled with deionized water circulating through the thyristor heat sinks. Deionized water is often used for power electronics converter cooling when the heat sinks are electrically live. The deionized water is passed through stainless steel coils cast into an aluminum body. Such a water-cooling technique clearly enables a compact thyristor module assembly. The ultimate heat rejection in the sea is achieved by water flowing between parallel plates.

The water cooling is used in many high-power variable-frequency drives (VFDs) for ship propulsion motors. It is generally configured in a closed loop consisting of primary and secondary sides separated by a plate-type heat exchanger. The primary side has a pump, heat exchanger, plumbing, and distribution manifolds for each side of the propulsion switchboard. The secondary side also has a pump, plumbing, three-way heat exchanger bypass valve, and the ultimate cooling medium (seawater in this case). The cooling water circuit is monitored with differential pressure flow meters in the primary pump discharge pipes, pressure sensors in distribution manifolds, and temperature sensors in both water circuits. Such complex monitoring is installed not only to alert the operator of cooling system problems but also to minimize the pumping power and maintain the temperature by varying the primary pump motor speed as required to meet the VFD load demand. The scheme often uses a pulse width modulated (PWM) drive for the centrifugal pump motor. The monitoring circuit also has shutdown functions to prevent damage due to out-of-range faults. In addition, leak detection is installed within VFDs under the cooling plates, which alters operators of either plate leakage or excessive condensation due to low temperature.

The typical cooling water is a 20%/80% mixture of propylene glycol and deionized water, respectively. Although propylene glycol has a lower specific heat and higher viscosity than plain water, and therefore reduced heat transfer performance, it raises the boiling point, lowers the freezing point, and contains many anticorrosive properties that extend the service life of components and reduce maintenance. The mixture is cooled by seawater circulating through a brazed plate heat exchanger. All plumbing components are stainless steel tubing.

With a range of seawater temperature typically encountered in practice, monitoring the VFD operating temperature and corresponding load adjustment is required. High water temperature would result in high drive temperature, poor performance, and possible damage if the drive load is not reduced. On the other hand, too low water temperature results in condensation in the VFD itself and possible damage. The recommended water temperature in the secondary side heat exchanger is +23°C with a minimum temperature difference of 5°C between the primary and secondary sides of the heat exchanger. Therefore, the primary side water is below 30°C. A three-way valve can bypass the cooling water around the heat exchanger to maintain the proper temperature range of the secondary cooling medium.

## PROBLEMS

*Problem 6.1*: A power electronics device has an on-state power loss of 30 W and a switching loss in watts equal to 1.5 times the switching frequency in kilohertz. The junction to case thermal resistance is 2.0°C/W, and the maximum allowable junction temperature $T_{jmax}$ is 180°C. If the device case is mounted on a heat sink to limit the case temperature to 60°C, determine the maximum allowable switching frequency for this device.

*Problem 6.2*: A transistor junction has power loss of 45 W and a junction-to-case thermal resistance of 1.1°C/W. The case is mounted on heat sink number 6 with 35-μm-thick mica insulation with thermal grease, resulting in a case-to-heat-sink thermal resistance of 0.7°C/W. The air temperature inside the converter cabinet can be as high as 65°C. Determine the maximum junction temperature in this device

*Problem 6.3*: The power electronics devices in a converter assembly are mounted on a metal base plate 15 cm high by 25 cm wide. The total switching power loss in the devices is 95 W. Determine the temperature rise of the plate due to convection and radiation cooling from both sides of the plate. The ambient air circulating inside the converter cabinet is at 55°C. Also, determine the effective thermal resistance of the plate in units of degrees centigrade per watt.

*Problem 6.4*: A power converter assembly with 1.2-kW power loss is mounted on a 30-cm square baseplate cooled by water. The water tubes are 2-mm internal diameter placed on a 25-cm square frame with four 90° bends. If the inlet and outlet water temperature difference is to be limited to 30°C, determine the water flow rate in liters per minute.

*Problem 6.5*: A piece of electrical equipment is cooled by a labyrinth of water tubes with an inner diameter of 1.0 cm and made of 12 straight sections, each 80 cm long, joined at ends by 90° connectors and 12-cm starlight bridging pieces. The cooling water enters the labyrinth from a large header and discharges to a large header such that the velocity at both ends is negligible. If the water velocity is 1.5 m/s, determine the pressure drop and heat removal capacity of the labyrinth with a 40°C water temperature rise.

*Problem 6.6*: A large vertical heat exchanger dissipates 10 kW per pair of cooling plates that are 30-cm wide and 95-cm high and separated by a 10-mm water gap. Determine the water flow rate and pressure drop to limit the water temperature rise to 30°C.

*Problem 6.7*: A 10-MW water-cooled power electronics converter is 97% efficient. Determine the water requirement in gallons per minute if the available cooling water inlet temperature is 30°C, and the outlet temperature must be limited to 60°C.

*Problem 6.8*: A three-phase, 3-cm outer diameter cable in air has a resistance of 2 mΩ/phase per meter and carries 100 A/phase load. Determine its outside surface temperature (rise + ambient) if the ambient air is 40°C.

## QUESTIONS

*Question 6.1*: Briefly summarize the mathematical analogy between the heat transfer and the flow of electrical current.

*Question 6.2*: How would you determine the thermal time constant of a piece of equipment merely by using two readings of a thermometer with the readings taken 1 minute apart?

*Question 6.3*: Why is passing the required quantity of cooling water not sufficient to limit the equipment temperature rise?

*Question 6.4*: Passing less water through a thinner tube can remove the same heat with a given length of tube. What is the downside of this?

*Question 6.5*: How can you reduce the pressure drop in the water cooling path by half?

*Question 6.6*: Discuss your experience (if any) in water cooling of shipboard or land-based electrical or power electronics equipment.

## FURTHER READING

Ellison, G. 2010. *Thermal Computations for Electronics*. Boca Raton, FL: CRC Press/Taylor & Francis.

Shabani, Y. 2010. *Heart Transfer, Thermal Management of Electronics*. Boca Raton, FL: CRC Press/Taylor & Francis.

Smith, S. 1985. *Magnetic Components*. New York: Van Nostrand Reinhold.

# Part B

## Electric Propulsion Technologies

Part B on the electric propulsion of ships is dominated by variable-frequency drive (VFD) technology based on power electronics converters. The VFD converts the fixed line frequency power to variable-frequency power at the motor terminals. The frequency which may vary to match with the required propeller speed during various phases of ship operations, such as approaching and leaving a port and while cruising at various speeds. Part B also covers at length various propulsion motors, including the permanent magnet and superconducting motors recently developed by the U.S. Navy for ship propulsion.

Electric propulsion in passenger cruise ships and navy warships offers minimum noise and vibration and maximum usable space for paying passengers or for combat weapons. It is fully developed in cruise ships, where it has opened up usable space for passenger cabins to bring greater revenue to owners. A common electric propulsion package fitted on many cruise ships consists of a 21-MW gas turbine driving a 21-MW generator that powers a synchroconverter, cycloconverter, or pulse width modulated (PWM) VFD with a 20-MW synchronous motor or an advance induction motor.

For warships, the electric propulsion technology that would meet military requirements on survivability and stealth is well under development. The U.S. Navy is committed to electric propulsion for surface ships and is investing additional efforts to develop a common technology that is adaptable to submarines as well.

It is in this light that electric propulsion technology has become increasingly important to shipbuilders and operators. The technology primarily involves high-power, high-voltage electric power systems, compact motors with high power and torque densities, and variable-speed power electronics motor drives. Part B covers the electric propulsion technology that exists today and the current research and development efforts that will bring changes in the coming years.

# 7 Electric Propulsion Systems

Electric propulsion technology changes the way a ship transmits power from the main engine (prime mover) to the propeller and the way it manages and distributes electrical power to both the propulsion and nonpropulsion loads. It does not change the primary power source of the ship, which remains the diesel engine, gas turbine, or steam turbine.

The ship propulsion system with mechanical drive and reduction gears was first developed in the United Kingdom. Following the development of the first large electric motor and generator in 1910, electric propulsion for ships was developed in the United States and elsewhere. Both systems competed against one another until 1920, when the British developed a lightweight, high-efficiency mechanical drive system, which dominated ship propulsion technology around the world for the decades that followed. The U.S. Navy revived electric propulsion technology during World War II for destroyer escorts because the U.S. gear-cutting capacity at the time was insufficient.

The first electric propulsion for ice-going vessels was introduced around 1939, when the Finnish icebreaker *SISU* was delivered with a Ward–Leonard dc electric propulsion system. Since then, various types of electric propulsion systems have been used for hundreds of icebreakers and ice-going vessels with propulsion power approaching 50 $MW_e$.

After World War II, mechanical drive technology continued to improve and remained dominant. Among warships, electric propulsion technology was widely adopted only for submarines, for which the diesel-electric power plant became the standard system. It permitted the submarine to propel itself submerged for limited periods of time on battery power, without access to the atmosphere for oxygen to burn diesel fuel onboard. Electric propulsion technology was also reexamined for use on navy ships other than small diesel-electric submarines. On commercial ships, electric propulsion was used in a few large cruise ships, such as *Normandie* in 1936 and *Canberra* in 1960.

These experiments, however, periodically confirmed that electric propulsion technology, while promising, was not competitive with mechanical drive technology for large submarines and surface ships until the 1980s, when the technological developments in motors, particularly in power electronics motor drives, made electric propulsion potentially more cost effective than the mechanical drive for large naval ships. In 1985, the United Kingdom began building the Duke class type 23 frigates, which used a combined diesel-electric and gas turbine-mechanical drive propulsion plant. It used a lower-power diesel-electric propulsion system for quiet sonar-towing operations at speeds up to 14 knots and a gas turbine-mechanical drive for higher speeds up to the maximum sustained speed of 28 knots for these ships.

In 1987, the *Queen Elizabeth II* (*QE2*) cruise ship underwent an overhaul when its onboard mechanical drive system was replaced with an integrated electric power (IEP) system. This system was operated successfully and set the stage for widespread adoption of electric propulsion technology for cruise ships. Today, most cruise ships in the world are being built with electric propulsion.

## 7.1  CURRENT STATUS OF ELECTRIC PROPULSION

The current status of electric propulsion in cargo, cruise, and navy ships, however, is quite different for each other, as described next.

### 7.1.1  COMMERCIAL CARGO SHIPS

There is not much current interest for electric propulsion in cargo ships hauling freight over long sea routes at one constant speed, with no passengers to please or weapons to launch with a high burst of power. Electric propulsion, however, is presently used almost exclusively in icebreakers and floating offshore oil platforms and is becoming more common in passenger and car ferries. Other kinds of commercial ships now being built with electric propulsion include shuttle tankers, pipe- and cable-laying ships, and research ships.

### 7.1.2  CRUISE SHIPS

Electric propulsion with a podded motor is a norm at present for cruise ships. As discussed further in this chapter, this provides great benefits for the internal layout and ship maneuverability. The pod propulsion motor is mounted in water beneath the ship. It is used extensively in cruise ships and is being considered for navy ships as well. Since the pod design is dominated by the motor and its cooling system, the newly developed compact motor technology is the main technology enabler for electric propulsion.

The electric propulsion systems used today in cruise ships and other commercial ships are generally made overseas, primarily in Europe. The two of the three primary European-based electric propulsion suppliers are Alstom and Asea Brown Boveri (ABB), which together account for most of the electric propulsion systems in operation today. The primary electric propulsion facilities of ABB are in Finland and Italy. Siemens of Germany has a smaller market share but has a lead in the permanent magnet motors and the advance motor drives.

### 7.1.3  U.S. NAVY SHIPS

Electric propulsion technology is already in use on a few U.S. government ships. They include the U.S. Coast Guard icebreaker *Healy*, several TAGOS-type ocean surveillance ships for Military Sealift Command, and a few AGOR-type oceanographic research ships operated by academic institutions under the University National Oceanographic Laboratory System (UNOLS) of the navy. The U.S. Navy in January 2000 selected electric propulsion technology for the DD-21 land-attack destroyer and considered it for other types of navy ships as well. The electric propulsion development effort of the navy centers on the IEP program. Several private-sector firms in the United States are now pursuing electric propulsion for the U.S. Navy. Electric propulsion offers significant benefits for navy ships in terms of reducing life-cycle cost, increasing ship stealth, survivability, and greater power available for weapons. The disadvantages include high development cost and technical risk, increased system complexity, and reduced efficiency in full-power operations.

Electric propulsion is installed on Virginia class (SSN-774) submarines procured in fiscal year 2010 and beyond. Other candidates for electric propulsion include the TADC(X) auxiliary dry cargo ships AND joint command and control (JCCX) ships planned for the navy, the second through fifth replacement of multipurpose amphibious assault ships, future aircraft carriers, and possibly the new cutters to be procured under the Coast Guard Deepwater project.

The interest of the U.S. Navy in electric propulsion technology is consistent with that of commercial ship operators (especially cruise ship operators) and other navies around the world. The British Navy in particular is also gradually moving to electric propulsion for its ships, although taking a somewhat different approach in motor drives. The strong interest in electric propulsion for naval ships taking different approaches suggests that electric propulsion not only offers several war-fighting and life-cycle cost advantages but also has multiple technical approaches to choose from to optimize power system design for a given military mission.

The U.S. Navy has established its integrated electrical power program for use in DD-21-type destroyer ships of the future. Under this program, the navy completed tests in 2000 on a full-scale, land-based electric propulsion demonstration system in Philadelphia. The prime contractor for this system was Lockheed Martin, but much of the actual equipment came from Alstom. In continuing work, five corporations with their own team members are involved in developing the electric propulsion systems for the navy:

The *Alstom* team, with marine operations in Britain, France, Germany, and the United States (Pittsburgh and Philadelphia), has developed an IEP that includes the synchronous motor, but the induction and permanent magnet motors are also being developed.

The team from *General Dynamics* has developed an IEP using the permanent magnet motor.

The IEP of the team from *Newport News Shipbuilding* also has permanent magnet motors.

*American Superconductor* has developed a compact superconducting synchronous motor that gives high torque and power densities suitable for applications in pod drives.

*General Atomics* is placing new effort in developing a superconducting homopolar motor that has been under development for many years by the U.S. Navy; Westinghouse built and tested one in the 1980s with some encouraging results and some reported shortcomings.

### 7.1.4 FOREIGN NAVY SHIPS

Until recently, the submarines were diesel-electric, in which the diesel engine uses the air drawn from the surface through a snorkel to generate electrical power, which is stored in large batteries. The battery power is then used to propel the ship during submerged operations. In a way, navies have been using electric propulsion for decades. Virtually all of the nonnuclear-powered submarines in the world have electric propulsion systems today. Some European countries are now introducing air-independent propulsion (AIP) that use fuel cells, sterling engines, or closed-cycle diesel engines as the prime movers. Regardless of the new types of prime movers, these submarines still have electric propulsion. The following are the current activities by foreign navies:

*United Kingdom:* The British Ministry of Defense (MoD) has funded development of advance electric propulsion technologies for possible use on future ships under their own IEP program. These technologies include a transverse-flux permanent magnet motor being developed by Alstom and Rolls-Royce. This motor departs significantly from the U.S. propulsion motors.

*France:* Electric propulsion technology with the permanent magnet motor is being developed by the French Navy for their nuclear-powered submarines.

*Germany:* A new type 212 German nonnuclear-powered attack submarine was designed in 2003 with an advance electric propulsion system that includes fuel cells as an alternative to the diesel engine and a permanent magnet motor, both made by Siemens. Germany is already using electric propulsion with permanent magnet motors in minesweepers, which are oceanographic ships that incorporate nonnuclear submarines.

*Italy:* Italy is building its own German-designed type 212 submarines. These ships will use Siemens-made fuel cells and permanent magnet motors. The ABB in Italy developed a prototype permanent magnet motor for the Italian submarine program, but the Italian Navy selected the Siemens motor.

*Netherlands:* The new amphibious ship *Rotterdam* of the Dutch Navy is an example of electric propulsion technology in a warship. Electric propulsion is also used in many Dutch ferries at present.

*Russia:* The electric propulsion technology in icebreakers is routinely used by Russian shipbuilders, who may also be developing advance electric propulsion technologies.

As stated, electric propulsion technology changes the way a ship transmits power from the main engines (prime movers) to the propellers, as well as the way it manages and distributes electrical power to both the propulsion and nonpropulsion loads. The transmission of power from the primary power source (diesel engine, gas turbine, or steam turbine) of the ship to its propellers is classified in the following ways:

*Mechanical propulsion:* Most commercial and military ships today transmit power from the main engine of the ship to the propeller using the mechanical drive system. The high revolutions per minute (rpm) of the engine are transmitted via a rigid shaft to reduction gears that reduce the rpm appropriate for the ship propeller. A second rigid shaft then transmits the low rpm from the reduction gears to the propeller as shown in Figure 7.1a. Ships with multiple propellers have multiple engines, reduction gears, and shafts. The electrical power for shipboard service loads is supplied by dedicated electrical generators as shown on the right-hand side of Figure 7.1b.

*Electric propulsion:* Most navy ships at present are powered by gas turbines that drive the propellers by means of heavy reduction gears and shafts. In electric propulsion, instead of turning gears, the turbines directly drive electrical generators at high rpm. The electrical power is then transmitted by cables toward the stern of the ship to the motor drive, which modifies the frequency and voltage for the propulsion motor of the ship to operate at a desired low speed. The electric motor then drives the propeller at low rpm. The ship with multiple propellers has multiple electric motors, motor drives, and cables, all located inside the hull as shown in Figure 7.2a. A damaged drive shaft can cripple the ship but not so with the cables, as multiple cables can be used for a desired reliability. If a missile or torpedo hits the ship, redundant cables can deliver the propulsion power. The ship service loads are powered by separate dedicated electrical generators as shown in Figure 7.2b.

*Podded electric propulsion:* In the electric propulsion discussed in the preceding paragraph, the propeller is driven by the electric motor located inside

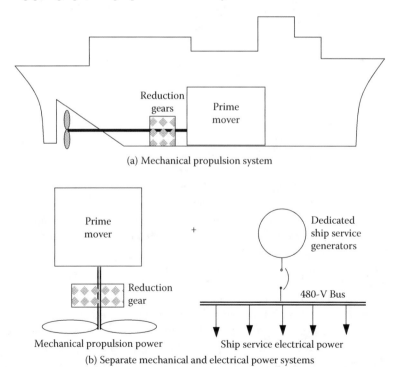

(a) Mechanical propulsion system

(b) Separate mechanical and electrical power systems

**FIGURE 7.1** Mechanical propulsion and separate electrical power system for service loads.

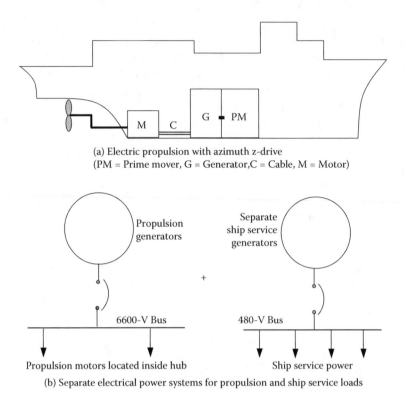

(a) Electric propulsion with azimuth z-drive
(PM = Prime mover, G = Generator, C = Cable, M = Motor)

(b) Separate electrical power systems for propulsion and ship service loads

**FIGURE 7.2**  Electric propulsion with motor in hull (z-drive) and separate electric power system for service load.

the hull and coupled to the propeller with a stern shaft. In podded propulsion, the propulsion motor is located in water outside the hull in the same housing that couples it directly to the propeller. The mechanical layout is shown in Figure 7.3a, which eliminates the stern shaft. It gives significant flexibility in placing the ship components where they offer the greatest benefits to the overall ship architecture.

*Integrated electric propulsion:* The propulsion of a ship typically requires 75–85% of the total power onboard. In electric propulsion, this power capability is devoted exclusively to ship propulsion and is not available for nonpropulsion use, even when the ship is stationary or traveling at low speed. The ship with IEP, in contrast, has all engines producing electrical power at a common bus that is used for both the ship propulsion and the nonpropulsion electrical loads as shown in Figure 7.3b.

*All-electric ships:* The all-electric ship program has the ultimate goal of converting all other loads that at present use hydraulic or compressed air power (e.g., steering gear, stabilizers, and deck machinery) on the ship to the electrical type, thus eliminating the hydraulic and air systems on board. For example, Rolls-Royce is currently developing electric steering gear for thrusters, which could also be applied to podded propulsion.

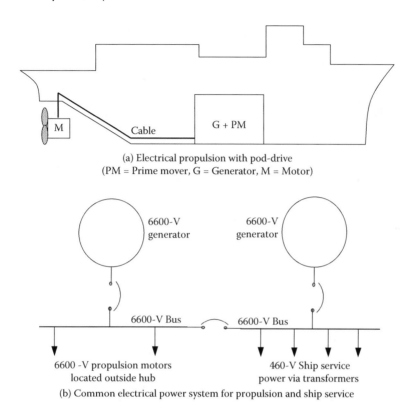

(a) Electrical propulsion with pod-drive
(PM = Prime mover, G = Generator, M = Motor)

(b) Common electrical power system for propulsion and ship service

**FIGURE 7.3** Integrated electric propulsion with motor in water (pod drive).

*Common electric propulsion:* This refers to the program of the U.S. Navy to design *all* ships with common components, which can be installed on various types of ships (e.g., submarines, surface combatants, amphibious ships, and auxiliary ships). The main benefits of the common electric propulsion system are reduced research, development, and procurement costs for a range of navy ships.

## 7.2 INTEGRATED ELECTRIC PROPULSION

In IEP, the electrical power produced by the engines and generators is delivered by cables to an electric switchboard that divides the electrical power into two distribution systems: one for propulsion of the ship and one for other electrical loads. The switchboard can alter the distribution of power between these two systems on a moment-to-moment basis to meet the propulsion and nonpropulsion requirements of the ship. The large amount of power needed to propel the ship at high speed is thus available for other uses if and when needed. Even when the ship is traveling at high speed, power can be momentarily diverted away from the propulsion system to a weapon system that needs a short burst of high power without appreciably slowing the ship. The flexibility of switching between the ship service power and the propulsion power inherently provides a higher degree of redundancy, survivability, and reconfigurability, which are the three major attractions of electric propulsion in warships. As a result, the IEP is of great interest to the U.S. Navy and other navies around the world.

In non-IEP systems, warships and cruise liners with relatively high service loads frequently leave the propulsion system operating at fractional loads at poor fuel efficiency. The IEP system offers the additional benefit of fuel saving by combining the total load of the ship on fewer prime movers running near full load at high efficiency. Moreover, fewer running hours are accumulated on each turbine and generator, requiring less maintenance and fewer spare parts.

For these reasons, the IEP has become the dominant propulsion power system in new cruise ships and is being rapidly adopted for warships following the development of enabling technologies. The IEP in such ships requires electrical power in the range of 50–100 MW. Fitting 50–100 MW of power generation, distribution, and utilization equipment, particularly in a warship with low displacement (for low propulsion resistance and high speed) poses a design challenge to naval architects. However, recent developments in both power electronics and propulsion motors with high power density allow fitting IEP systems to navy ships with smaller displacement. The IEP includes the following major subsystems:

*Prime movers:* The prime movers on commercial ships—particularly those with lower maximum speeds—are generally diesel engines. On U.S. Navy surface combat ships, they are gas turbines (modified versions of jet engines used on commercial airliners) that burn jet fuel. On U.S. Navy submarines and most U.S. Navy aircraft carriers, the engines are steam turbines obtaining steam from a nuclear reactor.

*Generators:* Convert *all* mechanical power of high-speed engines into electrical power.

*Propulsion power distribution system:* Distributes the electrical power to propulsion motors and other propulsion equipment.

*Propulsion motor drives:* Modify the frequency and voltage of electrical power as needed for the electric propulsion motors of the ship to operate at a desired speed during various operating phases.

*Propulsion motors:* Convert electrical power from the motor drives to low-rpm mechanical power suitable for the propellers of the ship.

*Propellers:* Run at low rpm and propel the ship through water.

*Nonpropulsion power distribution system:* Distributes the remaining electrical power to various nonpropulsion electrical loads on the ship. This system includes additional motor drives, cables, and switchgear.

## 7.3   AZIMUTH Z DRIVE

In the azimuth electric propulsion, the motor is located inside the hull of the ship, and the mechanical power of the motor is transferred to the propeller through shafts and gearboxes. The shaft from the motor to the propeller has two segments through the gears, forming a z shape as shown in Figure 7.2a. For its z-shaped shafts, this propulsion is also known as the *z drive*. The propeller can rotate a full 360° around its vertical axis to give propulsion thrust in any direction. It performs both the propulsion and the steering functions of the ship.

## 7.4  AZIMUTH POD DRIVE

In the azimuth pod drive, the propeller is connected directly to the motor shaft as shown in Figure 7.3a. The fixed-pitch propeller is driven by the variable-speed electric motor, which is in a submerged pod outside the ship hull, and the pod can rotate 360° around its vertical axis to give propulsion thrust in any direction. Thus, the ship does not need rudders, stern transversal thrusters, or long shaft lines inside the hull. The propeller of the pod usually faces forward in the pulling mode. This reduces the wake and hull resistance and results in higher fuel economy since there is no parasitic load from the flow of water over the propeller hull and rudder. Not having a traditional propeller shaft, the propeller can be located farther below the stern of the ship in a clear flow of water, providing greater hydrodynamic and mechanical efficiencies. The power rating of the motor fitted in azimuth pod drives ranges from 5 to 30 MW. The motor type can be 12-pole, 4000-V synchronous in the higher-power range or 640-V induction in the lower-power range that would run at 150 rpm at 15 Hz. The stator has either a single or a double winding-type motor as required for the reliability specification. The electrical power is transferred via cables and slip rings in the slip ring unit. Separate slip rings are provided for the propulsion motor stator windings, excitation, and control. The azimuth pod propulsion is used in cruise ships, icebreakers, navy ships, oil tankers, and offshore rigs.

The azimuth pod introduced by ABB in the 1990s with trade name Azipod® is a significant improvement over the azimuth z drive propulsion. It consists of one or more Azipod units (Figure 7.4), powered by a dry-type propulsion transformer, rectifier, cycloconverter, and synchronous motor, along with the usual propulsion control and monitoring systems. The pod propulsion systems have enabled ship designers to create new vessel types for icebreakers. For example, ABB has delivered a few dozen Azipod units for icebreakers with an ice class higher than or equal to 1A-Super according to the Finnish-Swedish ice rules.

A similarly rated azimuth pod assembly, Mermaid®, was developed by Rolls-Royce and Alstom. Figure 7.5 depicts the interior of Mermaid showing the motor directly coupled to the propeller. The electrical unit consists of a dry-type transformer, rectifier, load-commutated inverter (LCI), and synchronous motor. Both bearings of the motor are insulated, and the shaft is grounded at the drive end.

The pod drive offers the following benefits over the z drive:

- It eliminates complicated z-drive gears without giving up the 360° maneuverability.
- It gives better maneuvering capabilities with full thrust in all directions.
- Tugboat need is reduced and completely eliminated on some ships.
- At the stern end, it eliminates the need for a long exposed horizontal shaft leading to the propeller and rudder for steering the ship.

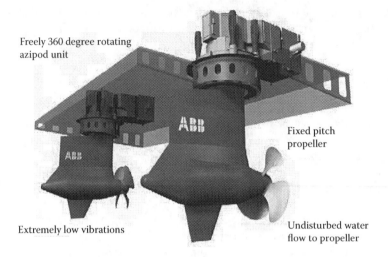

Freely 360 degree rotating
azipod unit

Fixed pitch
propeller

Extremely low vibrations

Undisturbed water
flow to propeller

**FIGURE 7.4**   Azipod® drive with 360° turning gear. (With permission from ABB Marine. Azipod® is a registered trademark of ABB Marine.)

- A flexible internal layout saves space in the engine and propulsion machinery rooms.
- It reduces maintenance and repair since the pod can be detached and quickly repaired or replaced by a like unit without need for cutting an opening into the hull of the ship and working around other equipment.

The pod design thus eliminates many additional systems, including (a) rudders and steering gear, (b) propeller shaft and line shaft from the engine, (c) rotating and stationary shaft bulkhead seals, (d) strut and line shaft bearings, and (e) redundant machinery, which saves space by replacing long shaft lines with electrical cables.

The layout for the total electrical power system developed by ABB using Azipods is shown in Figure 7.6. It starts from the prime movers driving the electrical generators, which feed power to the main switchboard. The power is then fed via propulsion transformers to the variable-frequency drives (VFDs) powering the Azipod motors. There is an independent control network that supports the propulsion power distribution network. The emerging trends in pod drives is the voltage source pulse width modulated (PWM) inverter with permanent magnet motor in lower-power ratings and advance induction motor in higher-power ratings.

Azipull® propulsion pod from Rolls-Royce is the new low-drag pulling thruster in the 1- to 5-MW power range. It combines the advantages of the pulling propeller with the flexibility of a mechanical drive and can be driven by a diesel engine, gas turbine, or electric motor. It is designed to offer efficient propulsion and maneuvering on higher-speed vessels, typically in the range of 20–25 knots. It is available with a controllable- or fixed-pitch propeller. A propeller is placed on the forward end of the pod, pulling rather than pushing. Figure 7.7 shows the exterior view, and Figure 7.8 shows the interior details of the Azipull system.

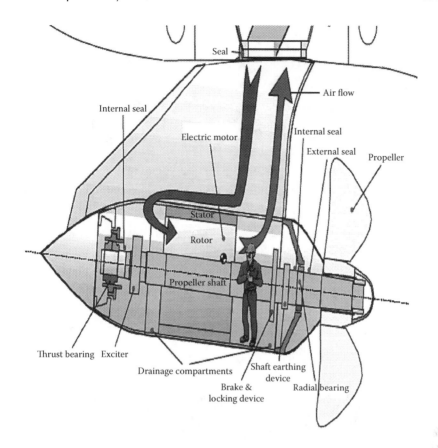

**FIGURE 7.5** Interior of Mermaid® pod with propeller. (With permission from Rolls-Royce plc. © 2012, all rights reserved by Rolls-Royce plc.)

## 7.5 ADVANTAGES OF ELECTRIC PROPULSION

Electric prolusion offers significant advantages to cruise ships and navy ships. However, merchant ships cruising at one speed over long hauls see fewer advantages today, but that may change in the future with new developments in the technology or in the market dynamics. One disadvantage of electric propulsion in merchant ships is lower efficiency than the mechanical-drive system at full-power operation due to the energy losses involved in converting the mechanical power into electrical power and the electrical power back into mechanical power. Naval ships, on the other hand, spend only a small fraction of their time at full power. Typically, about 80% of their time is spent at half speed, consuming about 1/8 of the rated propulsion power or less. Therefore, the power loss due to slightly lower efficiency at full power may be more than offset by the gain due to higher efficiency when operating at partial loads most of the time. For both the cruise and military ships, the common advantages of electric propulsion are as follows:

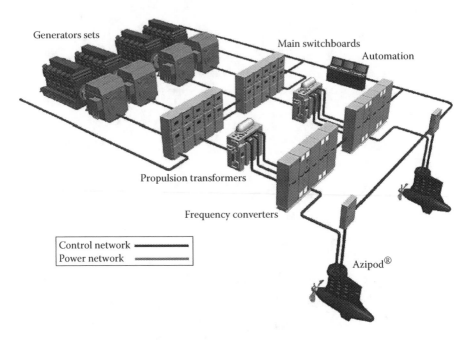

**FIGURE 7.6**  Electric propulsion power system layout with Azipods. (With permission from ABB Marine.)

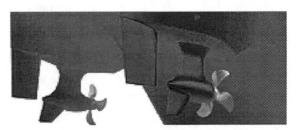

**FIGURE 7.7**  Azipull® propellers underneath ship's hull. (With permission from Rolls-Royce plc. © 2012, all rights reserved by Rolls-Royce plc.)

*Increased payload:* By eliminating the mechanical drive that requires the engines, reduction gears, shafts, and propellers to all be in long lines running at the bottom of the ship, electric propulsion makes it possible to install various components of the ship at locations that offer the most efficient packaging of the ship. For example, it may permit the turbine engines of the ship to be located higher in the ship, reducing the amount of interior space required for the ducts needed to take air down into the engine room and to carry exhaust gases away from the engine. In both these ways, electric propulsion may free up space aboard the ship that can be used to carry additional passengers, weapons, or sensors. It is estimated that a reduction of about 30% volume is possible compared to the conventional mechanical drive system.

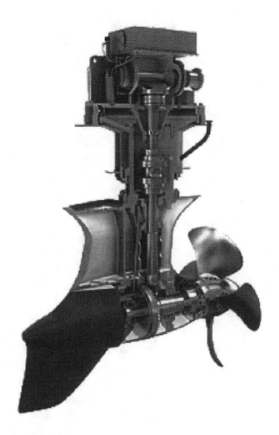

**FIGURE 7.8**   Azipull propeller assembly details. (With permission from Rolls-Royce plc. © 2012, all rights reserved Rolls-Royce plc.)

*Lower vibration:* Lower vibration and noise from the propeller machinery make the ship operate much quieter. This is of special interest to cruise ships for passenger comfort and to navy ships for stealth.

*Greater flexibility:* The electrical generators and cables to the propulsion motors can be located anywhere on the ship that is most advantageous instead of being restrained by bulky nonnegotiable mechanical shaft lines (often called the *tyrannical shaft line*). With an integrated electrical power system, the power distribution from the generators can be rapidly reconfigured in the event of damage to the ship to ensure a continued supply of electrical power to vital systems.

*Higher reliability:* The power can be transmitted by redundant cables, whereas the redundant shafts in the mechanical drive are not possible. Running as many multiple redundant cables as needed for greater reliability is possible at a small additional cost.

*Greater maneuverability:* The maneuvering and cruising operations are flexible since the ship speed can be varied continuously with infinitely variable motor speed as opposed to the discrete steps of mechanical gears. The podded propulsion can permit a tighter turning radius and give an ability to change the direction or orientation of the ship even at very low speeds. Moreover, the speed and direction of the prime mover need not be changed to affect the speed, direction, and rotation of the propeller. For icebreakers, ferries, tugboats, oceanographic vessels, and cable-laying ships that require frequent speed changes and direction reversals, electric propulsion offers a great advantage over mechanical propulsion. The speed and direction of rotation of the electric propeller can be changed rapidly at remote locations, making it possible to put the ship control directly in the hands of the vessel navigator or the dynamic-positioning computers.

*Reduced maintenance:* A major maintenance task on conventional mechanical drive ships is to keep the key components—the reduction gear, main shaft, and stern—well aligned. In electric propulsion, the gears are replaced by variable-speed motors for engine speed reduction, and shafts are replaced by cables for power transmission. Both of these eliminate the major maintenance work that is typical in the mechanical drive. With podded propulsion, the maintenance and repair are further reduced since the pod can be detached and quickly repaired or replaced by a like unit without need for cutting an opening into the hull of the ship and working around other equipment.

*Fuel savings:* For ships spending more time at low speed, fewer engines can run at full power, resulting in greater energy efficiency and hence less fuel consumption. Podded propulsion can reduce fuel consumption further by 5–15% due to the improved hydrodynamic efficiency. The fuel saving also comes from (a) fewer exposed components that create drag resistance to forward movement and (b) the propeller encountering a more uniform—less disturbed—water flow, increasing the efficiency. The reduced fuel consumption also translates into a reduced amount of space required for fuel storage. Depending on the ship type, the navy ship with IEP may save an estimated 15–25% in fuel compared to a similar ship with mechanical drive. The lifetime cost saving from reduced fuel consumption may exceed the higher initial procurement cost of electrical propulsion.

*Higher automation:* The electric propulsion system can be designed to be highly automated and self-monitoring, hence requiring less maintenance and fewer crew members to operate than with a mechanical drive system.

*Longer equipment life:* The smooth torque versus speed characteristic of the motor under continuously variable speed gives fewer mechanical transients—backlashes—in the drive motor shaft and couplings and fewer thermal transients in the motor, resulting in longer life for all equipment involved.

*Technology upgrade:* It allows replacement of today's prime movers—diesel engines, gas turbines, or steam turbines—and generators with more efficient power generation technologies in the future, including direct energy conversion devices such as fuel cells. This can save fuel and operating costs and help meet the evolving environmental regulations.

For navy ships, electric propulsion offers additional benefits as follows:

*Increased stealth:* Stealth is fundamental to the survivability and effectiveness of a submarine. Acoustic noise remains the most reliable method of detecting and tracking submarines at longer ranges. Being acoustically quieter than the mechanical drive, the ship equipped with electric propulsion is less detectable. Podded propulsion reduces the wake signature of the surface ship, further reducing the detectability by remote overhead sensors and improving the chances of defeating much-feared wake-homing torpedoes. The propulsion machinery vibration isolation has been extensively developed for navy ships, but much research at present is focused on noise reduction at the source, particularly the low-frequency (<100-Hz) hull vibrations. Several technologies are being applied, including active—dynamically controlled—magnetic bearings in this area.

*Increased survivability:* Ship survivability improves greatly by eliminating the possibility of one or more of the long shaft lines of the ship being thrown out of alignment and rendered useless by a nearby weapon explosion. Electric propulsion allows a wide separation of the propulsion system elements on the ship, making it less likely that a single weapon might disable the entire drive system. For example, with two redundant cables—one on the port side and the other on the starboard side—it is much more probable that at least one side of cables may survive an enemy strike and keep the ship fully functional.

*Increased power for weapons:* IEP can make a large amount of power available for nonpropulsion loads, such as powerful radar and sonar, laser weapons, high-power microwave weapons, electromagnetic rail guns, electrothermal guns, and electromagnetic aircraft launchers; or rapidly charging the batteries of unmanned air vehicles (UAVs), unmanned underwater vehicles (UUVs), and high-energy undersea sensor network. Some of these functions, particularly the weapons, may require peak power levels measured in tens of megawatts for a short time. Adding this much electrical power-generating capacity to a nonintegrated electrical power system would incur substantial additional cost.

*Other advantages:* The greater freedom in the center-of-gravity management, reduced propeller cavitation and noise, improved low-speed maneuverability, and attractive thruster propulsion are advantages in addition to those listed.

These advantages have attracted the U.S. Navy to fund research, development, and demonstration of full scale IEP for fast warships at various government laboratories and private corporations. The successful demonstrations of the newly developed technologies under these programs led the U.S. Navy to decide that the next class of combat ships, the DD-21 land attack destroyer, would use electric propulsion. This decision opened an enormous opportunity for redesigning the ship architecture for reducing the cost, noise, vulnerability, and maintenance demand while improving shipboard life. The decision has established a development budget of hundreds of millions of dollars to accelerate the development of the IEP, with Electric Boat of the General Dynamics Corporation taking a lead in demonstrating the key technologies. For merchant ships, the U.K. MoD study indicated a higher acquisition cost but lower operating and support costs with the overall reduction in the life-cycle cost. The ministry estimated a payback period of several years depending on the operating profile of the merchant ship.

Two disadvantages of electrical propulsion are (a) higher initial capital cost and (b) lower overall energy efficiency for ships running at full-rated speed all the time, such as in cargo ships and oil tankers.

## 7.6   ELECTRIC PROPULSION ARCHITECTURE

The power system architecture for ships with electric propulsion is shown in Figure 7.9. It has four 6600-V generators in two groups for redundancy required for the ABS-R2 class of ships. In some ships, watertight and fireproof separation is required between the two groups for ABS-R2S class. A dedicated emergency generator is not required in this architecture but may be provided if specifically required.

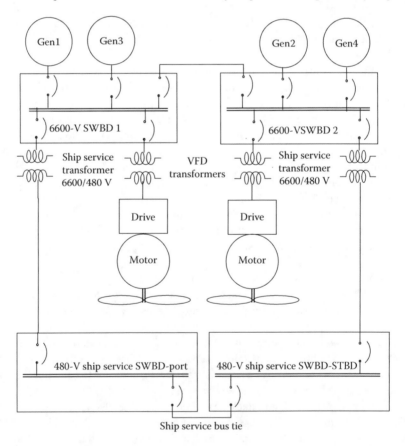

**FIGURE 7.9**   Integrated electric power system architecture with ABS-R2 redundancy. SWBD = Switch board, STBD = Star board.

A more detailed physical layout of the electrical power system distribution is shown in Figure 7.10.

In developing candidate architecture for a new ship line, the first major task for the system engineer is to trade-off the mechanical versus electric propulsion

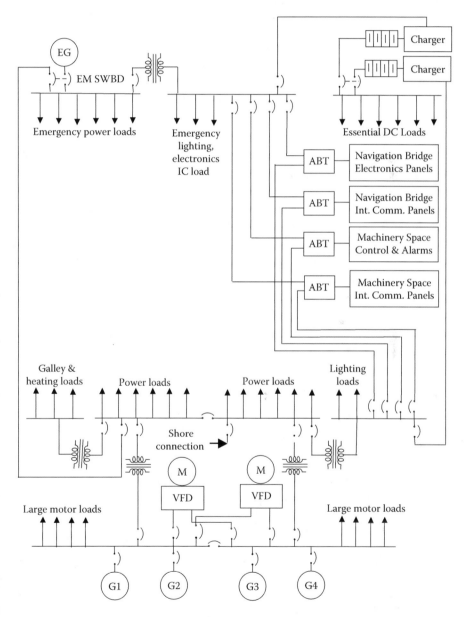

**FIGURE 7.10** Integrated electric power distribution system. ABT = Automatic bus transfer, EG = Emergency generator, EM SWBD = Emergency switch board, IC and Int.comm. = Internal communication.

option. A ship with the following requirements makes a good candidate for electric propulsion:

- High hotel or weapon load that is all electric
- High flexibility in major equipment placement
- More space for passengers or weapons
- Reduced vibration and noise
- High reliability, maneuverability, and reconfigurability

If the decision is for electric propulsion from technical, business, and other relevant considerations, then the next question is whether to use ac or dc power.

## 7.6.1  AC versus DC Power

Overall, the ac system today would give the least capital and life-cycle costs since much ac equipment is available off the shelf, and more manufacturers are in the business of making large ac generators, motors, transformers, circuit breakers, and so on. The use of dc has declined over the past few decades. It is hardly used now in new designs, except in special circumstances. As a result, few manufacturers offer dc equipment in their product lines. For this reason, ac is the most likely candidate for new ships and is generally the default choice unless a strong case for dc is made for a special reason. There is one advantage of dc over ac: It makes even less vibration and noise. The U.S. Navy has an ongoing interest in high-voltage dc power for special applications for which a ship has strict acoustic and vibration limits.

## 7.6.2  Voltage-Level Selection

The next question in the propulsion power architecture is which voltage level to select: low, medium, or high voltage. The main propulsion power rating ranges between 25,000 and 40,000 hp in the present merchant ships. High-speed merchant ships, large cruise liners, and navy combat ships require much greater propulsion power. Higher electrical power requires higher voltage to keep the current and $I^2R$ power loss low and efficiency high. Following are the general guidelines for selecting the voltage:

- 660 V are suitable for small ships with load less than 5 MW.
- 6600 V are suitable to power medium-size ships with loads in the range of 5–20 MW.
- 11 kV are used in the 20- to 70-MW power range.
- 13 kV may be needed for large high-speed ships with loads approaching 100 MW or higher, as some navy ships may need. Large cruise liners also need power around 100 MW. For example, the *Queen Mary 2* is a 80-MW ship.

Since power $P = V \times I$, and $I$ depends on $V$, the power varies with $V^2$. If the heritage data of the shipbuilding company indicates that the voltage level of $V_1$ has worked well for a power level $P_1$, then the new ship with higher power level $P_2$ would have the new optimum voltage level approximately equal to

$$V_2 = V_1 \sqrt{\frac{P_2}{P_1}} \qquad (7.1)$$

which is rounded to the nearest standard voltage. Regardless of the generator voltage rating, the hotel and small housekeeping loads are typically a 120/240-V, single-phase, three-wire system or 208-$V_{LL}$/120-$V_{LN}$, three-phase, $Y$-connected, four-wire distribution system via step-down transformers.

### 7.6.3 PROPULSION MOTOR

Selecting the propulsion motor has been the subject of much debate in the ship building industry since it involves many factors specific to each motor. Whichever motor is selected, sufficient torque margin must be provided to prevent stalling under the worst load conditions, steady state or dynamic. The propulsion motor can be dc, ac synchronous, or ac induction. The advent in power electronics drives for controlling the motor speed has attracted new interest in motor technologies, particularly for research ships, icebreakers, cruise ships, and most recently navy ships. For that reason, all potential candidates for ship propulsion motors—traditional and new—are covered in depth in the next chapter.

## 7.7  MOTOR DRIVES AND SPEED CONTROL

The basic architecture of the ac-dc-ac VFD drive was shown in Chapter 4, Figure 4.7. The 60-Hz ac power from the generator is first rectified to dc and then inverted into a variable lower frequency lower voltage, which is then applied to the induction or synchronous motor. The rectifier and inverter use semiconducting power electronics components (diodes, thyristor, and transistors). High-frequency harmonics produced by chopping currents in the converters are eliminated or minimized by an inductor-capacitor (L-C) filter at the VFD output. For navy ships with stringent vibration requirements at the motor feet, the motor must include an additional fine-tuned L-C filter network across its input terminals to minimize transmitted vibration and noise into the hull and the sea.

The converter may use the basic ac-dc-ac converter described or a cycloconverter, synchroconverter, or voltage source PWM converter, as shown in Figure 4.10. Many technical and commercial considerations are traded while selecting a system that can meet all the continuous and overload conditions for the specified duty.

The operating efficiency of large converters (>1-MW ratings) is 97–98% at full power. A 1-MW converter would have approximately 20–30 kW power loss turning into heat, which is removed by an air- or water-cooling system. The shipboard converter is a sealed unit. In an air-cooled converter, the cooling air is circulated through the converter through a series of internal ducts, passing through the principal heat-producing components: the filter reactor, thyristor stacks, and line reactors. The heat is ultimately dumped into the sea by a tube or parallel-plate air-to-seawater heat exchanger.

The ac motor speed is controlled by adjusting the output frequency and voltage of the power electronics drive. The synchronous motor speed is related linearly with the frequency, whereas the induction motor speed has a complex relation with the frequency as explained next.

The torque-speed characteristic of an induction motor working at frequency $f_1$ is shown by a heavy curve in Figure 4.6. If the frequency is changed to $f_2, f_3$, and so on, then the motor torque characteristic $T_{m1}$ shifts to the left as shown by thin curves $T_{m2}$, $T_{m3}$, Respectively. On the load side, the torque-speed characteristic of a propeller is similar to a pump load, as shown by the heavy dotted curve.

The propulsion system operates at the speed at which the torque developed in the motor equals the load torque on the propeller, that is, at the intersection point of the two curves where $T_{motor} = T_{load}$ in Figure 4.6. As the induction motor frequency is lowered, the operating point shifts from right to left, thus changing the motor speed. This is how the speed of the induction motor is changed by changing the frequency. But, along with the frequency, the voltage is also changed in the same proportion to keep the motor flux and torque constant at the rated values without saturating the magnetic core.

## 7.8 POWER ELECTRONICS CONVERTERS FOR VFD

### 7.8.1 SYNCHROCONVERTER FOR CRUISE SHIP

The synchroconverter (with current source invetrer) finds applications in electric propulsion of large cruise ships. Its first application was in the cruise liner *QE2* with diesel-electric propulsion. Each *QE2* converter rating is 11.5 MW, which is fed directly from the main propulsion bus of the ship at 10 kV. The thyristor selection in the power train is critical as it must withstand peak voltage that can be experienced under the normal, abnormal, and faulty operating conditions. The maximum repetitive forward voltage capability $V_{drm}$ of the thyristor is thus a significant characteristic of the definition of the device. Thyristors are available with ratings of 50 to 5000 V and with continuous current-carrying capacity ranging from 1 to thousands of amperes. For the *QE2* synchroconverter, the chosen device has a repetitive $V_{drm}$ of 3600 V and a continuous current rating of 1278 A. The power train has line and motor side bridges, each with 12 thyristors in series per armature phase. The rated dc link current of the converter is 1047 A. The surge suppression circuits at the input and output of the converter provide high transient voltage protection for the thyristors. The standard capacitor-resistor snubber circuits provide the *dv/dt* protection.

The peak torque requirement is specified at zero shaft speed when the motor input frequency is zero. Hence, one bridge in each antiparallel pair can operate in a dc condition for 30 s, and the device selection is dictated by this condition. For the *QE2* VFD, the device was chosen with the peak $V_{drm}$ of 5200 V and the peak current rating of 2900 A, compared with the continuous rated current of 1278 A.

### 7.8.2 LCI WITH SYNCHRONOUS MOTOR

The load-commutated inverter (LCI) with a synchronous motor (SM) is commonly used in large cruise ship propulsion, such as in some Carnival, Crystal, and Cunard cruise ships. The following are key features of these LCI-SM drives.

- Diesel engines run at constant speed (400 rpm, 60 Hz) where their efficiency is the highest.
- The generator power is converted into variable frequency, variable voltage
  to control the motor speed.
- The cross connection allows the motor to operate using either of the two
  converters in case one converter fails.
- The power electronics converters are basically used for both soft starting of
  the motor and low-speed operation.
- The motor starts with the inverter up to half the rated speed at no load,
  and then the load torque is applied by controlling the pitch angle of the
  propeller.
- When the inverter frequency rises to a full 60 Hz, the motor is switched
  directly to the generator to improve system efficiency.
- The machine excitation control maintains the motor at a slightly leading
  power factor near unity.
- The drive has speed control in the outer loop and torque control (proportional to current) in the inner loop. The drive is regenerative, with reversible
  speed.
- The propeller pitch angle adjusts the motor torque and varies the ship speed.
- The mode control allows operating in four modes: port, ready to sail, converter driven, and free sailing (direct on the bus).

### 7.8.3 CYCLOCONVERTER FOR ICEBREAKER

The icebreaker poses a severe torque duty on the motor, which must be accounted
for in both the motor and the VFD designs. A typical load characteristic of an icebreaker ship with electric propulsion is shown in Figure 7.11. It shows the motor
performance requirements during free running, operating in heavy ice, and the
worst-case condition towing requirement. The converter and motor design rating is
set at point C, the worst-case (bollard) pull condition. It defines the design values of
flux and current in the propulsion motor and correspondingly the current and voltage
of the converter.

In icebreakers, a severe overload requirement is specified at low shaft speed, such
as 1.75 times the rated torque pulse of 1 minute in a 5-minute cycle with 4 minutes
at 100% full-load torque. Such operation requires very high current at low voltage,
which defines the current-carrying capability of the VFD thyristors. Each thyristor
is protected with a fuse. The common mode of thyristor failure is a short circuit,
after which the thyristor continues to work as a diode. The protective fuse clears, but
full uncontrolled dc output is still available. This could give an unbalanced output
voltage. A monitoring circuit, such as the converter ripple detector, could indicate
such an event. Each thyristor in the power circuit must also be protected from the
transient voltage rate of rise ($dv/dt$) in the normal operation of the converter. This is
achieved by a simple resistor-capacitor snubber circuit connected across each thyristor. Since all devices do not simultaneously begin to conduct at turn on, interphase
reactors—as shown in Chapter 4, Figure 4.16—are fitted to ensure equal sharing of
current between the thyristor in each armature.

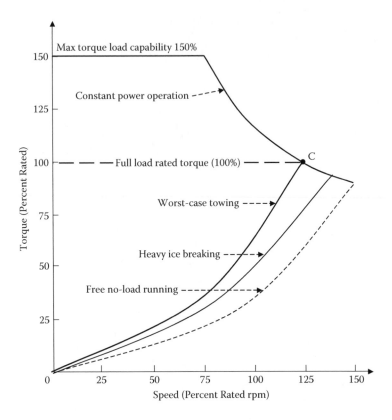

**FIGURE 7.11**   Load characteristic of icebreaker with electric propulsion.

The icebreaker cycloconverter is different from the synchroconverter in *QE2*. For example, the U.S. Coast Guard *Healy* icebreaker VFD is transformer-fed from the 6.6-kV main propulsion bus of the ship. The transformer secondary voltage has been determined to ensure that a single device per armature can satisfy the continuous and overload conditions specified for this application. Since the converter has a single device per arm, it cannot continue normal operation in case one fails in short circuit. The drive will trip within 15–20 ms, but the ship can still operate on the other propeller.

The arctic cargo ship *Norilskiy Nickel* is another example of a severe-duty icebreaker ship. It operates year around for Siberian exports along the severe northwestern Arctic sea route with solid ice thickness reaching 5.6 ft (1.7 m) and temperature falling to –50°C (–58°F). Its electric propulsion consists of three Wartsila 12V32 diesel engines, each with a maximum continuous rating of 6000 kW at 750 rpm, each driving an ABB generator with 8314-kVA capacity at the 6600-V, 50-Hz main switchboard. The ship has a single 13-MW Azipod drive with double-winding motor for redundancy in the propulsion unit. The ship also consists of propulsion transformers, two high-voltage distribution transformers, propulsion control and a remote control system, an additional 415-kVA emergency generator, and a 750-kVA shore generator. The overall parameters of the ship are listed in Table 7.1

---

**TABLE 7.1**

*Norilskiy Nickel* **Russian Icebreaker Dimensions and Propulsion Power Data**

| | |
|---|---|
| Classification | Russian Maritime Registry |
| Shipping ice class | LU7 Stern, LU6 bow |
| Length, overall | 554.6 ft (169 m) |
| Breadth, molded | 75.8 ft (23.1 m) |
| Draft at dwl | 29.5 ft (9 m) |
| Height to bulkhead deck | 46.6 ft (14.2 m) |
| Deadweight, ice draft 9 m | 14,928 dwt |
| Deadweight, at draft 10 m | 18,486 dwt |
| Number of 20-ft TEU containers | 648 |
| Gross tonnage | 16,994 |
| Icebreaking, sterns first | 1.5 m ice with 20-cm snow layer |
| Maximum speed in open water | 16.1 knots |
| Power plant | Diesel-electric |
| Main engines | 3 × Wartsila 12 V32 |
| Alternators | 3 × ABB 8314 kVA |
| Electrical propulsion | 13-MW Azipods |
| Propellers | 5.6 m diameter |
| Heavy fuel capacity | 1400 tons |
| Fresh water capacity | 200 ton |
| Crew cabins | 18 |
| Passenger cabins | 3 double berth |

*Source:* Adapted from *Marine Reporter and Engineering News,* October 2006, pp. 26–28.

---

Figure 7.12 is the diesel-electric propulsion system installed on an icebreaker operating in the St. Lawrence River in Canada. The river freezes in winters, but the ship operates throughout the year using the propellers as the ice cutters. It keeps the ship navigable all year around using the propellers as icebreakers in winter. The propulsion system—assembled by Canadian GE—basically uses a cycloconverter with a synchronous motor on a 4160-V, 60-Hz bus. The diesel engine efficiency remains high at constant engine speed, resulting in significant fuel saving. The VFD uses 36 phase-controlled thyristors that operate in six groups of six-pulse mode. The phase-splitting transformer on the input side makes it a 12-pulse converter, which significantly reduces the harmonic loading on the generator side. The motor field current is derived from the brushless exciter, and the motor is operated near unity power factor with no kVAR loading on the machine or the cycloconverter. The inverter is load commutated without additional commutation circuits. Direct stator flux control is used in the constant-torque operation, whereas a trapezoidal voltage is used in the field-weakening region. The cycloconverter works as a constant-current source in high-torque, low-speed operation and as a constant-voltage source in low-torque, high-speed operation. The instantaneous phase currents are controlled by feedback with injection of feed-forward counter emf to improve the performance.

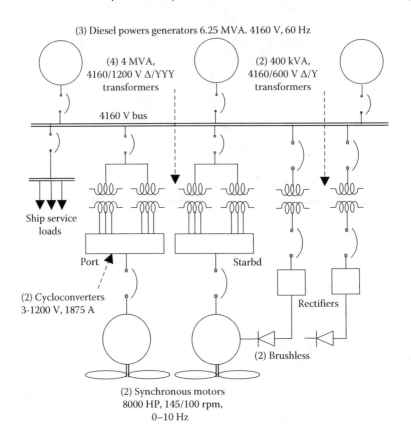

**FIGURE 7.12**   Electric propulsion with cycloconverter–synchronous motor for icebreaker.

### 7.8.4   DC Propulsion for a Small Ship

The dc motor speed control requires voltage and field current control as discussed in Chapter 4, Section 4.11. In the first approximation, the dc motor speed is related linearly with the voltage. Therefore, the dc speed drive needs the power electronics converter to supply variable-voltage output for the motor. This is generally obtained by the phase-controlled ac-dc converter using thyristors.

The training ship *Kings Pointer* at the U.S. Merchant Marine Academy is an example of a small ship with dc electric propulsion. It was used as a navy-military surveillance ship built in the early 1980s for collecting data on foreign ships. It is 224 feet long with 2250 tons deadweight and has a power plant with the following features:

- Diesel-electric propulsion designed for cruising at 10 knots
- Four diesel engines, each rated 970 hp
- Four synchronous generators, each 600 kW, 600 V ac, three phase, 60 Hz
- Two propulsion motors, each rated 800 hp, 750 V dc
- Two phase-controlled rectifiers to convert ac into dc for propulsion motors

- Two propellers, each with four blades, 8 ft diameter
- Emergency power system: 400-hp diesel engine with 250-kW generator
- Diesel fuel storage capacity 228,600 gallons
- Diesel fuel consumption 1850 gallons per day at 10 knots

## 7.9  PROPULSION POWER REQUIREMENT

The propulsion power required is basically determined by the hydrodynamic and aerodynamic resistances to the motion of the ship. At the blade front, the propeller load torque varies with $\alpha$-power of the propulsion speed, that is,

$$T_{prop.load} = k(Speed)^\alpha \tag{7.2}$$

$$Power = Torque \times Speed = k(Speed)^{\alpha+1} \tag{7.3}$$

where $k$ = constant and exponent $\alpha$ = 2 at low speeds and 3 at medium speeds and can be 4 or higher at high speeds when the fluid flow is highly turbulent with cavitations. This explains why designing a high-speed ship is a great technical challenge. Increasing the speed by 10% requires at least $(1.10)^3$ = 1.33 or 33% higher power. In the reverse, decreasing the speed by 10% saves about 27% fuel but takes 10% longer for the journey. This is a business decision for the ship owner to make, based on fuel cost and the per diem personnel cost of operating the ship.

The blade-level power Equation (7.3) is further broken down in its components as follows: The total propulsion resistance $R_T$ of the ship depends on its speed, displacement, and hull form. It comes primarily from three sources: (a) surface friction, (b) air resistance, and (c) wave (residual) resistance. The friction and wave resistances depend on the surface of the hull below the waterline, whereas the air resistance depends on the surface of the ship above the waterline.

A fluid with velocity $V$ and density $\rho$ exerts a force on a steady object that is given by

$$F = \tfrac{1}{2}\rho V^2 \times A \tag{7.4}$$

where $A$ = area of the object perpendicular to the fluid velocity. Equation (7.4)—known as Bernoulli's law—is used as a basis for expressing the propulsion resistances $R$ of the ship by means of dimensionless resistance coefficients $C$ as follows:

$$R = C \times \tfrac{1}{2}\rho V^2 \times Area \tag{7.5}$$

The total resistance of the ship to propulsion coming from three major sources is discussed next.

### 7.9.1  FRICTIONAL RESISTANCE

Frictional resistance is primarily a square function of ship speed. It also depends on the wetted area of the hull and on the frictional resistance coefficient $C_{friction}$. It represents a major part of the total resistance of the ship, typically 70–90% for low-speed ships such as bulk carriers and oil tankers and less than 40% for high-speed ships such as cruise liners. The frictional resistance is expressed as follows:

$$R_{friction} = C_{friction} \times \tfrac{1}{2}\rho\ V^2 \times Wetted\ surface\ area\ of\ hull \qquad (7.6)$$

The friction increases with fouling of the hull (i.e., by the growth of algae, sea grass, and barnacles). The fouling is minimized by painting the hull, which delays the growth of living organisms. The paints in the past typically contained tributyl tin, which was banned by International Maritime Organization (IMO) beginning in 2008 for its toxicity. New copper-based antifouling paints are now being used to replace the tributyl tin paints.

### 7.9.2  AERODYNAMIC RESISTANCE

The aerodynamic resistance (air drag) is proportional to the square of the ship speed and the cross-sectional area of the ship above the waterline that is perpendicular to the speed direction. The air drag typically represents a few percentage of the total resistance. For container ships in a headwind, the air resistance can be as much as 10%. The air resistance is expressed as follows, where a typical value of $C_{air}$ is 0.90:

$$R_{air} = C_{air} \times \tfrac{1}{2}\rho_{air}V^2 \times Cross\text{-}sectional\ area\ of\ ship\ above\ waterline \qquad (7.7)$$

### 7.9.3  WAVE (RESIDUAL) RESISTANCE

The ship in motion makes waves and eddies—particularly at the aft end of the ship—putting the water in motion as well. The kinetic energy dissipation rate in the water reflects the propulsion resistance on the ship. The wave resistance at low speeds is a square function of the speed but increases much faster at higher speeds. In principle, this means that a speed barrier is imposed, so that a further increase of the propulsion power of the ship will not result in a higher speed, since all additional power will be converted into making additional waves. The wave resistance normally represents 10–25% of the total resistance for low-speed ships and up to 40–60% for high-speed ships. The ship in shallow waters can have additional resistance as the displaced water under the ship will have greater difficulty in moving aftward due to the shear force with ground.

The wave resistance is expressed as follows:

$$R_{wave} = C_{wave} \times \tfrac{1}{2}\rho V^2 \times Cross\ section\ of\ ship\ below\ waterline \qquad (7.8)$$

### 7.9.4  TOTAL TOWING RESISTANCE

The total towing resistance $R_{Total}$ of the ship is the sum of the three components just described, that is,

$$R_{Total} = R_{friction} + R_{air} + R_{wave} \qquad (7.9)$$

The towing power necessary to move the ship through water at speed $V$ is $P_{Tow} = R_{Total} \times V$, which can be expressed as follows:

Towing power in SI units,

$$kW = \frac{R_{newtons} \times V_{m/s}}{1000} \qquad (7.10a)$$

Towing power in British units,

$$HP = \frac{R_{Lbf} \times V_{ft/sec}}{550} \qquad (7.10b)$$

The distribution of the total propulsion resistance of the ship is given in Table 7.2. The resistance due to fouling may increase by 25–50% throughout the lifetime of the ship. Resistance also increases because of sea, wind, and marine currents by approximately 20–30% depending on the shipping route. The total resistance when navigating in head-on sea may increase by as much as 50–100% of that in calm weather. For this reason, the ship speed is somewhat reduced in high waves.

Since all resistances in general are proportional to the square of the speed, we can write the towing power as the cubic function of the ship speed, that is,

$$P_{tow} = C_{tow} \times V^3 \qquad (7.11)$$

Equation (7.11) is valid at low-to-moderate speeds but does not hold true at high speeds when the wave resistance increases much faster. Consider, for example, a ship designed for 15 knots base speed, for which the propulsion power versus speed follows the cubic power relation. With the same hull design, the propulsion power for this example ship would be two times the base power at 18 knots, instead of increasing by the cubic power to $(18/15)^3 = 1.728 \times$ base power. The increase in this case actually follows the fourth power, and it would follow an even higher power at higher speeds, that is,

**TABLE 7.2**

**Distribution of Various Resistances in the Ship Total Propulsion Resistance**

| | % of Total Resistance | | |
|---|---|---|---|
| Resistance Type | High-Speed Ships (Cruise Ships and Ferries) | Medium-Speed Ships (Containers) | Low-Speed Ships (Bulk Carriers and Oil Tankers) |
| Friction $R_{friction}$ | 45 | 70 | 90 |
| Air resistance $R_{air}$ | 10 | 5 | 2 |
| Wave + eddy $R_{wave}$ | 45 | 25 | 8 |

$$P_{tow} = C_{tow} \times V^{\alpha} \tag{7.12}$$

where $\alpha = 3$ at low speed, 4 at high speed, and 5 or greater at very high speed.

As such, working from the power side of Equation (7.12), a further increase of the propulsion power beyond a certain level may only result in a small increase in the speed of the ship, as most of the extra power will be converted into wave energy. Thus, a speed barrier for the given hull design exists—known as the *wave wall*— shown by the asymptotic vertical dotted line in Figure 7.13. For this reason, a major modification in the hull design is necessary for high-speed ships.

The power delivered to the propeller $P_{prop}$, however, is greater than $P_{tow}$ due to the flow conditions around the propeller and the propeller efficiency $\eta_{prop}$, that is,

$$P_{prop} = \frac{P_{tow}}{\eta_{prop}} \tag{7.13}$$

The propeller efficiency depends on the ship speed, propeller thrust force, revolutions per minute, propeller diameter, and the design of the propeller (i.e., the number of blades, disk area ratio, pitch-to-diameter ratio, etc.). It typically varies from 0.50 to 0.80, the lower number for low-speed ships and the higher number for high-speed ships.

Total shipboard power system rating is the sum of the three components as follows:

*Total shipboard power = Propulsion power + Ship service power + Hotel power*
$$\tag{7.14}$$

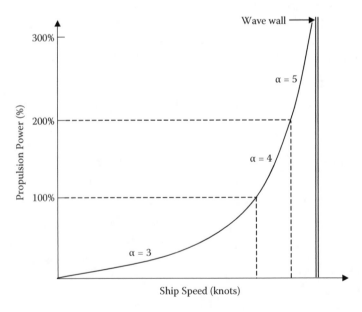

**FIGURE 7.13**  Propulsion power versus speed up to wave wall.

The relative proportion of these three powers can be quite different in cruise ships and navy ships.

## QUESTIONS

*Question 7.1*: List five benefits of electric propulsion in navy ships and place relative weights on them from 5 (most important) to 1 (least important). Prepare the list based on your own understanding or work experience and then discuss with others.

*Question 7.2*: Why is electric propulsion not economically viable at present for cargo ships?

*Question 7.3*: What percentage of cargo ships do you think may use electric propulsion in the next 25 years? Give top-level reasons to support your estimates.

*Question 7.4*: Explain the difference between the integrated electric ship and all-electric ship.

*Question 7.5*: List three major benefits of electric pod drives (e.g., Azipod and Mermaid) compared to the conventional mechanical drive.

*Question 7.6*: List three key requirements for the electric propulsion motor and VFD for the icebreaker ship.

*Question 7.7*: If cruise liner A with 70 $MW_e$ power has been operating satisfactorily at 11 kV, what voltage level would you recommend for a new cruise liner that will need 110 $MW_e$ power?

*Question 7.8*: Identify the components that make up the total resistance to ship propulsion.

*Question 7.9*: Explain the term *wave wall* as it applies to high-speed propulsion.

*Question 7.10*: Share your experience (if any) on spending more on fuel to cut the journey time short. Where does the limit on the ship speed really come from?

## FURTHER READING

Bender, B. 2000. U.S. Navy sets sights on electric attack submarine. *Jane's Defense Weekly*, July 26.

Bonner, K. 1999. Naval propulsion for the 21st century: The Azipod system. *U.S. Naval Institute Proceedings*, August, pp. 74–76.

Bowman, F.L. 1999. Submarines in the New World Order. *Undersea Warfare*, Spring, p. 7.

Finney, D. 1988. *Variable Frequency AC Motor Drive Systems*. London: Peter Peregrimis.

Institute of Electrical and Electronics Engineers. 2002. *Recommended Practice for Electric Installations on Shipboard*. IEEE Standard 45. New York: IEEE Press.

Islam, M.M. 2002. *Handbook to Electrical Installations on Shipboard*. New York: IEEE Press.

King, J., I. Ritchey, and C. Hodge. 2008. Marine propulsion: The transport technology for the 21st century. *Rolls Royce Report*, Rolls Royce, U.K.

Leonard, R.E., and T.B. Dade. 1998. The all electric ship: Enabling revolutionary changes in naval warfare. *Submarine Review*, October, pp. 43–53.

McCoy, J. 2000. Powering the 21st century fleet. *U.S. Naval Institute Proceedings*, May, pp. 54–58.

Murphy, M. Variable speed drives for marine electric propulsion. *Transactions of the Institute of Marine Engineers*, 108(Part 2), 97–107.

O'Rourke, R. 2000. *Navy Attack Submarine Program Background and Issues for Congress (on Virginia-class program)*. CRS Report RL30045.

O'Rourke, R. 2000. *Navy DD-21 Land Attack Destroyer Program: Background Information and Issues for Congress*. CRS Report 97-700 F.

Patel, M. R. 2011. *Shipboard Electrical Power Systems*. Boca Raton, FL: CRC Press/Taylor & Francis.

# 8 Propulsion Motors

The electric motors associated with electric propulsion for large ships can be divided into several basic categories listed in Table 8.1 with their long and short technical names. They can be ac or dc and can be traditional or newly developed motors. The construction and performance features of traditional motors can be found in many books, some of them listed at the end of this chapter. The key performance features of new motors specifically developed for electric propulsion on a ship are discussed in this chapter.

The new motors differ in terms of the technological maturity, power density, and potential applications in different navy ships. For that reason, much of the debate over electric propulsion today concerns selecting the electric motor since it is associated with specific corporations competing for the electric propulsion business of the navy. A preference for a certain type of motor can thus lead to a preference for the proposal from one company over the others.

With regard to motor performance, a large warship requires a motor that is reliable, compact, shock resistant, and quiet and can deliver high power and high torque at a slow speed suitable for ship propellers. Traditional high-power electric motors tend to be high-speed, low-torque machines rather than low-speed, high-torque machines. In the 1980s, large low-speed, high-torque electric motors were developed, but the horsepower was still inadequate for large warships. The conventional dc motor can be made in ratings up to 10–15 MW. This power range is large enough for small nonnuclear-powered submarines, for which dc motors have been widely used, but is not enough for large surface ships at high speeds.

From the motor design point of view, the limitation of traditional electric motors in developing high torque comes from both the electric current and the magnetic flux, which interact to generate torque competing for the same space in the magnetic core slots. The design that can provide adequate space for both results in a motor that is not compact. The stator winding conductors are embedded in slots in the iron core that carries the magnetic flux also. The torque developed at the motor shaft is the product of the current and the flux. Hence, the torque density of the machine can be increased only if the sizes of the conductors and the slots that accommodate them are increased. However, larger slots reduce the width of the teeth between the slots that carry the magnetic flux, reducing the flux from the stator core to the rotor poles. Moreover, all active torque-producing conductors are packed in a narrow annular space adjacent to the air gap. The rest of the machine volume remains empty of torque-producing current.

For this reason, much of the attention in electric propulsion is focused on compact motors that can yield high power and high torque densities per kilogram and per cubic meter of the motor. With significant research funding from the U.S. Navy, the industry has been somewhat successful in designing such new propulsion motors.

**TABLE 8.1**

**Candidate Motor Types for Large Electric Propulsion of Ships**

| Long Name | Short Name |
|---|---|
| ac wound-field synchronous motor | Synchronous motor |
| ac induction motor (or asynchronous motor) | Induction motor |
| ac permanent magnet synchronous motor | Permanent magnet motor or PM motor |
| ac superconducting synchronous motor | Superconducting synchronous motor |
| dc homopolar motor | Homopolar motor |
| dc superconducting homopolar motor | Superconducting homopolar motor |

High-power, low-rpm, high-torque electric motors have been developed that are sufficiently compact and quiet for use on surface combat ships and submarines. With regard to the motor drives, it has also been possible to develop power electronics converters that could handle high power efficiently and deliver high-quality (low-harmonics) power needed for the motor to operate at high efficiency with low noise and low vibration. The development of large power electronics devices is also enabling technologies for electric ships because these devices can efficiently convert large amounts of electrical power into different voltages and frequencies needed for propulsion and nonpropulsion loads alike. The performance comparison between major propulsion motors used on ships at present follows.

## 8.1 SYNCHRONOUS MOTOR

The synchronous motor is the most mature technology and is suitable for large ships. It has been successfully used for electric propulsion in most commercial ships, particularly cruise ships, for about two decades. It has better efficiency compared to the induction motor but may lose synchronism if a large load is applied in one step for quick maneuvering, as may be required in some navy ships. Also, if scaled up to high horsepower ratings needed for surface combat ships and submarines at high speeds (>30 knots), the motor would be large and heavy, unfit for use due to its relatively low power density compared to some other motor types. The volume and weight are critical design considerations in the submarine propulsion plant. While they may be somewhat negotiable in the surface combat ship interior, a high-power traditional synchronous motor may be too large to fit in the propeller pod. The motor can be made somewhat smaller and more power dense by using water cooling instead of air cooling. Large utility-scale synchronous generators in hundreds of megawatts do use water cooling in the stator. However, in the power range of 20–25 MW, the potential gain in power density using water cooling may not be large enough to justify the effort in providing the deionized cooling water system through the hollow stator conductors. The most likely electric propulsion system using the synchronous motor may be in large navy auxiliary ships, which are often designed for relatively lower speeds. Their internal space constraints may not be as great, and their propulsion system may not need to meet the low acoustic noise and high shock resistance standards of the navy. The synchronous motor could make auxiliary ships somewhat common with cruise ships and other commercial ships but may not be suitable for maintaining the commonality of electric propulsion technology on *all* U.S. Navy ships.

## 8.2  INDUCTION MOTOR

The induction motor is also a matured technology for use in large ships and is the most widely used motor in heavy industry on land for its rugged construction, low capital cost, and low maintenance. The induction motor was used in the full-scale, land-based electric propulsion demonstration tests of the U.S. Navy. It can be sufficiently power dense for use on U.S. Navy surface combat ships but perhaps is not suitable for navy submarines. Using the induction motor in electric propulsion might accelerate the integrated electric propulsion technology in surface ships but would preclude achieving motor commonality with the submarines. One possible approach to addressing the commonality issue may be to use an advance induction motor (AIM) for the surface ships and preserve the commonality with submarines in components other than the motor. Full commonality can be achieved by designing a DD-21-type system with the induction motor that could later be changed to a compact permanent magnet or superconducting motor.

An advance polyphase induction motor has been designed with significant reduction in machine size. The Alstom/Converteam's 20-MW, 180-rpm polyphase induction motor (presently 100 tons with a new target of 70 tons) is small enough to be considered for use in the type 45 frigate of the Royal Navy. However, using this induction motor on the DD-21 or other types of U.S. Navy ships can be a concern regarding the use of technology developed by a foreign firm. Alstom supplied the AIM (and other components) for the land-based electric propulsion demonstration system of the U.S. Navy. It has established U.S. subsidiaries in Pittsburgh and Philadelphia to support its electric propulsion efforts for the U.S. market. In parallel, U.S. firms have focused on developing compact permanent magnet and superconducting motors.

The traditional motors are designed for fixed-frequency operation for shipboard propulsion via heavy gears. However, the introduction of motor drives has resulted in relaxation of the constraints of a fixed 60-Hz or 50-Hz frequency. A direct gearless drive for a 150-rpm propeller, for example, needs a 48-pole motor in a 60-Hz system, whereas the same machine requires a 16-pole design at 20 Hz. The AIM design takes advantage of this to produce high torque density needed in the ship propulsion motor, as well as in the direct drive generator for wind energy farms, on land or in ocean.

The high-power, pulse width modulated (PWM)-fed AIM with custom-made variable-speed drive can result in a compact and efficient propulsion system. The optimally designed AIM with its own drive allows flexibility and optimization of the electromagnetic design. This in turn enables improved efficiency and power factor while allowing a larger air gap to provide a good shock-withstanding capability. The AIM system is now moving to a 15-phase machine driven by its custom-designed PWM converter. Recent developments in insulated gate bipolar transistors (IGBTs) have helped design more efficient PWM converters in combination with induction motors. The AIM can also be a 12-phase or 15-phase high-voltage motor that can be connected directly to the converter output without using a transformer, a major benefit for weight, volume, and cost reductions.

The AIM systems are being installed on the type 45 destroyers and the first two DDG1000 vessels of the U.S. Navy, some ships of the Royal Navy, and aircraft carriers of the French Navy. Major benefits of such an approach for naval propulsion are (a) direct connection of high voltage at the converter terminals without a transformer, (b) high power density, (c) high efficiency, (d) low noise and vibration, (e) high shock-withstanding capability, and (f) low life-cycle cost and maintenance.

## 8.3 PERMANENT MAGNET MOTOR

The next type of motor uses permanent magnets—generally on the rotor—that are made of high-energy product permanent magnet alloys such as samarium-cobalt (SmCo) or neodymium-iron-boron (NdFeB). These alloys exhibit linear demagnetization in the second quarter of the B-H (hysteresis) loop as shown in Figure 8.1. They have remnant flux density $B_r$ of about 1 Wb/m$^2$ and high coercive force $H_c$ of about 1000 kA/m. Therefore, the magnetic energy density $\frac{1}{2}B_r \times H_c$ is high in these magnets. The permeability of the magnet alloy is the same as that of air. Since the permanent magnet produces a constant flux, it can be represented as an excitation coil with a constant current source. Therefore, for analyzing the motor performance, the magnets are treated as a constant current source at its surface, equal to 1000 kA/m, which interacts with the stator current to produce the torque. The power- and torque-producing capability of the permanent magnet motor is proportional to its energy product, which is the area between its demagnetization line and the $x$ and $y$ axes, that is $\frac{1}{2}B_r \times H_c$ in Figure 8.1. The magnet with a larger energy product produces greater torque. The absence of rotor winding makes more space available to

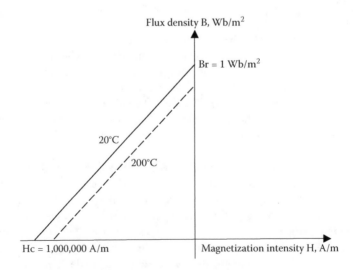

**FIGURE 8.1**  Permanent magnet demagnetization line in second quarter of B-H loop.

pack more high-energy-density magnets, resulting in high air gap flux, which also adds to the torque density. For these reasons, the permanent magnet motor has higher torque density, hence higher power density as well, and has high efficiency since there is no actual coil and no copper loss in the rotor.

The power density of the permanent magnet motor using various permanent magnet alloys is compared in Figure 8.2. The neodymium-iron-boron alloys have higher energy product, whereas the samarium-cobalt alloys have better thermal properties. The energy product of both groups is sensitive to the operating temperature. It degrades, as shown in Figure 8.1, from 20°C to 200°C. The energy product degrades faster at higher temperature and is totally lost at the curie temperature of the alloy. The mechanical shock is also known to degrade the energy product.

The NdFeB permanent magnet material has low cost and high strength and is able to operate up to 150°C (302°F), whereas the high-temperature SmCo permanent magnet material offers operation up to 290°C (554°F). These operating limits are taken into account in system design. For example, a high-temperature permanent magnet is needed in a permanent magnet generator directly connected to a high-speed gas turbine impeller operating at 900°C.

A performance risk with the permanent magnet motor exists in the event of a fault, such as a short circuit in the cables between the drive and the motor—a fault that can occur in any electrical motor. Unlike in other motor types, it is not possible in the permanent magnet motor to turn off the magnetic field of the rotor. Therefore, the motor with back emf starts working as a generator in the event of a fault, particularly if the ship is moving through the water at high speed, when the propeller would become a water turbine. The generated electrical energy would be fed by the machine—now working as a generator—back into the fault, aggravating the damage caused by the short. The back emf lasts as long as the rotor spins and cannot be ramped down quickly. But, in the synchronous motor, and more so in the superconducting motor, breaking the large dc field current is not easy due to a large

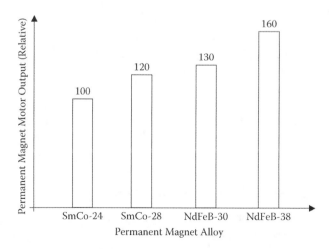

**FIGURE 8.2** Relative output power versus permanent magnet alloy used in large permanent magnet motors.

inductance of the field coil. An ability to turn off the magnetic field could be useful, but in the permanent magnet motor, the back emf has to be managed by a suitable control circuit. It can be diverted to a dummy resistor as we do in the dynamic braking employed in other motors, for which the faults are managed through the motor and the control system designs.

The permanent magnet motor gives power density with good acoustic performance, making it suitable for a broad range of ships. Surface-mounted and embedded magnet designs are robust and highly efficient at all speeds and loads. This motor offers about 30% weight reduction over the wound rotor synchronous motor. With a variable-frequency drive, it offers flexibility in the selection of the number of poles. More poles reduce the size and weight of the motors by reducing the required back iron and significantly reduce the motor core vibrations. The absence of $I^2R$ power loss in the rotor results in high efficiency, less heating, and a cooler motor. The production cost of permanent magnet motors is dropping with new magnets of high magnetic strength at increasingly high operating temperatures. The permanent magnet cost is now competitive compared with a hand-wound copper coil for the same magnetic field strength. With no windings on the rotor, permanent magnet motors are more shock tolerant and less susceptible to failures. Moreover, they can produce maximum torque at essentially zero speed, something the induction motor cannot do efficiently.

Although technologically less mature than the previous two types, the permanent magnet motor can be made quieter and more power dense for use in a common electric propulsion system for navy surface ships and submarines. As a result, the navy has tested it in quarter scale (6000 hp) for possible use on the Virginia-class submarine and other future submarines. It may soon be available in full ratings suitable for a common electric propulsion system. The funding for this motor by the navy, even at quarter scale, is taken by some in the industry as the confidence of the navy in permanent magnet motor technology. However, technical risk remains in scaling it up to needed power ratings for programs such as the DD-21 destroyer. If the permanent magnet motor is not mature enough to confidently install on DD-21-type ships, then it must be backfitted with an advance induction motor at some point in the future. This would have an adverse impact on the desired commonality in the electric propulsion system across the ship.

The permanent magnet motor design using the traditional radial flux topology competes well with the advance induction motor design by Alstom/Converteam Incorporated. However, the long-term stability of the permanent magnet materials used in large motors is not proven, which could lead to reliability and maintenance issues over the motor life.

## 8.4   SUPERCONDUCTING SYNCHRONOUS MOTOR

The superconducting synchronous motor today is even less-mature technology than the permanent magnet motor but could be even more power dense and quieter. It uses cryogenic equipment to cool the superconducting wire below its critical temperature when the wires become superconducting with precisely zero resistance for dc. Such wire can carry enormous current with no $I^2R$ power loss and achieve a significantly stronger magnetic field than the traditional synchronous motor. As a result, the superconducting synchronous motor can yield high torque and power densities, higher energy efficiency, and quieter operation compared to the permanent magnet motor.

The cost, reliability, and survivability of the cryogenic system of the superconducting synchronous motor may pose a technical risk issue in ultimate deployment. This risk has been reduced through the advent of new high-temperature (liquid nitrogen) superconductors, as opposed to the old low-temperature (liquid helium) superconductors. The much lower cryogenic (refrigeration) needed for the high-temperature superconductor (HTSC) has made it possible to achieve superconductivity without using the expensive liquid helium cooling system. Since American Superconductor Corporation (AMSC) introduced the HTSC technology in the late 1990s, the U.S. Navy funded a program to build and test a quarter-scale motor first and then a full-scale 25,000-hp HTSC motor, to have the motor possibly ready to enter the fleet in 2012.

Figure 8.3 depicts the circular cross section of a two-pole superconducting synchronous motor where the labeled parts are as follows:

- Rotor with HTSC dc field coil in liquefied nitrogen vapor (two-phase cooling) from stationary cryocooler through a rotary joint on the shaft.
- Aluminum shield that is a part of the rotor. It contains the vapor, provides centrifugal support to the rotor parts, works as the electromagnetic damper for rotor oscillations following sudden load changes, and shields the rotor from harmonic eddy current heating that may come from the stator flux harmonics.
- Air gap that provides clearance between the stator and the rotor.
- Stator armature coils of copper conductors in air (not in magnetic slots). The strong magnetic field produced by the superconducting rotor coils compensates for the absence of magnetic steel here. The armature carries ac and works at room temperature.
- Stator yoke (back iron) for return flux path for both the rotor flux and the stator flux (armature reaction).
- Stator enclosure; also provides casing for the cooling air circulated through the stator.

Table 8.2 summarizes key performance features of various motor candidates for electric propulsion. Figure 8.4 compares the motor efficiency versus load for various types of 20-MW motor. It shows that the superconducting motor offers significant efficiency advantage at full load and even greater advantage at partial loads. This is important for many navy ships, which spend a significant amount of time at reduced speed. Since the propulsion power depends on the speed cubed, the ship at 50% speed needs only $(\frac{1}{2})^3 = 0.125$ times the full-rated power. As for the weight and volume, Table 8.3 summarizes the design parameters of two superconducting motors (5 and 36.5 MW) designed by AMSC. Both motors were delivered to the U.S. Navy, the 5-MW motor in 2003 and the 36.5-MW motor in 2007. The potential applications of these motors include both military and cruise ships.

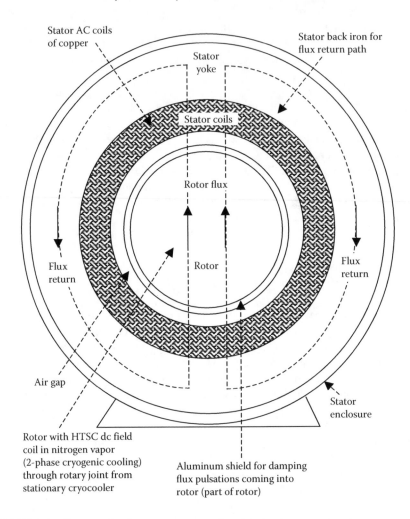

**FIGURE 8.3**  Superconducting motor construction showing key components.

Figure 8.5 compares the outer dimensions and weights of a 36.5-MW, 120-rpm, 6.6-kV superconducting motor with those of a 21-MW, 150-rpm, 4-kV traditional copper motor. The weight and volume are reduced to about a third for a comparable megawatt rating. The current density in the superconducting rotor coil is about 150 times that in copper coil, and the energy efficiency can be as high as 98%, which can significantly reduce the fuel cost over the life of the ship.

A concern remains that the superconducting wire in these motors may degrade over a long period of use. For this and other reasons, the superconducting motor technology may take longer to mature than the permanent magnet technology. When it does, it may prove suitable for use in a common electric propulsion system for use on navy surface ships and submarines. Since the superconductivity is already used in some land-based applications and in navy ships, the next chapter in this book covers it in further details.

**TABLE 8.2**
**Summary of Key Performance Features of Various Motors for Electric Propulsion**

| | |
|---|---|
| Induction motor (asynchronous motor) | Rotor energized inductively from the stator. Requires small precision air gap and hence is less shock tolerant. Large induction motor requires large air gap to allow for thermal growth, vibration, and shock. Large air gap decreases the power factor, efficiency, and torque density more than in other motor types. |
| Synchronous motor (wound rotor) | Requires a separate rotor excitation that increases the complexity. It has larger air gap than in comparable induction motor and hence is more shock resistant. Slip rings and brushes (if present) and rotor coil are prone to winding maintenance and failure-related problems. But, it has higher efficiency and power factor, which can be made unity with underexcited rotor coil. |
| Permanent magnet synchronous motor | Has high power density and low noise, hence is suitable for a broad range of ships. Surface-mounted and embedded magnet design is highly efficient at all speeds and loads. New permanent magnets at lower cost are available for increasingly high-energy product, magnetic strength, and operating temperature. This motor can produce maximum torque at essentially zero speed, something an induction motor cannot do efficiently. |
| High-temperature super-conducting synchronous motor | HTSC motor can be more power dense and quieter than permanent magnet motor but technology is not nearly as mature. It can carry high current density with negligible power loss in the rotor and can give more than double the power output of conventional motors with copper windings of similar size. It offers the highest energy efficiency among all motors. |

There is no such thing as an ac superconductor; superconductivity is only a dc phenomenon. Transformers and ac cables with superconductors have been built and tested with unattractive benefits.

## 8.5 SUPERCONDUCTING HOMOPOLAR MOTOR

The U.S. Navy worked on developing the homopolar motor for many years, starting in the early 1980s. Recent developments in electric propulsion have brought renewed interest in the homopolar motor. However, homopolar motor technology at present

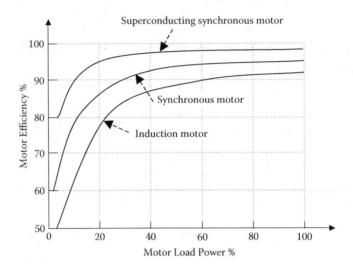

**FIGURE 8.4**   Efficiency of three types of 20-MW motors at various loading levels.

**TABLE 8.3**
**Superconducting Motor Specifications and Design Parameters**

**Specifications**

| | | |
|---|---|---|
| Output | 5,000 kW | 36,500 kW |
| Speed | 230 rev/min | 120 rev/min |
| Torque efficiency | 63% | 97% |
| Pole pairs | 3 | 8 |
| Voltage | 4.16 kV | 6 kV |
| Armature current | 722 A rms (3 Phase) | 1,270 A rms (9 phase) |
| Phase | 3 | 9 |
| Power factor | 1 | 1 |
| Frequency | 11.5 Hz | 16 Hz |
| Weight | 23 metric tons | 75 metric tons |
| Dimensions (L × W × H) | 2.5 × 1.9 × 1.9 m | 3.4 × 4.6 × 4.1 m |
| Stator cooling | Liquid | Liquid |
| Drive | Commercial marine | Commercial marine |

*Source:*   Office of Naval Research, courtesy of American Superconductors Corporation.

is even less mature than that of both the permanent magnet and the superconducting synchronous motors.

The conventional dc machine is *not a true dc machine*: It is ac inside and dc outside. The ac-to-dc conversion is done by the commutator. One reason the dc machine has fallen out of favor in modern times is the high manufacturing and maintenance costs and poor performance and reliability associated with the commutator. The homopolar machine *is a true dc machine* inside and out, with the need for the

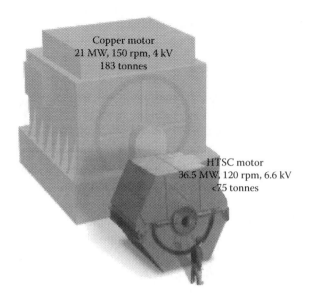

Copper motor
21 MW, 150 rpm, 4 kV
183 tonnes

HTSC motor
36.5 MW, 120 rpm, 6.6 kV
<75 tonnes

**FIGURE 8.5** Volume and weight comparison between 36.5-MW HTSC and 21-MW copper motors. (From The Office of Naval Research, photo courtesy of American Superconductor Corporation.)

commutator circumvented by the armature conductor cutting the magnetic flux of only one polarity as shown in Figure 8.6. All flux flows through the air gap in one direction from one side of the core to the shaft and then to the center section. The rotor conductor in the center section of the machine continuously cuts the flux in one direction only, making it a true dc machine. The dc current is transferred between stator and rotor via slip rings and current collectors made of solid carbon or liquid metal brushes. No commutator is needed in the homopolar machine. In addition, because the homopolar motor has only dc current, rather than the ac current in all other motor types discussed, it permits the motor drive to be less complex and thus less expensive than the motor drives associated with the other motor types.

Because only one conductor of the copper drum length generates the machine output voltage, the homopolar machine inherently has a very low terminal voltage. Therefore, high-power machines in tens of megawatt ratings must carry enormous current, posing a design challenge on the current collection with brush erosion and sparking. The prototypes built at the Westinghouse research and development center in the early 1980s for the U.S. Department of Defense for a multimegawatt pulse power electromagnetic gun system used liquid metal current collection to meet specific system requirements. The conducting liquid—mercury or sodium potassium—on a copper slip ring collected large currents. Such liquid metal *brushes* avoid the rubbing but introduce viscous drag to be managed. In addition, prototype testing revealed potential hydrodynamic instability of the liquid metal. A complex three-dimensional (3-D) finite-element magnetohydrodynamic analysis can predict such instability under load transients, when the liquid metal can expel from the well, collapsing machine performance and posing a personnel safety hazard. The subsequent

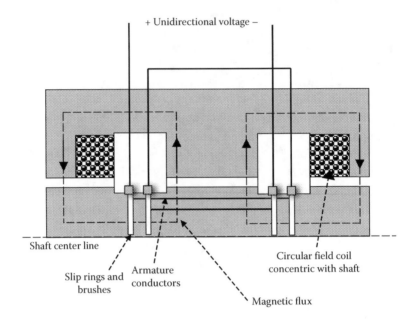

**FIGURE 8.6**  Homopolar machine construction features.

development was directed in developing solid brushes with silver fibers or pure graphite fibers to enhance slip ring performance. Since then, solid-fiber brushes have been developed that are superior to the liquid-metal current collectors previously used in homopolar motors. However, these new solid-metal brushes have been tested only at smaller scales. The successful collection of large currents needed for homopolar motors with tens of megawatts remains to be demonstrated.

Based on the room temperature homopolar motor development of earlier years as described, new efforts were targeted on developing the superconducting homopolar motor, which uses superconducting technology to achieve higher electrical current and a stronger magnetic field than the nonsuperconducting motor. As a result, the superconducting homopolar motor can be more power dense, quieter, and energy efficient than the permanent magnet motor. Such a motor can produce full torque at low speed that is suitable for direct coupling with the propeller. The U.S. Navy demonstrated a 400-hp superconducting homopolar motor and 3000-hp superconducting dc motor on the test ship *Jupiter II* in the early 1980s. Two difficult design aspects of these early motors were that (a) both used a low-temperature superconductor in a liquid helium bath and (b) high current collecting brushes made of liquid metal showed an inherent tendency to become hydrodynamically unstable under load transients. The superconducting homopolar motor with greater power density poses the same challenge in current collection between the stationary parts to the rotating parts, only more severe than the room temperature homopolar motor.

Advances in HTSC and in solid silver fiber current-collecting brushes could eliminate these problems in the next generation of homopolar motors. Such efforts have continued in recent years by General Atomics. It is estimated that a quarter-scale

(6000-hp) homopolar motor can be built and demonstrated at sea in a few years at a cost of about $25 million in 2010. However, designing, building, and testing a full-scale (40,000-hp) homopolar motor may require additional time and significant funding, which does not appear to be on the horizon at present.

## 8.6   OTHER MOTOR TYPES

All motors can come in three versions based on the direction of the magnetic flux: (a) the conventional radial flux, (b) the axial flux, and (c) the transverse flux. These versions differ in the design and orientation of their fixed and rotating elements—the stator and the rotor—and consequently in how the magnetic flux and the electrical current in the motor interact to create the mechanical torque. The conventional radial flux motor has a cylinder rotating within another annular cylinder, the axial flux motor has a disk-shaped wheel spinning between two other fixed disks, and the transverse flux motor has a rimmed disk whose rim spins inside slotted rings.

The limitations of the conventional radial flux machines are (a) excessive nonuseful current in long overhangs at the ends, (b) bulky yokes to carry flux between poles, and (c) a limit on the number of poles without smearing the magnetic flux pattern between adjacent north and south poles. A motor with more poles that runs at lower speed matches better with the propeller speed that can eliminate bulky gears.

Axial flux motor: In the conventional motor, the flux flows radially through the air gap between the rotor and the stator. In the axial flux motor, the flux flows parallel to the axle of the motor. The rotor—often referred to as a pancake rotor—can be made much thinner and lighter; hence, this motor is often used for applications requiring quick changes in speed. The novel topologies being developed by Jeumont Industrie in France and Kaman Electromagnets in the United States are shown in Figure 8.7. The design uses a permanent magnet on a disk-type wheel.

Transverse flux motor: The transverse flux machine uses topology that provides relief from the competition for space between the electric and magnetic circuits, thus allowing higher power and toque densities. One topology of the transverse flux motor is shown in Figure 8.8, which gives a single slot per pole. The rotor is basically a solid conductor and works like the induction motor, as opposed to the axial flux synchronous motor described.

Figure 8.8 illustrates the stator winding and magnetic topology in a transverse flux motor that was developed by Rolls-Royce with support from the U.K. Ministry of Defense (MoD). It offers a potential fourfold increase in power density compared with a conventional synchronous motor. One of the early Rolls-Royce designs gave 20-MW power at 180 rpm with 1.8-m diameter, 2-m length, and 40-ton mass. On the other hand, a 1.5-MW, 100-rpm dc propulsion motor used in the type 23 frigate of the Royal Navy occupies the same space. As the development of the transverse flux motor has proceeded, better conversion efficiency has allowed the electromagnetic design to be somewhat relaxed. A new optimized 20-MW, 180-rpm motor is

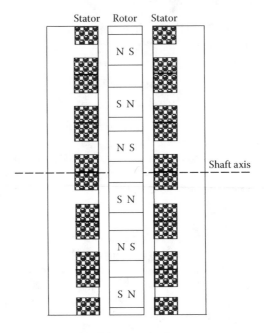

**FIGURE 8.7**  Axial-flux motor with disk-type wheel.

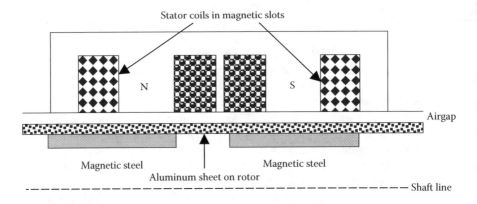

**FIGURE 8.8**  Transverse-flux motor construction features.

estimated to occupy 2.2 m³ with a 70-ton mass. The transverse flux design has a potential for higher efficiency than the radial flux design but poses greater development risks. An obstacle at present is poor efficiency of the power electronics converters suitable for this machine.

Comparing various propulsion motors is justifiably debatable due to the difficulty of agreeing on the relative importance of various performance criteria. However, the figure of merit—called the *goodness factor G* by Laitewaite—can be a valuable figure for the motor design engineer in comparing the conventional and unconventional machines. In the first approximation, the goodness factor is given by

$$G \approx \left\{ \frac{Equivalent\ solid\ thickness\ of\ rotor\ conductor}{rotor\ resisitvity \times airgap} \right\} \times \quad (8.1)$$

$$stator\ pole\ pitch \times frequency$$

## 8.7 OTHER COMPONENTS

The debate on motor types at present appears to be focused on the merits of competing motor designs, although the motor is merely one component of a much larger electric propulsion system that contains several other major components, such as the power electronics converters, cooling design, and cryogenic equipment for the superconducting machines. A motor that is best when viewed in isolation may not lead to the best overall system design because of its effects on other subsystems, particularly on the motor drive. The final decision should be based on the best overall electric propulsion system incorporating all components.

For example, developing a newer and more compact and lightweight generator would reduce total system weight and space requirements and make it easier to place the gas turbine and generator higher in a surface ship, thus reducing the amount of internal ship volume occupied by the large air intakes and exhaust ducts of the gas turbines—one of the potential architectural advantages of electric propulsion. Similarly, there may be potential for developing motor controllers even more compact than the PWM controllers now being developed. The shipwide electrical distribution system can evolve and change in terms of key characteristics, such as the type of current (dc, main frequency ac, or low- or high-frequency ac) and the higher voltage levels.

## 8.8 NOTES ON PROPULSION MOTORS

Some noteworthy aspects of electrical propulsion are as follows:

- Some old submarine and cruise ships used dc motors below 8-MW ratings, which is their design limit due to commutation and other reasons.
- Propulsion motors driving propellers at both ends—fore and aft—have been unsuccessfully tried to improve propulsion.
- Permanent magnet motors are used up to 10-MW ratings.
- The advance induction motor is designed to give optimum performance when integrated with its own variable-frequency drive.
- Axial flux motors give a lighter machine but have low shock resistance.
- The transverse flux machine, as developed by the British Royal Navy, gives better power density.
- Power electronics-induced harmonics are best handled by active filters at the main power bus that can be tuned with the changing bus load.
- The integrated electrical power system of the U.S. Navy uses a variable-frequency drive with a PWM converter with advance induction motor. It is rated 19 MW, 115 rpm, 4-m diameter, with a stator designed for 18$g$ impact in the vertical direction.

## QUESTIONS

*Question 8.1*: List the three most important performance features desired for electric propulsion of U.S. Navy combat ships.

*Question 8.2*: Based on what you read in this chapter, which motor can produce the highest power and torque densities at low speed? If you have additional knowledge based on your work experience, share it with the class.

*Question 8.3*: The power and torque densities of a permanent magnet motor depend on the energy product of the permanent magnet alloy. What is this energy product?

*Question 8.4*: How, by focusing only on the motor, can the engineering team be distracted from making a decision that is optimum for the entire power system on a ship? Share your experience, if any.

*Question 8.5*: A high number of poles results in a slow-running motor that can directly drive the propeller without reduction gears. What makes it difficult to design such motors?

*Question 8.6*: Make three key statements on the superconducting motor.

*Question 8.7*: Make three key statements on the homopolar motor.

## FURTHER READING

Ahmed, A. 1999. *Power Electronics for Technology.* Upper Saddle River, NJ: Prentice Hall.

Alger, P.L. 1981. *The Nature of Polyphase Induction Motor.* New York: John Wiley & Sons.

Buck, J.B., et al. 2007. Factory testing of a 36.5 MW high temperature superconducting propulsion motor. *American Society of Naval Engineers,* June, p. 25.

Edward, J. 1999. Transforming shipboard power. *Sea Power,* October, pp. 50–52.

Gieras, J.F. 2009. *Permanent Magnet Motor Technology: Design and Applications.* Boca Raton, FL: CRC Press/Taylor & Francis.

Gieras, J.F., and R.J. Wang. 2000. *Axial Flux Permanent Magnet Brushless Machines.* Berlin: Springer.

Krishnan, R. 2010. *Permanent Magnet Synchronous and Brushless DC Motor Drives.* Boca Raton, FL: CRC Press/Taylor & Francis.

O'Rourke, R. 2000. *Navy Attack Submarine Program Background and Issues for Congress on Virginia-Class Program.* CRS Report RL30045. Conqressional Research Service, Washington, D.C.

O'Rourke, R. 2000. *Navy DD-21 Land Attack Destroyer Program: Background Information and Issues for Congress.* CRS Report 97-700 F. Conqressional Research Service, Washington, D.C.

Say, M.G. 1983. *Alternating Current Machines.* New York: John Wiley & Sons.

Walters, J.D., et al. 1998. Reexamination of superconductive homopolar motors for propulsion. *Naval Engineers Journal,* January, pp. 107–116.

Whitcomb, C. A. 2000. Commercial superconducting technology for ship propulsion. *The Submarine Review,* April, pp. 62–73.

# 9 Superconductors in Navy Ships

The superconducting technology is fully developed on ground for many applications. The U.S. Navy has invested significant funds in developing high-temperature superconducting technology for use on ships. The developments for the navy include (a) a superconducting synchronous motor with high torque and high power densities for electric propulsion in destroyer ships and (b) superconducting degaussing coils for combat ships. The superconducting technology may also find applications for storing a large amount of energy for delivering an intense burst of power for weapons on combat ships. In commercial ships carrying supercooled liquefied natural gas (LNG) at 110 K (−163°C), the LNG tankers are good candidates for yet-to-be-developed high-temperature superconductor (HTSC) applications using the same onboard cryogenic equipment.

## 9.1  SUPERCONDUCTIVITY

The electrical resistance of a conductor is temperature dependent, decreasing with decreasing temperature as shown in Figure 9.1. If the temperature is reduced toward absolute zero in the cryogenic range (the range of liquid nitrogen to liquid helium), the dc resistance of certain conductors abruptly drops to precise zero at some critical temperature $T_{cr}$. Below this temperature, the superconducting coil requires theoretically zero voltage to produce enormously high current and the resulting high magnetic flux. In practice, a negligibly small voltage is needed to overcome the lead transitions from the coil to the room temperature voltage source. The coil itself needs no voltage to circulate a steady-state dc current in the coil, and the coil terminals can be shorted to continue circulation of the current forever. The circuit time constant $L/R$ is now infinite, meaning that the current will continue to flow in the coil indefinitely. The coil is said to have attained the superconducting state with zero resistance. The energy stored in the coil then *freezes* and remains stored indefinitely.

The superconductors used until the 1980s worked around 4 K in a liquid helium bath (boiling point 8 K). This is mighty cold, considering that the lowest background temperature of our universe is 3 K. The high temperature superconductor (HTSC) was invented in the late 1980s that works at liquid nitrogen temperature (boiling point 77 K). This reduced the cryogenic cooling requirement of the superconductor by severalfold and accelerated the applications of superconductivity in the industry. The critical current density of an HTSC can be as high as 1000 A/mm² (depends somewhat on the actual working temperature and the magnetic field density also), whereas that of copper in conventional electrical equipment is about 1 A/mm², which is lower by a factor of 1000. Recently discovered materials become superconducting even at higher temperatures, some exceeding 100 K. These new HTSCs are leading

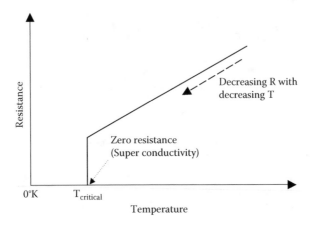

**FIGURE 9.1**    Conductor resistance versus temperature approaching absolute zero.

the industry to new technological breakthroughs. They can be easily operated in a liquid nitrogen bath at 75–80 K, which has a number of advantages:

- Liquid nitrogen costs much less than liquid helium.
- It has greater heat capacity, which means less liquid is needed.
- It can be transported in a portable container.
- It has an ideal temperature range for optimum operation of semiconductor devices, making hybrid circuits of semiconductors and superconductors attractive.

Some of the applications of HTSCs in navy ships are discussed in the following sections.

## 9.2   DEGAUSSING COIL

The ferromagnetic materials used in ship construction, once magnetized during construction or under the magnetic field of Earth, retain permanent magnetism due to the hysteresis effect. This *permanent* magnetism is not really permanent in a strict sense; it is influenced by the history of the magnetization, mechanical stress, and temperature. On a macroscopic scale, this implies that even after demagnetizing these materials (called *degaussing* or *deperming*), the magnetization of a steel hull gradually builds up. Such magnetization has a unique pattern in each ship, known as the *magnetic signature* of the ship. The magnetization of the ship, moving or not, disturbs the local magnetic field of the earth. The magnetic disturbance, if measurable, poses the following threats to the ship:

- Detection and classification of the ship by the enemy becomes easy.
- The performance of magnetic sensors and smart signal processors widely used in modern ships is disturbed.

- It poses a risk of detonation of sea mines triggered by a sensor that detects variations in the ambient magnetic field when the massive hull of a ship passes overhead.
- It poses an even greater danger if the ship itself has additional magnetism due to heavy electrical currents flowing in and out of the electrical power system or the weapons system.

Naval mine strikes have been the cause of about 77% of U.S. Navy ship casualties since 1950. Most recently, three modern warships—USS Samuel B. Roberts, USS Princeton (CG-59), and USS Tripoli (LPH-10)—were severely damaged by mine warfare during Persian Gulf conflicts. With increased operations in the coastal areas of the world's oceans, U.S. Navy ships potentially face an increasing threat from naval mines.

Minimizing the magnetic signature of the ship to an acceptable level is therefore necessary to protect the ship from these threats. This is accomplished by proper choice of materials and by using *degaussing coils* to counter the magnetic field pattern of the ship. The degaussing coils are used to erase the magnetic signature of the ship, and the compensating coils are used to reduce the inductive effect of the enormous ferromagnetic mass (hull) moving through the magnetic field of Earth. The degaussing coil design starts with establishing the theoretical model that can approximate the behavior of the ship in the magnetic field and then determining the optimum degaussing coil configuration to achieve the desired reduction. The magnetic signature study is made at a certain measurement depth. This depth is different for each class of ships and depends on the latitude of the navigation area and on the geometry of the ship.

For continuous degaussing of the ship, simulation studies have shown that the degaussing coils with high currents in the aft and bow coils are more useful for reducing the magnetic field strength of the ship. Demagnetization is done using three coils in three axes as follows:

1. Longitudinal coil for the vertical component of the longitudinal magnetization
2. Transverse coil for the vertical component of the transverse magnetization
3. Vertical coil for the vertical component of the vertical magnetization

The magnetic lines generated by the permanent or induced magnetic field of the ship play an important role in the degaussing process that nullifies their effect on the magnetic field of Earth. The following are various types of coil arrangements (Figure 9.2) used onboard, with each coil set drawing a certain amount of power:

*Main coil (M):* It is installed in a horizontal plane at the waterline. When current is passed, the coil compensates the induced and permanent vertical components of the magnetic field of the ship. When the ship changes the hemisphere, the coil polarity is to be manually adjusted.

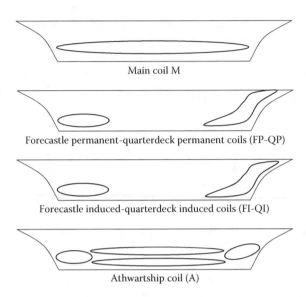

**FIGURE 9.2**   Degaussing coil locations in navy ships.

*Forecastle permanent and quarterdeck permanent (FP-QP) coils:* They compensate the longitudinal permanent components of the magnetic field. The FP coil is installed horizontally forward of the ship and encircles approximately one-third of the ship at the main deck, while QP coil encircles one-third of the aft region of the ship.

*Forecastle induced and quarterdeck induced (FI-QI) coils:* With location almost the same as for FP-QP coils, they compensate for the longitudinal magnetic components of the magnetic field. The polarization is automatically changed by converting a gyro input signal to a magnetic heading. The current in the coil is directly proportional to Earth's magnetic field.

*Athwartship (A) coil:* Installed in the vertical position, it compensates the athwartship's permanent and induced components of the magnetic field of the ship.

There are two general ways to camouflage the ship against magnetic detection. Ships have onboard degaussing coils that are supplied with a current to generate their own magnetic field that opposes Earth's field at the ship's location. In conjunction with degaussing coils, ships are also routinely bulk demagnetized in a process called *deperming.* For example, *Jimmy Carter* is a Seawolf-class attack submarine based in Bangor, Washington. Its mission is to seek and destroy enemy submarines and surface ships, collect intelligence, and engage in antiship and strike warfare. It underwent a complete hull deperming (Figure 9.3) before it was commissioned in 2003.

The U.S. Navy has developed superconducting degaussing coils that are significantly lighter and more efficient than the conventional copper-based coils. They operate at much lower voltages and give a wider range of neutralizing the magnetic signature. Following an in-depth study in 2008, prototype degaussing coils using

**FIGURE 9.3**  *USS Jimmy Carter*, Seawolf-class attack submarine, undergoing deperming at Naval Base Kitsap Bangor.

high-temperature superconducting ceramic materials were installed aboard the *USS Higgins* (DDG 76) to counter underwater mines. They were developed by the Office of Naval Research (ONR) and the Naval Surface Warfare Center Carderock Division (NSWCCD) at the Ship Engineering Station in Philadelphia. The HTSC degaussing coils installed on *Higgins* completed a pass over the U.S. Navy Magnetic Silencing Range in San Diego, California. This was the first-ever measurement of HTSC degaussing coils installed on a naval combat ship. Following the successful demonstration, HTSC degaussing coils were deployed onboard the *USS Higgins* (DDG 76) in 2009.

The HTSC coils cooled to 33 K (− 240°C or −400°F) by a cryogenic compressor are operated at current densities that are 100 to 200 times higher than those in room temperature copper coils. They produce magnetic flux for degaussing at a fraction of the weight of room temperature copper coils. The weight savings is estimated in the 50–80% range, offering significant potential for saving fuel or adding payloads. The HTSC degaussing technology provides new options to the naval architects in designing future degaussing systems that will use less energy to achieve even greater degaussing performance.

## 9.3  SYNCHRONOUS MACHINES

The superconducting synchronous machines (generator and motor) have been under research and development since the mid-1970s. Highly efficient generators and motors developed by companies like General Electric, Westinghouse, Siemens, and American Superconductors may become commercially viable products. The superconducting machine gets rid of the magnetic core in the stator armature, giving much higher (four- to fivefold) power density (excluding the cryogenic overhead). Toshiba Corporation of Japan and GEC-Alstom, along with Electricite de France, have also been actively perusing developments in this field. The Massachusetts Institute of Technology, General Electric, and Westinghouse Electric had active programs funded by the Department of Energy (DoE), the Department of Defense (DoD), and the Electric Power Research Institute (EPRI) to design and demonstrate large utility-scale machines. A 300-MVA prototype superconducting generator was fully designed and key components tested successfully at the Westinghouse Research and Development Center under funding by the EPRI, which was terminated in 1987 due to funding cuts. Table 9.1 lists some of the design features of that machine.

For ship propulsion, the U.S. Navy has ongoing interest in the superconducting synchronous motor up to 50,000 hp for an integrated electric power system for warships and submarines. Under NAVSEA (Naval Sea Systems Command) funding, the American Superconductor and Northrop Grumman Corporations in 2009 completed full-power testing of a 36.5-MW (49,000-hp) high-temperature superconducting ship propulsion motor at the Land-Based Test Site of the U.S. navy in Philadelphia. The electric propulsion motor was designed for navy combat ships and submarines. The full design specifications of this motor were given in Table 8.3 in Chapter 8 but are briefly summarized below.

- An ac synchronous motor, 36.5-MW (49,000-hp) shaft output, 120 rpm, 2.9-million n-m (2.2-million lb-ft) torque, 97% efficiency, 70-ton motor, 75-ton total system
- Three-phase, 60-Hz, 6-kV line-to-line voltage

### TABLE 9.1
### Design Parameters of 270-MW Superconducting Generator

| | |
|---|---|
| Rating | 300 MVA, 0.90 power factor |
| Voltage | 24 kV, 60 Hz, 3 phase |
| Rotor speed | 3600 rpm |
| Rotor cooling | Liquid helium |
| Stator cooling | Gaseous hydrogen |
| Efficiency | 99.5% |
| Active length | 78 inches |
| Rotor outer radius | 18.6 inches |
| Armature inner radius | 23.5 inches |
| Armature outer radius | 32.5 inches |
| Core inner radius | 35.5 inches |
| Core outer radius | 52.0 inches |
| Rotor inertia constant | 0.62 second |

*Source:*  From Patel, M. R. 1990. Dynamics of rotating electrical machines on space platforms. In *AIAA Proceedings of the 25th Intersociety Energy Conversion Engineering Conference*, pp. 990–996.

- Motor with nine-phase stator coils, class F insulations, 1275 A, suitable with variable-frequency drive
- Rotor made of HTSC field coils operating at 30 K; cryogenic refrigeration system with two gaseous helium loops to extract heat from the HTSC coils

The full-scale version of this motor in navy ships may potentially make the ship about 200 tons lighter and more fuel efficient with additional space for war-fighting payload.

## 9.4   SUPERCONDUCTING ENERGY STORAGE

The energy stored in the capacitor's electric field $E$ in insulation of permittivity $\varepsilon$ is $\frac{1}{2}\varepsilon E^2$ J/m$^3$ between the plates or $\frac{1}{2}CV^2$ Joules in terms of its capacitance $C$ and voltage $V$. Most capacitor liquids break down around 10 kV/mm. An oil-filled capacitor operating at 5 kV/mm—an upper limit on design stress under normal operating voltage—has the energy storage density of $\frac{1}{2}(8.85 \times 10^{-12}) \times (5 \times 10^6)^2 = 110$ J/m$^3$ between the electrode plates. This is several orders of magnitude lower than that in the magnetic field of an inductor, as discussed next.

The energy stored in the magnetic field $B$ of the coil in medium of permeability $\mu$ is $\frac{1}{2}B^2/\mu$ J/m$^3$ in the hollow volume of the coil. In terms of the coil current $I$ and inductance $L$, the energy storage is $\frac{1}{2}L \times I^2$ joules. The energy stored in the magnetic field in air at a flux density of 2 Wb/m$^2$—the upper design limit at the saturation point of the magnetic steel—is $\frac{1}{2}2^2/(4\pi \times 10^{-7}) = 1.6$ MJ/m$^3$. This is 15,000 times greater than the energy storage density in the capacitor. A coil wound in air without the magnetic steel core has no saturation limit. If it can carry much higher current, it can store even higher energy density in the current square proportion. The conventional copper coil current is limited by the $I^2R$ heating, but the superconducting coil with $R = 0$ has no theoretical limit on the current. Commercial superconducting coils (often called the *superconducting magnets*) are available for a variety of applications at present to produce up to 15 Wb/m$^2$ field in one liter volume in air that stores 90 MJ/m$^3$. That is a million times greater energy density than that in the capacitor. For this reason, the superconducting coil finds increasing applications in large-scale energy storage.

To understand the energy storage in the superconducting coil, we consider a coil of inductance $L$ and resistance $R$ supplied by dc voltage $V$ as shown in Figure 9.4. When the circuit breaker is closed with the switch open, the coil would draw current $i$ and charge the coil with energy according to the basic circuit law, that is, Kirchhoff's voltage law,

$$V = R \times i + L \frac{di}{dt} \tag{9.1}$$

We know from the basic R-L circuit analysis that the inductor current during charge rises exponentially with time constant $\tau = L/R$. After a long time (>5$\tau$), the current becomes steady-state dc equal to $V/R$. The voltage required to circulate the dc current is then $V = R \times I$.

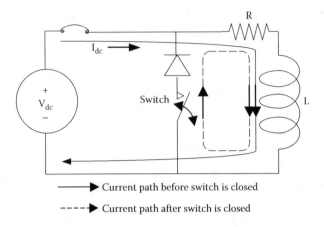

Current path before switch is closed

- - - - ▶ Current path after switch is closed

**FIGURE 9.4**   Inductor coil under charge and discharge of energy.

If the coil with initial current $I_0$ at $t = 0$ is shorted by closing the switch, it will discharge the stored energy, with the current exponentially decaying as

$$i(t) = I_0 e^{-\frac{t}{\tau}} \tag{9.2}$$

where the time constant $\tau = L/R$. If $R$ is made zero (by making the coil superconducting), the time constant $\tau$ becomes infinitely long, that is, the current—and the stored magnetic energy—would take forever to decay to zero. This can also be explained by the constant flux linkage theorem. A shorted coil with zero resistance will keep its flux constant and freeze the energy—will keep its energy constant—forever. The superconducting coil makes this possible.

The superconducting coil can store energy for numerous applications, small and large. On combat ships, it can provide an extremely high burst of power for launching weapons without significantly increasing the generator megawatt rating, except for a small amount of power needed to charge up the system between two launches. On land, the grid power company can charge up the system at night and meet the daytime peak demand. Such large-scale energy storage can eliminate brownouts and reduce the peak generating capacity of the power plant, thereby reducing the cost of electrical energy delivered to consumers. Many other industry applications are described in Sections 9.5 and 9.6.

A typical superconducting energy storage system is shown in Figure 9.5. The coil is charged by an ac-dc converter in the power supply. Once fully charged with the required current and cooled below the critical temperature, the switch is shorted, and that makes the superconducting coil a *permanent* energy storage coil. The converter continues providing small voltage needed to overcome losses in the room temperature parts of the circuit components. This keeps the constant dc current flowing (frozen) in the superconducting coil. In the storage mode, the current is circulated through the normally closed switch. The power supply for charging can be a room temperature ac-dc converter or a superconducting dc generator, such as a Faraday disk. The system controller shown in the figure has three main functions:

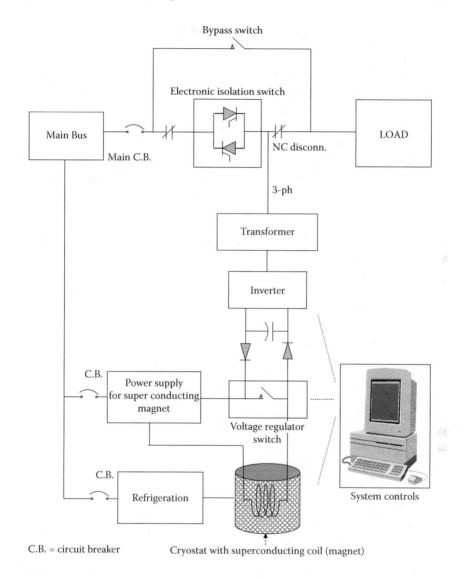

**FIGURE 9.5** Large-scale energy storage system using superconducting coil (magnet). CB = circuit breaker.

1. Control the solid-state isolation switch
2. Monitor the load voltage and current
3. Interface with the voltage regulator that controls the dc power flow to and from the coil

If the system controller senses the line voltage dropping, it interprets it as the system being incapable of meeting the load demand. The switch in the voltage regulator opens in less than 1 ms. The current from the coil now flows into the capacitor bank until the system voltage recovers to the rated level. The capacitor power is fed directly to the load or after inverting into 60 or 50 Hz ac power, if necessary. The bus voltage drops as the capacitor energy is depleted. The switch opens again, and the process continues to supply energy to the load continually. The system is sized to store sufficient energy to power the load for a specified duration.

The superconducting energy storage has the following advantages over other energy storage technologies that are discussed in Chapter 15 of this book.

- The round-trip efficiency of the charge-discharge cycle is high, around 95%. This is higher than that attainable by any other technology.
- It has much longer life, up to about 30 years.
- The charge and discharge times can be extremely short, making it attractive for supplying large power for a short time if needed.
- It has no moving parts in the main system, except in the cryogenic refrigeration components.
- With no moving parts, it has high reliability.
- The stored coil energy could be switched on in less than 30 ms, compared to 15 minutes to prepare and start a conventional emergency generator.

## 9.5   INDUSTRY APPLICATIONS

Although superconductivity was discovered decades ago, industry interest in developing practical applications started in the early 1970s. One of the most developed application of the HTSC is in building powerful and yet lossless superconducting coils without an iron core for large utility-scale energy storage, dc field coils of large motors and generators, power transmission lines such as high-voltage dc links in the power grid, magnetic levitation for trains, magnetic resonance imaging (MRI), and more. These applications are briefly discussed next:

*Energy storage:* A grid-connected 8-KWh superconducting energy storage system was built with funding from the DoE and was operated by the Bonneville Power Administration in Portland, Oregon. The system demonstrated over 1 million charge/discharge cycles, meeting its electrical, magnetic, and structural performance goals. Conceptual designs for a large superconducting energy storage system up to 5000-MWh energy for utility applications have been developed. The Wisconsin Power Company built and tested a large superconducting ring for stabilizing the grid.

The superconducting coil is also a potential candidate for energy storage for space-based missile defense weapons to produce brief pulses of extremely high power needed to drive antimissile lasers, particle beams, and electromagnetically propelled canons. The defense department has also funded programs at utility companies to develop a large superconducting energy storage ring. In normal use, it would store energy generated at low cost during off-peak hours for the use by the utility during peak hours. However, the Pentagon could turn to them in the event of a missile attack and instantly divert the stored energy for the defense need of the moment to launch high-energy weapons.

*Power cable:* A power cable made with HTSC wire can conduct up to 100 times more power than the same-diameter cable made with copper wire. By replacing copper cables with superconducting cables in cities using existing underground tunnels and ductwork, utilities can avoid digging up city streets while relieving grid congestion and increasing the reliability and security of the power network.

Several superconducting cable deployments in the U.S. power grids had been partially funded by the DoE. In 2006, National Grid and American Electric Power (AEP) energized distribution-voltage superconducting cable systems in their commercial power grid in Albany, New York, and Columbus, Ohio. The first transmission-voltage cable system in the world was energized on Long Island in 2008. This 138-kV system is a permanent part of Long Island Power Authority's primary transmission corridor to and from Connecticut. At full capacity, this power cable system is capable of transmitting up to 574 MW of electrical power. Another superconducting cable project is now ongoing with Consolidated Edison in Manhattan, with partial funding from the U.S. Department of Homeland Security.

The LS Cable Company and KERI of the Republic of Korea completed testing of a 100-m, 9.9-kV superconducting cable. American Superconductor Corporation (AMSC) delivered the HTSC wire to Korea's LS Cable, which then stranded the wire into a superconducting cable system capable of carrying 50 MW of power. The cable is about a half mile in length, making it the longest distribution-voltage superconducting cable system in the world; it was installed in Seoul and energized in 2011.

*Power electronics:* Although superconductivity does not apply to semiconducting devices, power electronics converters working at cryogenic temperature are being developed for Air Force applications. They are either dc-dc converters or ac-dc converters depending on the power source. With reduced power losses at low temperature, this technology results in high conversion efficiency of about 99%, coupled with high power density, around 1 kW/kg for small dc-dc converters and 5 kW/kg for ac-dc converters in large power ratings typical of defense weapons. These power densities are about double of what can be achieved with conventional designs.

## 9.6   RESEARCH APPLICATIONS

### 9.6.1   Particle Accelerator and Collider

Over 1600 superconducting magnets (coils), with most weighing over 27 tons, are installed in the Large Hadron Collider (LHC), which lies in a tunnel 27 km (17 m) in circumference and 175 m (574 ft) deep at the French–Swiss border near Geneva. Approximately 96 tons of liquid helium are needed to keep the magnets—made of copper-clad niobium-titanium—at their operating temperature of 2 K (−271°C), making the LHC the largest cryogenic facility in the world at liquid helium temperature. Some 1232 dipole magnets keep the particle beams on their circular path, while an additional 392 quadrupole magnets are used to keep the beams focused to maximize the chances of interaction between the particles in four intersection points, where the two beams will cross. The 3.8-m (12-ft) wide, concrete-lined tunnel, constructed between 1983 and 1988, was formerly used to house the Large Electron–Positron Collider. It crosses the border between Switzerland and France at four points, with most of it in France. Surface buildings hold ancillary equipment such as compressors, ventilation equipment, control electronics, and refrigeration plants.

The LHC is expected to address some of the most fundamental laws of physics governing the interactions and forces among the elementary particles, the deep structure of space and time, and in particular the intersection of quantum mechanics and general theory of relativity, for which current theories and knowledge are unclear or break down altogether. It was built by the European Organization for Nuclear Research (CERN) for testing various predictions of a large family of new particles. Over 10,000 scientists and engineers from over 100 countries, as well as hundreds of universities and laboratories around the world, participated in its design and construction.

Superconducting quadrupole electromagnets are used to direct the beams to four intersection points, where interactions between accelerated protons take place. Once or twice a day, the protons are accelerated from 450 GeV to 7 TeV as the field of the superconducting dipole magnets is increased from 0.54 to 8.3 Wb/m$^2$. The protons will each have an energy of 7 TeV, giving a total collision energy of 14 TeV. At this energy level, the protons will have a Lorentz factor of about 7,500 and move at about 0.999999991 times the velocity of light, or merely 3 m/s slower than the speed of light.

### 9.6.2 COMPUTERS

Improved superconductors may be used in supercomputers to replace the wires between transistors and chips. This could increase computing speed while reducing the amount of power computers dissipate in heat. Superconducting interconnects between semiconducting devices could increase circuit speed as devices become smaller. Critical current densities around 1 MA/cm$^2$ are needed for interconnects between devices on chips, and 100,000 A/cm$^2$ densities are necessary on printed circuit boards between chips.

### 9.6.3 MANY OTHER APPLICATIONS

Magnetic bearings and levitating trains suspended on a magnetic field have moved from prototype to commercial use. In medicine, diagnostic machinery that uses superconducting equipment to probe the soft tissue of the body could become more sensitive and less expensive. In space, superconducting switches—sensitive to certain kinds of radiation—could be used by the military to locate enemy satellites by detecting minute amounts of infrared radiation.

## 9.7 CRITICAL *B, J, T* OF THE SUPERCONDUCTOR

The superconductor is a metallic alloy that conducts electrical current with precisely zero resistance below a certain critical temperature $T_{cr}$. The superconductor also has a critical current density $J_{cr}$ and a critical magnetic flux density $B_{cr}$. If any one of the three operating parameters—temperature $T$, current density $J$, or flux density $B$—exceeds its respective critical value, the conductor ceases to be superconducting and reverts to the normal conducting state with normal resistance. For a given alloy, the value of $J_{cr}$ on the vertical axis in Figure 9.6 depends on the magnetic flux density and the operating temperature as shown in the figure. It shows that the critical values of

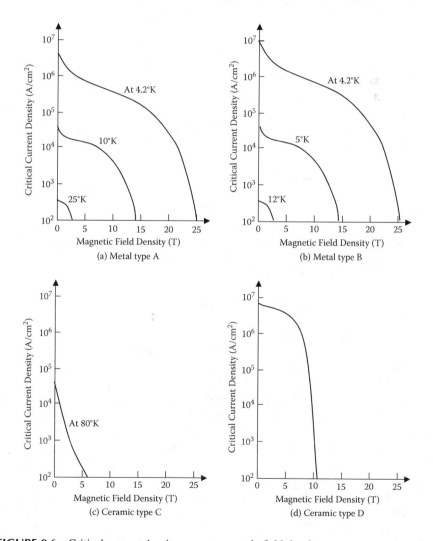

**FIGURE 9.6** Critical current density versus magnetic field density versus temperature for various superconductors showing approximately constant ($B_{cr} \times J_{cr} \times T_{cr}$) product or surface envelope.

these three parameters—$B_{cr}$, $J_{cr}$, and $T_{cr}$—are mutually related in the inverse way; if one increases, the other two decrease. In a way, one can think that the superconductor has a critical $B_{cr}$, $J_{cr}$, and $T_{cr}$ volume of operating space in which it remains superconducting. For a given superconducting alloy, the $(B_{cr} \times J_{cr} \times T_{cr})$ product, which can be viewed as the critical volume bounded by the critical surface, is constant in the first approximation. Thus, each superconductor has a certain approximately constant operating $(B_{cr} \times J_{cr} \times T_{cr})$ product, beyond which the superconductivity ceases. Two noteworthy features of the superconductor are

1. The superconductor has precise zero resistance only for dc. In ac applications, the superconductor has some small resistance, which adds substantially to the cryogenic cooling requirement and makes the ac applications much less attractive.
2. Above the critical temperature, critical current, or critical magnetic flux density, it loses the superconductivity and generates extremely high $I^2R$ loss (heat) from that point on with normal resistance.

Thus, the three operating parameters—the temperature, the current density, and the magnetic flux density—jointly determine when the alloy attains precisely zero resistance. The critical value of each parameter—marking the transition to superconductivity—varies with changes in the other two. For example, as temperature and magnetic field increase, the critical current density decreases. Hence, the superconductivity thrives below but vanishes above a three-dimensional critical surface, which is unique to each superconductor. Among the early superconductors, niobium-titanium (NbTi) and niobium-tin (NbSn) had been extensively used. Both have the critical temperature around 9 K, which requires liquid helium cooling to 4 K to allow some operating margin. The maximum flux density of around 9 Wb/m$^2$ can be produced using NbTi at 4 K and around 16 Wb/m$^2$ using NbSn at 4 K. The critical current density is about 1000 A/mm$^2$ at 10 Wb/m$^2$, which is orders of magnitude higher than that in the conventional copper coil (5 A/mm$^2$). NbSn is mechanically weak and brittle compared to NbTi, posing some difficulty in manufacturing.

The 1986 discovery of HTSCs has made three new types of superconductors available, all made from bismuth or yttrium-cuprate compounds. Current density and field requirements for various metal and ceramic superconductors are shown in Figure 9.6. In the case of yttrium barium copper oxide, the critical surfaces for thin film and bulk material made with the melt-textured growth process of AT&T extend past 80 K, the liquid nitrogen temperature. The critical surfaces of the earlier superconductors—niobium titanium and niobium tin—are much smaller, which forces them to operate near 4 K, the liquid helium temperature. Referring again to Figure 9.6, when superconducting metals (A and B) and ceramics (C and D) operating at 4 K and 80 K, respectively, are compared for current density and flux density relationships, the melt-textured bulk ceramic (C) is inferior, which would barely show up on the graph if normally processed as opposed to receiving the melt-textured growth process.

## 9.8    COIL DESIGN AND COOLING

The superconducting coil is wound with wires made of numerous superconducting filaments in a copper or aluminum matrix. In their nonsuperconducting state, copper and aluminum have lower resistance than the new superconductors. Thus, if any of the critical values were accidentally exceeded in operation, the electrical current would switch from the superconductor fibers to the surrounding metal matrix, resulting in enormous ohmic heating. The goal of the present development programs funded by the DoE and DoD are to produce superconducting wires with the design current-carrying capability of 1000 A/mm$^2$ based on the entire cross-sectional area of the wire, including the copper matrix and the insulation. This compares with 2–3 A/mm$^2$ in the traditional copper coil in standard atmosphere. The manufacturing process includes the coating of noble metals, like silver, to turn brittle, high-temperature superconducting ceramic into flexible wires. It applies not only to the powder-based wires made of bismuth-strontium compound, but also to the next generation of wires using yttrium barium copper oxide, which many companies are developing in the United States, Europe, and Japan. The current density and magnetic field requirement for various superconducting applications are shown in Figure 9.7.

The strong magnetic field coming out of a conventional solenoid coil would interfere with ship control and other neighboring equipment sensitive to the magnetic field. For large energy storage, the coil must be air cored in a toroidal configuration shown in Figure 9.8 to contain the magnetic field within itself so it does not disturb

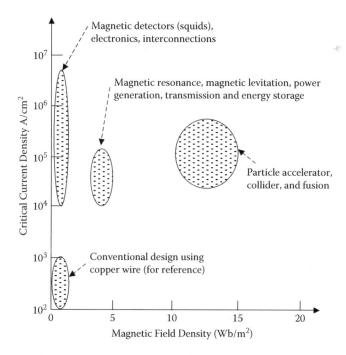

**FIGURE 9.7**    Current density and magnetic flux density requirements for various superconducting applications.

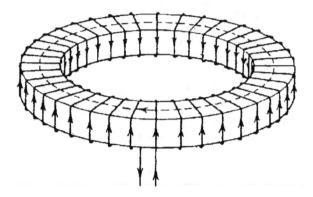

**FIGURE 9.8** Self-contained magnetic flux of toroidal coil.

the working of the surrounding equipment. The toroidal coil geometry keeps the flux self-confined to a closed ring-shape volume as shown in Figure 9.8. An air-core toroid with major radius $R$ and the number of turns $N$ carrying current $I$ produces the internal magnetic field $B_\varphi = \mu_o N \times I/(2\pi \times R)$, which varies inversely with the major radius. The circular current produces hoop stress in the conductor, and the circumferential current produces repulsive force between the two sides of the toroid. The resulting mechanical stresses vary around the circumference, thus causing strong hoop and bending stresses. Bending stress needs a center support structure. Once it is supported, the remaining force produces a pure hoop stress, which is internally supported by the tensile strength of the conductor. The superconducting coil is generally manufactured with high prestress, so that cooling to the operating cryogenic temperature does not produce damaging stress. However, the mechanical creep during room temperature storage and high-temperature excursions during forced warmup may be of concern.

The old NbTi and NbTi alloys stay stable at high $B$ and high $J$ without quenching (i.e., without losing superconductivity). They usually operate in a liquid helium bath at 4 K at 1 atmospheric pressure. Attaining 4 K is a challenge as heat removal becomes progressively difficult. Removing 1 W of heat from 4 K to room temperature costs about 500 W of refrigeration power, even with perfect Carnot cycle efficiency. Therefore, heat leakage from the leads connecting to the external circuit must be controlled by proper thermal design as it determines the capital and running cost of the cryogenic equipment. In the superconducting energy storage system, the main running cost is to keep the coil below the critical superconducting temperature.

On the other hand, the room temperature magnet loses much more power in the magnet itself and little or none in the cooling power. The room temperature power loss in the conventional magnet coil amounts to more than 100 times that in the superconducting magnet. Thus, the design trade is highly favorable even for the liquid helium superconducting magnet. The new superconductors with the critical temperature around 100 K are cooled by liquid nitrogen, which needs orders of magnitude less refrigeration power.

The copper matrix avoids local thermal transient heating and subsequent loss of superconductivity (also called *quenching*). If the coil quenches, the copper matrix provides an alternative path for the normal currents to flow with high-resistance heating. During quench, the stored magnetic energy is converted into heat. Generally, a quench starts from a small local hot spot due to friction under vibration or another reason. The spot may become overheated more than the bulk surrounding. The hot spot temperature can be controlled by rushing coolant to avoid local burning. If the coil is quenched, a large potential of hundreds—or even thousands—of volts is induced in the coil, which may cause arcing between the turns. Switching an external dump resistor or a short-circuited coil across the superconducting coil that inductively transfers part of the stored energy can provide quenching protection.

The semiconducting devices, inductors, transformers, capacitors, resistors, and the structural materials used in other power system components in the superconducting coil system behave differently at liquid nitrogen temperature (80 K) than at the conventional operating temperature around 100°C (372 K). These differences must be accounted for in designing the superconducting coil system.

## QUESTIONS

*Question 9.1*: How do the inductor and capacitor compare in their relative energy storage density per liter of volume?

*Question 9.2*: What is the half-life of current in a shorted superconducting coil having inductance $L$?

*Question 9.3*: What is the magnetic signature of a ship, and what is degaussing?

*Question 9.4*: Why are degaussing coils needed on the ship?

*Question 9.5*: What is the difference between the old superconductor and the new HTSC?

*Question 9.6*: Explain the critical $B \times J \times T$ volume or bounding surface of a superconductor.

*Question 9.7*: What is the quenching of a superconductor, and how and when can it occur?

*Question 9.8*: If you have come across any superconducting application other than those mentioned in this chapter, share it with the class.

*Question 9.9*: As a design engineer, identify a potential product for which you would like to explore using a superconductor in your design.

## FURTHER READING

Balachandran, U. 1997. Super power, progress in developing the new superconductors. *IEEE Spectrum*, July, pp. 18–25.

DeWinkel, C.C., and Lamopree, J.D. 1993, Storing power for critical loads. *IEEE Spectrum*, June, pp. 38–42.

# 10 Fuel Cell Power

Ships have long been using the cheapest oil in the world—the bunker fuel—which is the residue of the oil refineries after all cleaner fuels have been extracted from the crude oil. At about 60% cheaper than clean oil, it is still widely used under the corporate pressure on profits. However, bunker fuel is high in sulfur—the main component of acid rain and haze—as well as particulate matter consisting of minute sooty specks directly linked to lung cancers, heart attacks, and respiratory diseases. The United Nations has estimated that emissions of sulfur dioxide alone from ships was about 16 million tons in 2010 and is expected to increase 40% to 23 million tons per year by 2020. The forecast is in sharp contrast to land-based industries and vehicles, which have been forced by successive clean air laws to reduce emissions and use higher-quality fuels. Current European Union rules limit the sulfur content of fuel in cars to 15 parts per million, against 45,000 parts per million for ships. It is in this light that hybrid propulsion using fuel cell power may find application in ships—small and large—to meet the rising environmental regulations, particularly near busy ports.

To reduce emission from the electrical power generation on the ship, especially when in ports, new shipboard power systems with fuel cells for base load and battery for peak maneuvering power are being investigated by the U.S. Maritime Administration (MARAD) in partnership with the industry. Such a power system is achievable in the near future, especially in ships transporting liquid natural gas (LNG), which is a natural fuel for the fuel cell. Such a ship, when powered by fuel cells, would emit significantly less $CO_2$ than current container ships using bunker fuel. The fuel-cell-powered electrical propulsion for a totally clean ship is feasible to design and operate, although at high cost at present.

The fuel cell was first developed for space power applications and was first used in a moon buggy. It was routinely used in NASA's space shuttles for two decades. Today, it is increasingly used in some industrial and commercial applications. The U.S. energy policy has recently placed emphasis on hydrogen-based economy, and that may accelerate fuel cell use for broader applications. The fuel cell has a potential of replacing the combustion engine in (a) automobiles and metro buses, (b) emergency power generators, (c) commercial buildings and hospitals, (d) ships and military applications, and more. On small-scale levels—but with a large market potential—it may one day replace the battery in consumer electronics, including cell phones, eliminating the inconvenience of frequent battery charging. Sharing a couple of drops of alcohol with our cell phone at dinnertime will keep us connected for the next day. Table 10.1 lists various applications of the fuel cell and specific benefits for each application.

**TABLE 10.1**

**Fuel Cell Applications and Their Benefits**

| Application | Major Benefits |
|---|---|
| Space shuttles | Greater power capability and range than the battery |
| City cars and buses | Replace diesel engine for emission reduction |
| Emergency power generation in commercial buildings and hospitals | Replace diesel engine for reducing emission, vibration, and noise |
| Submarines, submerged rescue vehicles, remote rescue vehicles, etc. | Greater range of operation, reduced vibration and noise |
| Ship approaching a port | Replace diesel engine for emission reduction near ports |
| Portable electronics | Eliminates inconvenience of frequent battery charging |

## 10.1 ELECTROCHEMISTRY

The fuel cell converts the chemical energy of fuel directly into electrical energy in dc form. It is a static electrochemical energy converter. However, unlike the battery, it does not run down in energy and does not need to be recharged. It keeps producing electrical power as long as the fuel is continuously supplied. Typical fuel is pure hydrogen gas or a hydrogen-rich mixture and an oxidant. Some call the fuel cell the *gas battery*.

The working principle of the fuel cell is the reverse of the electrolysis of water. In electrolysis, electrical power is injected between two electrodes in water to produce hydrogen and oxygen. In the fuel cell, hydrogen and oxygen are combined to produce electrical power and water. The energy conversion is direct from chemical to electrical. Since the process is isothermal, conversion efficiency is not limited by Carnot efficiency. This is unlike the thermodynamic energy converter using steam or an internal combustion engine as the prime mover. The fuel cell efficiency, therefore, can be about twice that of the thermodynamic converter. It can be as high as 65% in some low-cost designs and 75–80% in solid metal oxide fuel cells developed for large utility-scale power plants. No noise and high reliability with no moving parts are additional benefits over thermodynamic power generators.

In construction, the fuel cell consists of anode and cathode electrodes separated by a liquid or a solid electrolyte and an external load circuit as shown in Figure 10.1. Hydrogen or a hydrogen-rich mixture is fed to the anode. The hydrogen fuel is combined with oxygen of the oxidant entering from the cathode port. The hydrogen does not burn as in the internal combustion engine. It splits into hydrogen ions ($H^+$) and electrons ($e^-$) and produces electrical power by an electrochemical reaction. Water and heat are the by-products of this reaction if the fuel is pure hydrogen. With natural gas, ethanol, or methanol as the source of hydrogen—as in some fuel cells—the by-products include some carbon dioxide, traces of carbon monoxide, hydrocarbons, and nitrogen oxides, all of which are negligible.

Thus, the fuel cell is a static electrochemical device that generates electrical power by chemical reaction without altering the electrodes or the electrolyte material. The

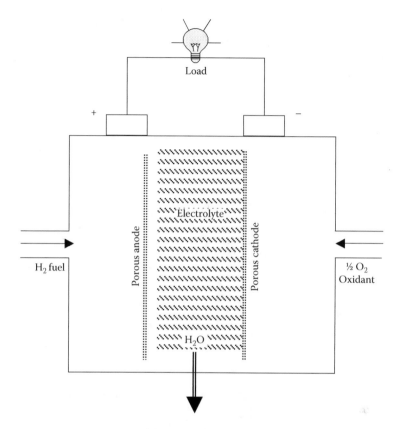

**FIGURE 10.1**    Fuel cell construction features.

chemical reaction is $2H_2 + O_2 = 2H_2O$, and, of course, the energy is released in the process. Unlike the conventional battery, the fuel cell has no electrical energy storage capacity. Hence, it requires a continuous supply of reactants and removal of the reaction by-products during operation.

The fuel cell runs on hydrogen, the simplest element and most plentiful gas in the universe. Hydrogen is colorless, odorless, and tasteless. Each hydrogen molecule has two atoms of hydrogen, which accounts for the $H_2$ we often see in chemical equations. Hydrogen is the lightest element, with a density of 0.09 grams per liter at standard pressure, yet it has the highest energy content per unit weight among all fuels, about 52,000 Btu/lb, which is three times the energy of a pound of gasoline. With molecular weight of 1 for hydrogen and 16 for oxygen, 1 lb of hydrogen needs 8 lb of oxygen to form $H_2O$ and release 52,000 Btu energy in the process.

The functions of the major fuel cell components (Figure 10.1) are as follows:

- The anode (fuel electrode) provides a common interface for the fuel and electrolyte, catalyzes the fuel oxidation reaction, and conducts electrons from the reaction site to the external circuit.

- The cathode (oxygen electrode) provides a common interface for the oxygen and the electrolyte, catalyzes the oxygen reduction reaction, and conducts electrons from the external circuit to the reaction site.
- The electrolyte transports the ions involved in the fuel and oxygen electrode reactions, while preventing the conduction of electrons to avoid a short circuit.

Other components may also be necessary to seal the cell, to provide for gas compartments, and to separate one cell from the next in a fuel cell stack. Figure 10.2 depicts a commercial assembly of a 5-kW, 240 V dc fuel cell from Plug Power

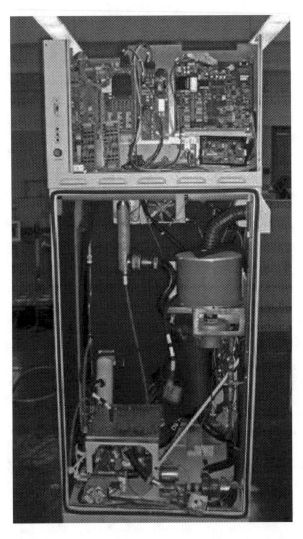

**FIGURE 10.2**    A 5-kW fuel cell assembly with 60-Hz inverter at top. (From U.S. Merchant Marine Academy/Plug Power Corporation.)

Corporation. The bottom compartment is the fuel cell stack, and the top compartment is the power electronics inverter, which converts the dc output of the fuel cell into 60-Hz or 50-Hz power for ac loads.

## 10.2 ELECTRICAL PERFORMANCE

The fuel cell basically works as a voltage source with an internal resistance. The theoretical value of the elementary fuel cell voltage is 1.25 V, about the same as in NiCd and NiMH battery cells. Multiple fuel cells are stacked in series-parallel combinations to obtain the desired stack voltage and current. The fuel cell voltage drops significantly due to various losses—more than in the battery—when the current is drawn. The drop increases with increasing current as follows:

$$V_{drop} = \alpha + \beta \, Log_e J \qquad (10.1)$$

where $J$ = current density at the electrode surface, and $\alpha$ and $\beta$ are constants that depend on the temperature and the electrode surface. The electrical performance of the fuel cell is represented by the electrode voltage versus surface current density, commonly known as the polarization curve or the $v$-$i$ curve, as shown in Figure 10.3. An ideal $H_2$-$O_2$ fuel cell produces about 1.25 V dc at ambient conditions. Undesirable ions and products of the intermediate irreversible reactions

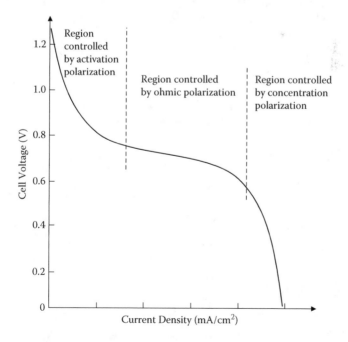

**FIGURE 10.3** Fuel cell polarization curve showing voltage versus current in different regions.

decrease the cell voltage even in an open circuit. As seen in Figure 10.3, further voltage drop under load current results from the following three distinct irreversible polarizations in the cell:

- *Activation polarization,* that is, the reluctance of the fuel and oxidant to undergo reaction at each electrode. It causes energy loss associated with the reaction.
- *Ohmic resistance polarization,* which is due to the energy loss in the electrodes, contacts, and ionic impedance of the electrolyte.
- *Concentration polarization,* that is, the accumulation of ions and reaction products and the depletion of consumed ions and reactants in the electro-lyte near the electrode surfaces. It causes energy loss associated with mass transport.

With the three polarization-related voltage drops, a practical fuel cell produces only 0.5 to 1.0 V dc at current densities of 100 to 400 mA/cm$^2$ of the electrode area. The fuel cell performance improves by increasing the cell temperature and reactant partial pressure. A design trade-off exists between achieving higher performance by operating at higher temperature and pressure and confronting the material and hardware problems faced at higher temperature.

The practical operating range of the fuel cell is influenced by the ohmic resis-tance. The voltage-versus-current (*v-i*) characteristic in this region is similar to that of the battery, except that the average discharge voltage is lower. The voltage also drops with time measured in operating hours. Figure 10.4 shows the terminal

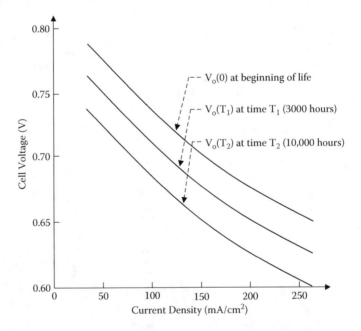

**FIGURE 10.4**   Fuel cell voltage versus current at different operating hours.

voltage-versus-current curves at three different operating times. The voltage is highest at the beginning of life and then progressively decreases as the operating hours are accumulated. At any given time, the terminal $v$-$i$ relationship can be approximately expressed as

$$V = V_o - kI \tag{10.2}$$

where $V$ and $I$ = terminal voltage and current, respectively; $V_o$ = open circuit voltage; and $k$ = constant. The value of $k$ increases and $V_o$ decreases with time. Using Equation (10.2), we can express power $P$ delivered at any operating point as

$$P = V \times I = (V_o - k \times I) \times (V_o - V) \div k \tag{10.3}$$

The maximum power that can be delivered by the fuel cell can be found from equation $dP/dI = 0$, which occurs at $V = \frac{1}{2}V_o$, leading to

$$P_{max} = \frac{V_o^2}{4k} \tag{10.4}$$

Since the open circuit voltage $V_o$ degrades with time, so does the $P_{max}$. The value of $V_o$ at time $t$ can be expressed as the initial voltage less $K_o$ times the hours of operation, where $K_o$ is a constant, that is,

$$V_o(t) = V_o(0) - K_o \times \text{Hours of operation} \tag{10.5}$$

Operating the fuel cell to deliver $P_{max}$ all the times is not of practical interest as it would also consume maximum fuel. For this reason, it is rather operated at the maximum fuel efficiency until the end of operating life. The fuel cell is operated at $P_{max}$ level only at the end of life.

To supply a constant-rated voltage to the load, a voltage-regulating converter is necessary between the fuel cell and the load to boost the diminishing output voltage of the fuel cell. However, the voltage regulator would not be able to boost it up to the rated load voltage when the fuel cell voltage has dropped below a certain level. The fuel cell is then defined to have come to the end of operating life. At this point, the $P_{max}$ of the fuel cell, which decays with the voltage, is defined to be the rated power of the fuel cell. Thus, the $P_{max}/P_{rated}$ ratio of the fuel cell can be as high as two in the beginning of life but would reach one (by definition) at the end of life. The fuel cell life can be estimated from the relation of $V_o$ versus time, from which the $P_{max}$-versus-time line is established. Then, the expected life of the fuel cell when the $P_{max}/P_{rated}$ ratio drops to 1.0 is determined as shown in Figure 10.5.

One performance parameter of interest to the electrical power engineer is the transient electrical response of the fuel cell to a step load change. Such response is complex and is still under study. It includes at least four influence factors that affect the electrical output: (a) thermal, (b) electrochemical, (c) mass flow transient, and (d) the electrical circuit response with the internal inductance and capacitance. None of these parameters is well understood at this time, and they are the subject of advanced research studies.

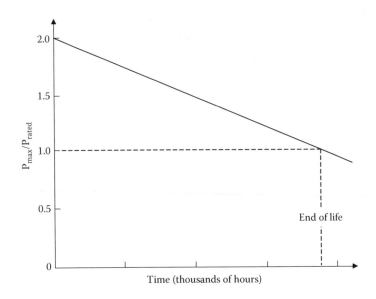

**FIGURE 10.5** Fuel cell $P_{max}/P_{rated}$ ratio degradation versus hours accumulated in operation.

## 10.3 COMMON FUEL CELL TYPES

### 10.3.1 ALKALINE FUEL CELL

Large alkaline fuel cells have been built with 20- to 35-kW output ratings for NASA's space shuttles and deep submergence rescue vehicles (DSRVs). The alkaline fuel cell of NASA combines hydrogen and oxygen to generate electrical power with water as a by-product. Oxygen enters the cell through the sintered-nickel cathode, where a catalyst produces OH ions and delivers them to the potassium-hydroxide electrolyte of the cell. The ions drift through the alkaline electrolyte to the anode, where they combine with hydrogen atoms to form water molecules and release electrons, and the electrical power is delivered to the external load. The primary energy released by 1 kg of hydrogen combined with 8 kg of oxygen is about 34,000 Wh. That is 3780 Wh/kg of total fuel ($H_2 + O_2$). At 50% energy conversion efficiency of these fuel cells, this amounts to 17,000 Wh electrical output per kilogram of hydrogen or 1890 Wh per kilogram of total fuel ($H_2 + O_2$). In NASA's space shuttles, the by-product water was used for the space crew.

### 10.3.2 PROTON EXCHANGE MEMBRANE

The proton exchange membrane (PEM) is another fuel cell technology that is widely used in commercial, industrial, and marine applications. The PEM fuel electrical assemblies come in many voltage and power ratings. One such 50-kW PEM fuel cell made by Proton Motor GmbH for a 100-passneger public ferry for ATG Alster-Touristik GmbH of Germany is shown in Figure 10.6. It has about 55% efficiency and an operating life of about 6000 hours.

**FIGURE 10.6**   A 50-kW proton exchange membrane-type fuel cell. (Photo courtesy of ATG Alster-Touristik GmbH/Proton Motor GmbH.)

The PEM fuel cell uses a thin, permeable polymeric membrane as the electrolyte; hence, it is also called a polymer electrolyte fuel cell. It was first developed by General Electric in the early 1960s for the U.S. Navy's Bureau of Ships and the U.S. Army Signal Corps. The membrane is very light. Platinum electrodes used on either side of the membrane catalyze the reaction, Oxygen in the form of air is supplied to the cathode and combines with the hydrogen ions to produce energy and water. The PEM fuel cell operates at low temperatures of about 80°C, allowing rapid startup. At present, the electrical power output ratings are up to 250 kW, and efficiency is in the 40–60% range. PEM fuel cells are often compact and lightweight compared to other fuel cell types. Because the electrolyte is a solid rather than a liquid, there is easier sealing of the anode and cathode gases, making these cells less costly to manufacture than some other types of fuel cell. Furthermore, the solid electrolyte can lead to a longer cell and stack life as it is less prone to corrosion than some other electrolyte materials. For these reasons, the PEM fuel cell makes a best candidate for use in cars, ships, and buildings.

NASA is currently developing advance PEM technology for a variety of space applications, which would offer enhanced safety, longer life, lower weight, higher reliability, and higher peak power capability. The goal of NASA's current PEM programs is to achieve three times the power and four times the life compared with the alkaline fuel cell of the same mass and volume.

### 10.3.3   HIGH-TEMPERATURE FUEL CELLS

The high-temperature class of fuel cells has higher energy conversion capabilities per kilogram of weight, but costs relatively more. Solid oxide, solid polymer, and molten carbonate fuel cells fall in this category. The Fuel Cell Commercialization Group in the United States has field tested a molten carbonate direct-fuel cell for a 2-MW utility-scale power plant.

The German car manufacture Daimler-Benz and Ballard Power Systems of Canada have developed a solid polymer fuel cell for automobiles as an alternative to battery-powered vehicles. Other automakers around the world have also accelerated

**TABLE 10.2**
**Performance of Various Fuel Cells Compared**

| Fuel Cell Technology | Specific Power (watts/kg) | Life (hours) |
|---|---|---|
| Alkaline | 100–150 | About 50,000 |
| Solid polymer | 100–150 | About 50,000 |
| Proton exchange membrane | 150–350 | About 6000 |
| Alkaline (used in space shuttles) | 300–400 | 3000–5000 |
| Lightweight cell for space applications | 600–700 | Under development |

research funding for fuel cell cars. For the German Navy, Howaldtswerke-Deutsche Werft AG of Germany has built four 212A class submarines that are powered by fuel cells. The design is air independent and produces no noise or exhaust heat. It carries a crew of 27 with displacement of about 1450 tons.

Solid oxide fuel cells of several different designs are being investigated worldwide. Most success to date has been achieved by Westinghouse-Siemens Corporation in the United States and Mitsubishi Heavy Industries in Japan. The cell element in this geometry consists of two porous electrodes separated by a dense oxygen ion-conducting electrolyte. It uses a ceramic tube operating at 1000°C. The fuel cell is an assembly of numerous such tubes. It is developed for large combined cycle gas turbine and fuel cell plants up to 60-MW capacity. The cell voltage degrades less then 0.1% per 1000 hours of operation and has an operational life in tens of thousands of hours. The prototype has been tested for over 1000 thermal cycles with zero performance degradation and 12,000 hours of operation with less than 1% performance degradation.

For ground use, other low-temperature (250°C) fuel cells are now commercially available from several sources. They use phosphoric acid as the electrolytic solution between the electrode plates. The performance parameters of various fuel cells available today are compared in Table 10.2.

## 10.4    PRESENT AND FUTURE USE

Power solutions based on fuel cell technology are expected to offer significant benefits in power generation applications as well as in the shipping industry, for which international emission regulations are becoming increasingly stringent. For these reasons, economic and political leaders worldwide have shown increasing interest in adopting hydrogen-based energy technologies, especially in the transportation sector, which has benefits that are more transparent. The investments being made in this direction will also contribute to solving clean energy challenges faced by many nations. For example, Iceland has set a goal to become the world's first hydrogen economy supporting all its energy needs, including transport vehicles and shipping, by 2050. This means a total elimination of fossil fuels that should reduce its greenhouse emissions by 50%. The hydrogen will be derived using the electrical energy from its geothermal power plant. Considerable progress has been made by Iceland toward this goal. Many of the country's public buses are already converted to use hydrogen; the buses are refueled by filling stations on the outskirts of towns. In addition, there are plans to convert the country's entire fishing fleet for hydrogen use. The natural geothermal energy resources and waterfalls of Iceland certainly give it an advantage over countries less endowed with such natural energy sources. In other countries, some current developments in fuel cell technology are discussed next.

### 10.4.1  Hydrogen-Powered Cars

The hydrogen-powered car can be about twice as efficient as the current petroleum-based cars. For this reason, making fuel cell cars has become a truly international effort. For example, in the United States, engineers at General Motors have developed the vehicle chassis as well as the engineering and electrical system integration for a hydrogen-electric vehicle. German engineers worked on the integrated fuel cell propulsion system, which was first shown at a Frankfurt motor show. The Italian and American engineers worked together closely to build the car body, and the SKF Group in Sweden developed the electric motor and wire technology.

Today, some fuel stations in Germany offer hydrogen as part of the Clean Energy Partnership. Bus refueling depots in a number of small European cities are providing hydrogen for public service vehicles as part of the Clear Urban Transport program. In British Columbia, Canada, a seven-node hydrogen refueling station network from Victoria to Whistler was developed to coincide with the 2010 Winter Olympic Games, and fuel stations in California have been opened by the California Fuel Cell Partnership.

### 10.4.2  Navy and Military Use

Type 212 submarines of the German and Italian navies use fuel cells to remain submerged for weeks without the need to surface. Unmanned aerial vehicles (UAVs), undetectable from the ground, are widely used by the military to scan terrain for possible threats and intelligence. New fuel-cell-powered UAVs are developed by the U.S. Office of Naval Research (ONR) and Naval Research Laboratory (NRL) for gathering critical information more efficiently and more quietly. A hydrogen-powered fuel cell can travel farther and carry heavier payloads than earlier battery-powered designs. A fuel cell propulsion system can also deliver potentially twice the efficiency of an internal combustion engine, while running more quietly and with greater endurance, exceeding the duration of previous flights by severalfold. The UAV with fuel cell technology can be used for similar marine expeditionary missions for meeting power needs afloat. For fighting wars, future fuel-cell-powered UAVs like Ion Tiger offer stealthy characteristics due to small size, reduced noise, low heat signature, and zero emissions.

The power for some small UAV fuel cells comes from hydrogen stored in the form of sodium borohydride, a white-grayish powder used in manufacturing pharmaceuticals and other things. Hydrogen is released when this powder is mixed with water and exposed to a catalyst and heat. Essentially, adding water and shaking the mixture makes the fuel ready for the fuel cell. As a built-in control, the fuel cartridge controls the rate of hydrogen release as needed by the fuel cell to produce electrical energy. The by-product water can be used in many ways or recycled back to the fuel cartridge. What is left in the used cartridge after extracting all hydrogen is sodium borate solution (essentially borax), which is a relatively benign chemical used in making detergents and fire retardants. The cartridge can be recycled.

### 10.4.3 Fuel Cell in Merchant Ships

Marine transportation is incorrectly perceived to have lower emissions than most other forms of transportation. However, it is estimated that 17% of domestic $NO_x$ in Japan is due to coastal shipping lines using diesel engines. With shipping emissions, the issue intensified after the 2009 Copenhagen Climate Change conference, and the world's shipping lines are under increasing pressure to introduce *greener ships*. MARAD, in partnership with the industry, is looking into the ways to reduce emission from the ship power generation, especially when in ports, using fuel cells for base load and batteries for peak maneuvering power.

Japanese carrier NYK Lines and other major shippers, such as Toyota and Mitsui OSK Lines, are working to develop a low-emission SuperEcoShip™ by 2030. NYK has teamed with design partners Monohakobi Technology Institute, marine consultants Elomatic of Finland, and ship designers Garroni Progetti of Italy. The ship will produce 70% less $CO_2$ emissions than current container ships by powering the ship by LNG fuel cells that produce 30% less $CO_2$ than comparable marine diesels; plus, it will also have 31,000-m$^2$ solar panels in addition to retractable sails. The design also features a bow-mounted bubble projector to reduce friction as the ship moves through the water. It is proposed that the vessel will actually break into segments in port to facilitate freight handling and thus aid efficiency. As a further emission reduction goal, NYK plans to have a zero-emissions fleet by 2050.

Wärtsilä, a provider of marine power products, and Versa Power Systems (VPS), a developer of high-power solid oxide fuel cells (SOFCs), agreed in 2010 to integrate Versa Power's SOFC technology into Wärtsilä products, especially for products with a larger power range. A key goal is to develop fuel cell products that generate power and heat for various commercial and marine applications for clean and efficient electrical energy. The solid oxide fuel cell has low emissions and high power density, both in kilowatts per kilogram and kilowatts per liter. Some of Wärtsilä's pilot projects include a fuel cell unit that operates on landfill gas and produces electricity and heat for the city of Vaasa in Finland. In 2010, the company installed a 20-kW fuel cell unit onboard a car carrier owned by Sweden's Wallenius Lines, and a 50-kW unit has undergone internal validation tests.

Among other companies addressing the emission issue, Rolls-Royce is developing fuel cell systems, such as the solid oxide system for land-based applications. Such systems could ultimately be applied in electric ship propulsion, subject to the enabling reforming technology using typical marine fuels. Fuel cells with virtually constant efficiency up to 60–70% at all loads—full or part loads—could result in 40–50% reduction in fuel consumption and $CO_2$ emissions compared to diesel engines of comparable rating.

### 10.4.4 REGENERATIVE FUEL CELL

The types of fuel cells presented above convert energy one way, from fuel to electricity. They are not designed for the reverse recharging operation. Recharging the fuel cell requires electrolyzer to decompose the water back into hydrogen and oxygen. The electrolyzer is generally a separate unit from the fuel cell stack and the two cannot operate simultaneously. The assembly that has the fuel cell to convert fuel energy into electricity and the electrolyzer to convert electricity into fuel is known as the *regenerative fuel cell* (RFC), which works like a rechargeable battery. It stores energy in the form of hydrogen when excess electricity is available. Compared to the rechargeable battery, the RFC is less efficient (Table 10.3), but it has a long life. The battery needs total replacement after several years, whereas the RFC can last 20–25 years with only a few components requiring maintenance.

### 10.4.5 RFC FOR RENEWABLE ENERGY FARMS

Both wind and solar energies, although renewable, are intermittent and nondispatchable sources of energy. On some days, the wind may not blow and the sun may not show. For this reason, the stand-alone large wind and solar energy farms not connected with the power grid need some form of energy storage to maintain power availability. Among many competing alternatives, the RFC is certainly one on the horizon. The energy farm can charge the RFC with hydrogen when the energy is in excess and discharge it when short of electrical power. With poor round-trip energy efficiency of less than 50%, the RFC would perhaps need some initial tax incentives. For example, in Germany, the hydrogen produced using wind or solar energy is tax free, but that produced by conventional energy sources is subject to the same high energy tax as other fuels.

### 10.4.6 RFC FOR SPACE COLONIES

Long-term missions to establish a human colony on the moon or Mars will need high power. The colony based in space would use large solar-photovoltaic (PV) arrays to generate electricity, and the power system will use the RFC that performs like the electrochemical battery. A part of the solar energy will be used for the electrolysis

### TABLE 10.3
### Round-Trip Energy Efficiency of RFC and Battery

| Energy Technology | Round-Trip Energy Efficiency |
|---|---|
| Nickel-cadmium battery | 70–80 |
| Lithium ion battery | 75–85 |
| Mass-optimized hydrogen–oxygen RFC design | 50–60 |
| Efficiency-optimized hydrogen–oxygen RFC design | 60–70 |

of water to produce hydrogen and oxygen as energy storage. Proton Energy Systems has been developing a 1- to 100-kW RFC for space applications. The company has tested 25–35% dc-dc round-trip energy efficiency, which is very low, but the specific energy and specific power are high compared to the electrochemical battery. At stack level, an energy density of 300–400 Wh/kg and a power density of 70–100 W/kg are possible, including 10% for hydrogen storage. This compares with about 200-Wh/kg energy density with the lithium-ion battery.

When the human exploration of space grows into colonies on the Moon and Mars, high-efficiency RFCs and other power system components with life in decades would be required. The colonies may have other alternatives for energy storage, but the RFC provides a significant advantage. It coincidentally furnishes a backup supply of oxygen and water, which are valuable to the inhabitants in the event of emergencies and unforeseen circumstances. A simple architecture of such a power system for human colonies in space is shown in Figure 10.7. On the moon, it would include the following:

- PV solar array for power generation during the cyclic daytime periods of approximately 14 Earth days
- Electrolyzer that uses dc electricity from the array and electrochemically converts water into hydrogen, oxygen, and low-grade heat
- Fuel cell stack that electrochemically converts hydrogen and oxygen into dc electricity, water, and low-grade heat during the cyclic night period of approximately 14 Earth days
- Storage tank for pure water
- High-pressure hydrogen and oxygen tanks
- Automated controls

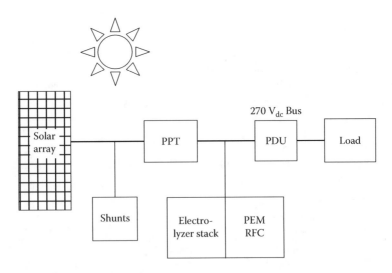

**FIGURE 10.7**  Power system architecture for a human colony on Moon or Mars. PPT= Peak power tracker PDU= power distribution unit.

On the moon, compared to that on Earth, the environment is very demanding, while some aspects are simpler for operating an RFC. The moon operation requires storage of hydrogen, oxygen, and water. The terrestrial operation does not need to store oxygen since the natural air can be fed through the cathode on a one-pass basis. It is easy to maintain the PV power generation schedule on the moon as the sunlight and dark times are steady and nearly constant in the lunar month. On Earth, the seasonal variations and the clouds can make the sunlight vary over a wide range.

## 10.5   PRESENT DEVELOPMENT ISSUES

Hydrogen, abundant on Earth, combines readily with other elements and is usually found as part of some other substance, such as water, or in hydrocarbons, including those that make up all plants and animals. Any hydrogen-rich material can serve as a possible fuel source for the fuel cell. Many hydrocarbon fuels—methanol, ethanol, natural gas, petroleum distillates, liquid propane, and gasified coal—can yield hydrogen in a process called reforming. Hydrogen can also be extracted from landfill gas, anaerobic digester gas from wastewater treatment plants, biomass technologies, or hydrogen compounds containing no carbon, such as ammonia or borohydride.

Hydrogen is made, shipped, and used safely today in many industries worldwide. Hydrogen producers and users have had a good safety record over the last half-century. Liquid hydrogen trucks on the nation's highways have carried an average 70 million gallons of liquid hydrogen per year without major incidents. Hydrogen has been handled and sent through hundreds of miles of pipelines with relative safety for the oil, chemical, and iron industries.

Yet, safety concerns are still widespread, some actual and some perceived. Compressed hydrogen gas must be stored in a pressure vessel that meets the American Society of Mechanical Engineers (ASME) pressure vessel code. People may be concerned to carry such a pressurized cylinder in the trunk or under the hood of their car. At a practical level, there are also issues in terms of how we store and transport hydrogen at present. Hydrogen is a very light gas, making it far more difficult to work with than the gasoline we are accustomed to using. For example, compressing the gas for easy transportation and storage requires significant amounts of energy, and the resulting compressed hydrogen contains far less energy than the same volume of gasoline. However, many companies have started looking at how to store hydrogen in a solid form. This involves storing the hydrogen in a chemical called sodium borohydride. This chemical is created from borax, and with the simple addition of water, it releases hydrogen and turns back into recyclable borax.

Because hydrogen is such a very light gas, it is difficult to store a large amount in a small space. That is a challenge for auto engineers who want to use it in today's 300- to 400-mile range vehicle. With the support of the U.S. Department of Energy, automobile engineers are examining several hydrogen storage options. Today's prototype fuel cell vehicles use compressed gaseous hydrogen tanks or liquid hydrogen tanks. New technologies, such as metal hydrides and chemical hydrides, may become viable in the future. Another option would be to store hydrogen compounds—methanol, gasoline, or other compounds—onboard and extract the hydrogen when the vehicle is operating.

The cost of producing hydrogen is another issue; it is still relatively high. Small systems currently cost around $0.08 per kilowatt-hour of hydrogen, while larger systems, which favor liquid hydrogen, cost $0.04 per kilowatt-hour versus $0.06 per kilowatt-hour for compressed gaseous hydrogen. This is a lower running cost compared to traditional gasoline or diesel but is offset by the currently higher initial capital cost of the fuel cell. The industry will eventually need to move from the natural gas-based hydrogen, which is being used during the market development phase, to industrial-level hydrogen production using renewable resources on a large economical scale. Hydrogen is not a *green fuel* in itself; it merely works as an energy storage medium. It still has to be made with either fossil or renewable fuels. Iceland is a special case as it has an enormous amount of geothermal energy to tap for producing hydrogen.

## QUESTIONS

*Question 10.1*: Explain the difference between the working of the fuel cell and the electrolyzer.

*Question 10.2*: If you have seen any fuel cell being used in the industry or anywhere, describe why in particular the fuel cell was adopted for that specific application.

*Question 10.3*: In your judgment, what is the future of the fuel cell in the shipping industry, and how would it be used?

*Question 10.4*: Use of fuel cell power, although fully developed, is a negligible percentage at present. With the emphasis of the U.S. government on hydrogen economy, how far in the future do you think we may see at least 10% of the U.S. energy demand met with the fuel cell? Share your judgment with reasons.

*Question 10.5*: List and discuss three roadblocks for fuel cell applications at present and the expected solution that may come in the future.

## FURTHER READING

Allen, D.M. 2001. Multi-megawatts specifications power technology comparison. *ASME Proceedings of the 36th Intersociety Energy Conversion Engineering Conference*, 1, 243–249.

Babasaki, T., T. Take, and T. Yamashita. 1999. Diagnosis of fuel cell deterioration using fuel cell current-voltage characteristics. *SAE Proceedings of the 34th Intersociety Energy Conversion Engineering Conference*. Paper No. 01-2575.

Barbir, F., L. Dalton, and T. Molter. 2003. Regenerative fuel cells for energy storage— Efficiency and weight trade off. *AIAA Proceedings of the 1st International Energy Conversion Engineering Conference*. Paper No. 5937. August 2003. Portsmouth, VA.

Burke, K. 2003. Fuel cells for space science applications. *AIAA Proceedings of the 1st International Energy Conversion Engineering Conference*. Paper No. 5938. August 2003. Portsmouth, VA.

Gou, B.K., A. Woon, and B. Diong. 2009. *Fuel Cells Modeling, Control, and Applications*. Boca Raton, FL: CRC Press/Taylor & Francis. August 2009. Denver, CO.

Hall, D.J., and R.G. Colclaser. 1998. Transient modeling and simulation of tubular solid oxide fuel cell. *IEEE Power Engineering Review*, July, Paper No. PE-100-EC-004. IEEE, Piscataway, NJ.

Hoberecht, M., and W. Reaves. 2003. PEM fuel cell status and remaining challenges for manned space flight applications. *AIAA Proceedings of the 1st International Energy Conversion Engineering Conference.* Paper No. 5963.

Oman, H. 2002. Fuel cells power for aerospace vehicles. *IEEE Aerospace and Electronics System Magazine,* 17(2), 35–41. IEEE, Piscataway, NJ.

Palmer, D., and W. Sembler. 2009. Environmentally friendly very large crude carriers. *Journal of Ocean Technology, Green Ships*, 4(3), 73–90.

Palmer D., W. Sembler, and S. Kumar. 2009. Fuel cells as an alternative to cold ironing. *ASME Journal of Fuel Cell Science and Technology*, August, Vol. 6, Paper No. 031009-1.

Perez-Davis, M.E., et al. 2001. Energy storage for aerospace applications, *ASME Proceedings of the 36th Intersociety Energy Conversion Engineering Conference, 2,* 85–89.

Voecks, G.E., et al. 1997. *Operation of the 25-kW NASA Lewis Research Center Solar Regenerative Fuel Cell Testbed Facility.* NASA Technical Report No. 97295. NASA Lewis Research Center. Cleveland, OH.

# 11 Hybrid Propulsion

*Hybrid propulsion* in this chapter refers to a combination of conventional propulsion and electric propulsion, with usually one or the other working at a time, but not simultaneously. It is used in a manner that optimizes the overall size, energy efficiency, and emission reduction of the power plant during one or more segments of ship operation. Hybrid propulsion is becoming more important in view of the environmental regulations being imposed on ships and ferries in busy ports and sea routes around the world in response to the rising public concern about emissions. Before addressing the hybrid power system architecture, we first review the gravity of the environmental issue faced by the shipping industry today.

## 11.1 ENVIRONMENTAL REGULATIONS

The shipping industry has been reducing its fuel consumption in response to rising energy prices, even more so now to cut emissions as well. Large cruise ships with 2000-passenger capacity move about 40 feet per U.S. gallon of diesel oil (3.3 m per liter), compared to 20 feet per gallon (1.65 m per liter) some 20 years ago. This fuel efficiency is partly achieved by recycling waste heat. Even with the most energy-efficient diesel engine, a large oil tanker emits more than 300,000 tons of $CO_2$ per year, equivalent to a medium-size coal power plant on land. And, there are more than 90,000 large ships in the world today. The industry has grown rapidly since industrial production has shifted away from the United States and Europe to China and south Asia, which means larger cargo traveling longer distances. Ships now carry more than 90% of the world's trade by volume, and have tripled the total tonnage capacity since the 1980s.

The International Council on Clean Transportation has estimated that ships produced more $SO_2$ soot that is associated with acid rain than all of the world's cars, trucks, and buses combined. The International Maritime Organization (IMO) and the U.N. Intergovernmental Panel on Climate Change estimated that the annual emissions from the world's merchant fleet have reached 1.2 billion tons of $CO_2$, or nearly 5% of all global emissions of the main greenhouse gas, about twice as much as all airplanes combined. The U.N. agency also estimated that $CO_2$ emissions from ships will rise an additional 30% by 2020, with other pollutants from shipping rising even faster than $CO_2$ emissions. Sulfur and soot emissions, which give rise to lung cancers, acid rain, and respiratory problems, are expected to increase more than 30% by 2020. The health implications of shipping emissions are most acute for Britain and other countries bordering the English Channel, one of the world's busiest shipping lanes.

The shipping emissions at present are not taken into account by the European targets for cutting global warming. The aviation industry is now provisionally included following public pressure, but shipping has so far escaped the publicity. However, the pressure is now increasing on the European Union to include the shipping industry in the total emission accounting that will require ship owners to switch to cleaner fuels.

The IMO in 2010 agreed to designate waters off the North American coasts as an Emission Control Area (ECA). Canada and France joined the United States in this North American ECA, implementing a coordinated geographic emissions control program. This will force large ships operating in North America to use much cleaner fuel and technology. The large commercial ships that visit the nation's ports—oil tankers, cruise ships, and container ships—currently use bunker fuel, which has a high sulfur content. These ships, mostly under foreign flags, make more than 57,000 calls at more than 100 U.S. ports annually. More than 30 of these ports are in metropolitan areas that fail to meet federal air quality standards. In total, nearly 127 million people currently live in such areas. Ships burning dirty fuels add harmful levels of particulate matter and nitrogen oxide that can travel hundreds of miles inland, causing severe respiratory symptoms in children and adults.

Enforcing the stringent ECA standards will reduce sulfur content in fuel by 98%, slashing particulate matter emissions by 85% and nitrogen oxide ($NO_x$) emissions by 80%. To achieve these reductions, tougher sulfur standards will phase in starting in 2012, ultimately reaching no more than 1,000 parts per million by 2015. Also, new ships must use advance emission control technologies beginning in 2016 that will help reduce $NO_x$ emissions.

The Marine Environment Protection Committee in 2010 addressed greenhouse gas reductions for the maritime transport sector. It agreed to a preliminary draft that mandates requirements for an Energy Efficiency Design Index (EEDI) and the Ship Energy Efficiency Management Plan (SEEMP). The EEDI is derived from emission factors associated with the fuel consumed by the main engine emissions, nominal auxiliary engine power, and auxiliary generator power, all of which are adjusted by a factor for any innovative energy-efficient technologies used onboard. On the other hand, the SEEMP establishes a mechanism to implement environmental management, energy monitoring, and efficiency improvement systems of the ship's operation. The ship emission reduction efforts will require the following:

- Large ships switch to cleaner fuel before entering a port
- Ships to shut off the onboard power plant and use land-based power on the port (cold ironing)
- Tugboats to use multiple smaller engines and a hybrid power system

## 11.2   HYBRID TUGBOAT

Hybrid propulsion in tugboats with both mechanical and electric power sources can be effective in meeting environmental regulations. The hybrid tugboat can use the electrochemical battery, hydrogen fuel cell, or other sources of clean power, such as wind and solar photovoltaic (PV).

Figure 11.1 depicts the data collected by the Foss Company for Dolphin-class tugs. They show that these boats spend up to 60% of their time at less than 20% power, and fully 95% of their time is spent at less than 67% power. The average load on the main engines during the 22-day monitoring period was a remarkably low 16%

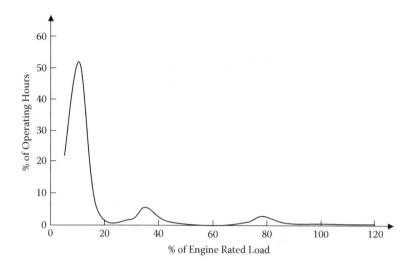

**FIGURE 11.1** A 5080-hp tugboat load profile in Los Angeles and Long Beach harbors. (Adapted from Faber, G., and J. Aspin, *The Foss Hybrid Tug from Innovation to Implementation*. Foss Maritime Company, Aspin Kemp & Associates, and ABR Company Ltd., Seattle, WA.)

overall. Although the tugboats operate at substantially lower loads most of the time, the main diesel engines must be sized to satisfy the maximum power of the tug for the peak thrust requirement and emergency situations.

Since the main engines are typically designed to have the highest fuel efficiency and lowest emissions at full-rated output, the engines at partial loads run way below their peak performance. The fuel efficiency typically starts to falls off around 60% load, but the harbor tugs generally operate at much below 30% load most of the time. A hybrid tug design can use one small engine to meet the average load and a large battery bank to meet the peak load demand. However, the drain on the batteries for several minutes of full bollard may make the battery size unmanageable in most tugs. A hybrid power plant with two diesel generators—one or both can be used as needed—can make the design more manageable to meet the power requirement of the tugboat within the space and weight constraints.

The hybrid tugboat designed by Foss has two diesel generators to supply electrical power for propulsion and boat services and to charge the batteries as needed. The electrical configuration of the Foss hybrid design is shown in Figure 11.2. One or both diesel generators can provide propulsive power. The number of generators in operation is determined by the demand of the propulsion system and the service loads. If the loads exceed the capacity of the generators for a period longer than a predetermined limit, the system would automatically change the configuration and start the additional generator to deliver the required power. Energy for a peak demand is provided by the battery.

Such a hybrid system has the dual benefit of engines running at peak efficiency and at reduced emission. It offers fuel saving of about 25% and reduces the life-cycle costs of major components. For example, because the engines are

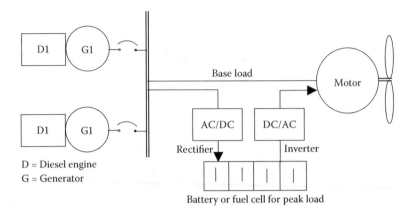

**FIGURE 11.2**   Hybrid tugboat propulsion power system with battery or fuel cell.

off-line most of the time, the time span between engine overhaul is increased. However, the added hybrid machinery and the development cost may partially or fully offset the benefit of the hybrid technology at present. The learning curve may bring the future costs down after a few such tugs are made and operated. At ports where emission reduction is mandated, hybrid tugs may be required even at a higher cost. The emission reduction in a hybrid tug compared to the conventional tug is estimated to be 44% $NO_x$ and 44% particulate matters. The other pollutants, such as $CO_2$ and $SO_x$, should also be reduced in proportion to the fuel consumption.

Tugboats with short transit runs in harbors like San Pedro, California, are prime candidates for hybrid tugs. While performing tug-assist jobs, these tugs spend a considerable amount of idle time at low power and little time at full power. For this reason, the ports of Long Beach and Los Angeles in 2008 gave financial support to Foss Company to build and operate a hybrid tug in San Pedro Harbor for a period of 5 years. The aim is to build the necessary technology and experience base in hybrid propulsion power.

## 11.3   HYBRID FERRY

A German tourist ferry, *Alsterwasser* (Figure 11.3), is an example of a fuel cell and battery hybrid vessel that uses two 50-kW fuel cells powering a 100-kW (134-hp) hybrid electric propulsion system with lead acid batteries. A consortium of seven German organizations and a Czech research institute together built the vessel, which entered service in 2008. The hull has an overall length of 25.5 m (84 ft), a 5.2-m (17-ft) beam, a draft (laden) of 1.3 m (4.2 ft), and displacement of 72 metric tons (79 short tons). It is a 100-passenger tourist vessel that cruises the canals. The hydrogen is stored at a pressure of 350 bar (5075 psi) in cylinders, and the 50-kg (110-lb) capacity is sufficient for 3 days of normal operation. Refueling with compressed hydrogen is carried out at a station in Hamburg and takes approximately 12 minutes. Its liquid-cooled proton exchange membrane (PEM)-type fuel cell system is made

**FIGURE 11.3** Fuel cell-battery-powered 100-passenger hybrid ferry operated by ATG Alster-Touristik, Germany.

by Proton Motor GmbH, modified from cells used in commercial vehicles. When the propulsion motor is not running at full power, excess electrical energy is stored in the batteries. Compared to a comparable diesel-powered ship, it is estimated that the annual net reductions in environmental emissions are 1,000 kg (2200 lb) of $NO_x$, 220 kg (485 lbs) $SO_x$, 40 g (1.4 oz) of particulate, and 70 metric tons (77 short tons) of $CO_2$. The project cost of €5.2 million ($7.4 million) is shared by a European Union grant and investment from the consortium members.

Two hybrid ferries went into operation in 2006 to carry passengers for Alcatraz Tours under a 10-year contract. The electrical power system of the ferries uses solar and wind energy (Figure 11.4) with zero emission at the wharf. The power system is designed in a manner similar to the *Solar Sailor* ferry in Sydney, Australia, to accommodate 600 passengers and operate at 12–15 knots speed. The ferries are fitted with a rigid solar panel covered with PV cells. The same panels also capture the wind. Large batteries onboard store the electrical energy that powers electric motors for propulsion. The batteries are charged using the excess power collected from the sun and the wind by the sail panels or the diesel generators burning low-sulfur diesel fuels. Under favorable conditions, the ferries can sail back and forth using mostly the panel power (solar or wind). In extreme weather, the power panels automatically fold down flat like a roof above the deck, and the battery power is used before turning on the diesel engines. While loading and unloading passengers, the diesel engines are turned off, and the serviced load is met solely by the battery energy. The ferries can also be plugged into shore power to charge the batteries. It is estimated that the Alcatraz hybrid ferries cut total emission by 70–80% compared to conventional ferries.

The solar irradiation on a PV panel normal to the sun in the earth's atmosphere is about 1000 $W/m^2$, or 1000 cos $\delta$ $W/m^2$ if the panel is off-normal to the sun by angle $\delta$. The electrical power output of the panel is then

$$P_o = 1000 \cos \delta \times \text{Panel area} \times \text{PV conversion efficiency} \qquad (11.1)$$

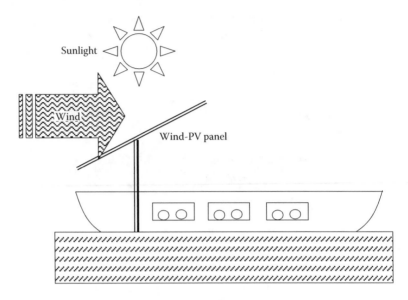

**FIGURE 11.4**   Hybrid ferry with wind-PV-battery power sources.

On a cloudy or overcast day, the irradiation changes from 1000 to about 500 W/m². The sun-to-electrical power conversion efficiency varies from 10% to 20% depending on the type of PV cell, with 10%-efficient cells costing about half of the 20%-efficient cells. The market prices per watt of electrical power generation capacity are approximately constant. Therefore, other factors (e.g., available area, structural requirement, etc.) determine the selection of cell type.

For wind power, air of density $\rho$ moving at speed $V$ m/s has the incoming power given by $\frac{1}{2}\rho V^3$ watts per square meter of the flow area. Accounting for the wind-to-electrical conversion efficiency, the electrical power output of the generator is given by

$$P_o = \frac{1}{2}\rho V^3 \times \text{Swept area of wind turbine rotor blades} \times \text{Rotor efficiency} \quad (11.2)$$

where swept area = $\pi \times$ (Turbine rotor blade radius)², and rotor efficiency varies from 35% to 40% in modern two- and three-blade rotors.

With 7-m/s (15.75-mph) average wind and 1.225-kg/m³ air density, the incoming wind power is $\frac{1}{2} \times 1.225 \times 7^3 = 210$ W/m² of the blade swept area. With 40% conversion efficiency, the generator output is about 84 W/m². However, because of the cubic power relation, the increase in power at high wind is much more than the decrease in power at low wind. This almost doubles the yearly average power output to $2 \times 84 = 168$ W/m². Therefore, 10-m diameter blades will generate $168 \times \pi \times 5^2 = 13,195$ W in 7-m/s average wind.

For a fixed-voltage output, the additional energy coming in when the sun is bright and normal and wind is high must be wasted. Modern renewable power plants, therefore, generate variable voltage and variable frequency to capture the maximum possible energy from the wind or the sun and convert it to a fixed voltage and fixed frequency for users. This is done by using the power electronics converters, which have the same design as the variable-speed drives for the motors, except they work with the reverse power flow.

## 11.4   HYBRID AUTOMOBILE

Because of their dominance in the society, some cars and trains already use gasoline–battery hybrid propulsion if fuel saving and emission reduction are important. The hybrid electric vehicle (HEV) has an internal combustion engine, an electrical machine that works as either the motor or the generator, and an energy storage battery that charges while braking and discharges during acceleration. During braking, the kinetic energy of the moving vehicle is recovered by regenerative braking and converted into electrical energy by the generator, which is rectified into dc by the power electronics to charge the battery. During acceleration, the battery energy is converted into ac by the power electronics inverter, which then drives the motor. Thus, the power electronics has a large contribution in today's hybrid vehicle in recovering the braking energy, which would have otherwise gone into friction at the brake pads, wasting gasoline and also wearing the brake pads and dumping the dust and heat in the environment. At low speed during starting, the HEV runs entirely from the battery. It starts drawing gasoline only after depleting the battery energy. The battery is typically nickel-metal hydride (NiMH) but can be any electrochemistry. Supercapacitors can also serve this purpose of storing the energy. Toyota Prius is one such top-selling hybrid car today.

For the HEV, the best option overall for the electrical machine is the internal permanent magnet synchronous motor. The motor is designed such that the reluctance torque is higher than the magnet torque (about 60/40% split) to have reduced flux. Such design has the following benefits: (a) lower back voltage at high speed, so that the fault current and pulsating torque are reduced; (b) lower iron loss, particularly at high speed; (c) higher efficiency than the induction or reluctance motor, particularly at low-speed, low-torque operation; and (d) with high energy density NdFeB permanent magnet, the power density of the motor is high with lower weight and inertia. On the negative side, the NdFeB magnets are expensive at present, but the cost is falling.

The hybrid train—similar to the hybrid car in principle—has been developed in Japan for routes with frequent starts and stops. It has no electrical supply and runs on diesel engine power augmented by battery power when needed. The engine runs at a constant speed for maximum efficiency. The battery boosts the engine power while accelerating and absorbs the regenerated energy during braking. The depleted battery is charged by the diesel engine en route or by the grid power in the yard.

There is also the pure electric vehicle (EV), which is not to be confused with the HEV. The EV runs only on a large battery, while the HEV runs primarily on gasoline and uses a small battery only when needed. For reference, the EV, like GM Volt, running only on the electrical energy uses a 16-kWh lithium-ion battery and a 120-V electric motor that delivers 320 n-m peak torque and gives an all-electric range of about 65 km (40 miles), which is adequate for about 75% of users to meet their daily commute. Longer-range EVs are being developed at present. For example, the 2011 Chevy Volt uses a small gas engine that extends the range to 350 miles. When the battery runs out of the initial charge, the gas engine drives the generator to charge the battery. The car has a 110-kW electric motor and 84-hp gas engine. The Volt can be fully charged in 10 hours from a 120-V plug or 5 hours from a 240-V plug. In both EVs and HEVs, the kinetic energy during braking is converted into electrical energy using regenerative braking and is reused during acceleration. This results in further fuel savings.

## 11.5  MAGNETOHYDRODYNAMICS PROPULSION IN SHIPS

Although not a hybrid of gasoline and electrical power for propulsion, the propulsion alternative presented here blends the electromagnetic and hydrodynamics principles to generate the propulsion power for a ship without using the propeller. It has been fully designed and tested in a small prototype, although with no encouraging results. But, sometimes a long-past failed experiment is worthy of study with a fresh start because the materials and other supporting technologies may have substantially changed since it was built and unsuccessfully tested. Magnetohydrodynamics (MHD) is one such technology for ship propulsion. Its last experiment was based on the old liquid helium superconducting technology, which has now been improved by severalfold by high-temperature liquid nitrogen superconductors. New engineers with new information can perhaps design, develop, and test it again to improve the MHD technology for ship propulsion. Until then, it may be a good graduate research topic.

In the working principle of MHD propulsion (Figure 11.5), electric voltage $V$ is applied between two fixed plates—located under the ship—separated by distance $d$ in seawater of electrical conductivity $\sigma$. This produces a current density in the liquid that is given by Ohm's law $J = \sigma(V/d)$. If the magnetic flux density $B$ is also applied perpendicular to the current, the mechanical force is developed that is given by $F = J \times B$ newtons per cubic meter of the liquid. This force propels the ship like a jet plane. Thus, the MHD propulsion converts the electrical energy to directly increase the momentum of the working fluid, which is seawater for ships.

Two theoretical advantages of the MHD propulsion are as follows: (a) There is much quieter and maintenance-free operation as there are no propellers or other moving parts, and (b) the speed-limiting phenomenon of cavitations at the propeller blades is completely eliminated.

The MHD principle was demonstrated in a small model ship built by Steward Way at the University of California in 1965, while on leave from the Westinghouse Electric Corporation. He assigned his senior year undergraduate students to develop a submarine with this new propulsion system using conventional room temperature magnets with copper coils, and they did.

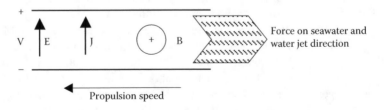

**FIGURE 11.5**  Magnetohydrodynamic principle for ship propulsion.

Another prototype was designed in Japan by the Ship and Ocean Foundation (later known as the Ocean Policy Research Foundation) using liquid helium superconducting magnets. The ship was first tested in Kobe Harbor in June 1992. It was propelled by two MHD thrusters, which run without any moving parts. The Japanese 1.1-ton, 11.5-ft long MHD boat (Figure 11.6) achieved a speed of 2.5 ft/sec (1.5 knots). A follow-on model, Yamato-1 (Figure 11.7), was also tested in Kobe, Japan. It produced a magnetic flux density of 4 Wb/m$^2$ and attained a speed of 6 knots with energy efficiency of a mere 4%. Steering the ship straight was difficult due to the nonuniform magnetic field; the problem was later solved by incorporating a dynamic control.

For commercial viability of an MHD propulsion system, it is necessary to have a superconducting magnet that can produce over 10-Wb/m$^2$ flux. A saddle-type dipole superconducting electromagnet (Figure 11.8) was built by Los Alamos Laboratory that can produce high flux density. The Argonne National Laboratory in Illinois, Newport News Shipbuilding in Virginia, and the Naval Underwater Systems Center in Newport, also have jointly demonstrated this principle on a 21-foot-long model using liquid helium superconducting magnets.

However, the commercial viability of MHD propulsion for ships has not been demonstrated as yet due to the following technical problems that remain to be resolved:

- An increase in the ship speed to 15–20 knots is needed. With the normal value of seawater conductivity, the Yamato-1 study projected that a 20-knot speed can be achieved only by having the magnetic field around 20 Wb/m$^2$. This is possible using the present-day high-temperature superconducting technology but is difficult to achieve in a large volume.
- The primary energy efficiency of around 40–50%, equivalent to that in the present-day cargo ships, must be achieved.
- MHD propulsion would work well at high sea, but not so in areas having a significant amount of freshwater or gray water, such as in many docks around the world.

**FIGURE 11.6**   First MHD-propelled demonstration boat with liquid helium superconducting coils.

**FIGURE 11.7** Improved MHD-propelled demonstration boat Yamato-1 tested at Cobe, Japan.

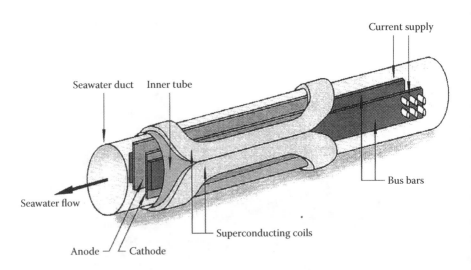

**FIGURE 11.8** MHD propulsion duct with saddle-type dipole superconducting magnet. (From Los Alamos National Laboratory.)

- Saltwater would quickly corrode the electrodes, making them a high-maintenance item.
- Solving the cryostability or quenching problem in a large superconducting magnet, which can lose superconductivity and thermally destroy the coil if even a tiny spot exceeds the critical temperature limit. A great deal of the design details must be worked out and numerous sensors must be continuously monitored to ensure a stable superconducting operation under the steady-state and the worst-case transient conditions.

The development cost of technically and economically viable MHD propulsion is expected to be high. As a result, no significant research is under way at present to develop MHD propulsion for ships or submarines. But, with continuing improvements in superconducting technology, it may gain research interest again in the future.

## QUESTIONS

*Question 11.1*: Why are tugboats good candidates for using hybrid propulsion? Identify the major design limits.

*Question 11.2*: Identify a potential candidate for a hybrid ferry between two ports familiar to you.

*Question 11.3*: Identify and discuss the ship stability problem if a conventional wind turbine on a tall tower is used on a ship. How would you quantify this problem?

*Question 11.4*: How does the hybrid car work, and why does it give higher mileage per gallon of fuel?

*Question 11.5*: Explain the working principle of MHD propulsion as it applies to ships.

## FURTHER READING

Faber, W.G., and J. Aspen. 2008. *The Foss Hybrid Tug, from Innovation to Implementation.* Report by Foss Maritime Company, Aspen Kemp & Associates, and ABR Company Ltd., Seattle, WA.

Patel, M.R. 2005. *Spacecraft Power Systems.* Boca Raton, FL: CRC Press/Taylor & Francis.

Patel, M.R. 2006. *Wind and Solar Power Systems.* Boca Raton, FL: CRC Press/Taylor & Francis.

# Part C

---

# *Emerging Ocean Energy Technologies*

Part C of this book is not related to the shipboard power systems; it covers the emerging technologies for extracting renewable energy from the ocean. Exploiting these technologies will require marine engineers' deep involvement in the construction, operation, and logistic support from the shore to the renewable energy site in the ocean. Moreover, the ocean energy extraction technologies have a strong connection with power electronics, which is a dominant part of this book. The ocean power generated at randomly varying frequencies of the ocean waves and varying wind speed in offshore and farshore wind farms—spreading rapidly in Europe and many countries—needs to be converted into fixed 60-Hz or 50-Hz power before connecting to the loads or to the land-based grid. Here, the power electronics are involved as much as in ship propulsion. As such, the electrical power system for marine current power generation is the exact reverse of the variable-frequency drive (VFD) for ship propulsion. If the ship is standing still in marine current, the propeller would work as a water turbine, the propulsion motor as a generator, and the VFD would convert the variable-frequency power of the generator into a fixed-frequency power feeding the main switchboard and then to the shore if so connected. There is no need for additional equipment for extracting the marine current energy in this way.

The Energy Information Agency (EIA) of the Department of Energy estimated that the present contributions of various sources of fuel for generation of electrical power in the United States are (a) 48% coal; (b) 21% natural gas; (c) 20% nuclear; (d) 6% hydroelectric; (e) 4% renewable (wood, biomass, solar, geothermal, and wind); and (f) less than 2% oil. However, the worldwide investment in recent years for building new renewable energy sources of electricity have been more than for new fossil fuel power plants. The current U.S. administration has set a goal of generating

20% of all electrical power from renewable sources by the year 2025. This is a moving target because the demand for electrical power is expected to increase by 25% between now and the year 2025, perhaps much more if a large number of electric cars enter the road.

For increasing the present 4% of the renewable energy to 20% by 2025, the virtually limitless wind, photovoltaic, and ocean energies will have to be tapped. However, both solar and wind are inherently intermittent energy sources, unavailable when the sun does not show and the wind does not blow. In addition, they have huge footprints compared with conventional power plants. A new coal-fired or nuclear plant of 1,000-MW$_e$ capacity would occupy less than 1 mi$^2$ (2.6 km$^2$). A wind farm of equivalent capacity would require about 200 mi$^2$ (256 km$^2$).

The oceans, covering two-thirds of the earth's surface, offer virtually unlimited renewable energy resources in open space where new energy technologies can be developed without significant interference with the environment or normal human activities. The ocean energy sites with smaller footprints and fewer down days offer an estimated exploitable global potential of over 100,000 TWh/year (T = tera = 10$^{12}$), about six times the present electrical energy consumption in the world of 16,000 TWh/year. The exploitable potential with the technology of today is about 45,000 TWh/year from wave energy; 2000 TWh/year from tidal and marine current energy, 20,000 TWh/year from salinity gradient energy, and 33,000 TWh/year from ocean thermal energy—a total of about 100,000 TWh/year from all forms of ocean energy.

Part C of this book covers three sources of generating electrical energy from the ocean (a) wave energy, (b) tidal and marine current energy, and (c) offshore and farshore wind energy. The first two of these technologies, although still in their infancy compared to other renewable energy sources, are rapidly developing. After several encouraging demonstration projects, commercial projects for tapping the ocean energy are now under way. This part of the book introduces the renewable ocean energy technologies to marine engineering students and professionals who may become involved in providing the offshore and shore logistic support needed to tap this vast energy potential of the ocean.

# 12 Ocean Wave Power

Ocean waves bring an enormous amount of energy to the shore, day and night, all year around, in most countries on the earth. If we can tap only a couple of percent of this energy, a small fraction of developable ocean shores can provide all the energy for the world.

Wave power is really wind power. A wind blowing for a long time (about 10 hours) across a very large area of the ocean causes the surface water to flow at about 2% of the wind speed, with water piling up in the wind direction. Gravity tends to pull the water down the pile of water. Thus, the wind acting on the ocean surface generates waves. The winds are driven by the uneven solar heating at different latitudes and the spin of Earth on its axis. The wind in turn drives the surface currents and waves of the oceans. The global wind patterns are the powerful westerlies and the persistent easterlies, commonly known as the trade winds (Figure 12.1). Big waves are generated by big windstorms. Wave heights of 70 to 100 feet are possible in storms of 70 to 100 miles per hour. The wave energy, either extracted from the wave or not, is replenished by the continuous interaction of the wind and ocean surface. The wave energy is thus a renewable source that is more continuous than other renewable energy sources.

The wave-generation mechanism is depicted in Figure 12.2 and is as follows:

(a) The waves are generated by the wind blowing against the surface of the ocean. The stronger and longer the wind blows over a wide stretch of water, the greater the wave height.
(b) As the waves spread far beyond the storm area, they become rolling swells. The water appears to be moving a distance, but it really just goes up and down in circles. The energy of the waves, however, moves forward in a falling domino pattern.
(c) When the wave hits an underwater obstruction, such as a reef or a seamount, the shear friction from the obstruction distorts the circular motion of the water, and the wave breaks up (trips over itself.)

The water particles under the waves actually travel in orbits that are circular in deep water, gradually becoming horizontal-elliptical (flat-elliptical) near the surface as in Figure 12.3.

## 12.1 WAVE POWER ESTIMATE

The power that can be generated by an ideal sinusoidal water wave of certain width is estimated next. The power is the rate of change in potential energy as the water level rises above and falls below the average sea level at the site. Referring to Figure 12.4, the mass of water in one-half sine wave that is above the average sea level is given by

## Global Wind Patterns

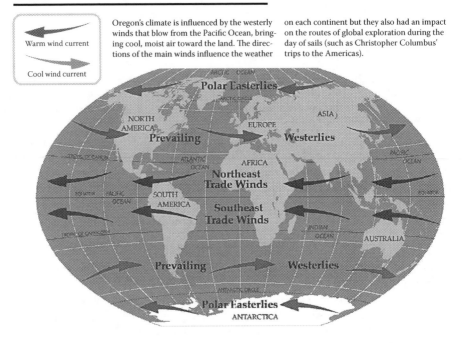

Warm wind current

Cool wind current

Oregon's climate is influenced by the westerly winds that blow from the Pacific Ocean, bringing cool, moist air toward the land. The directions of the main winds influence the weather on each continent but they also had an impact on the routes of global exploration during the day of sails (such as Christopher Columbus' trips to the Americas).

**FIGURE 12.1**    Global wind patterns showing westerlies and easterlies (trade winds).

$$mass = w\rho\left(\frac{\lambda}{2}\right)\frac{h_w}{2\sqrt{2}} \tag{12.1}$$

where $w$ = wave width, $\rho$ = density of seawater, $\lambda$ = wave length, and $h_w$ = wave height (trough to crest).

The center of gravity of mass in the wave crest is $h_w/(4\sqrt{2})$ above the average sea level, and that of the wave trough is $h_w/(4\sqrt{2})$ below the average sea level. The total change in the potential energy $PE$ during one wave cycle is therefore given by (Mass × Gravity constant × Change in height of center of gravity), that is,

$$\Delta(PE) = w\rho\left(\frac{\lambda}{2}\right)\frac{h_w}{2\sqrt{2}} \times g \times \left\{\frac{2h_w}{4\sqrt{2}}\right\} \tag{12.2}$$

Equation (12.2) reduces to

$$\Delta(PE) = g\,w\rho\lambda\left(\frac{h_w^{\,2}}{16}\right) \tag{12.3}$$

The frequency of waves in deep ocean is ideally given by

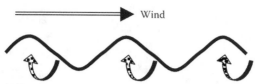

Wind

(a) Storm generates waves by friction of wind against the water surface.

(b) As waves move farther, they become rolling swells. Water looks like it is moving forward, but it just goes around in circles.

(c) An underwater reef or seamount breaks the wave, distorts the circular motion, and the wave basically trips over itself.

**FIGURE 12.2** Ocean wave generation mechanism in three steps.

$$f = \sqrt{\frac{g}{2\pi\lambda}} \text{ Hz} \tag{12.4}$$

The power developed by the ocean wave is then simply $\Delta(PE) \times$ Frequency, which gives the wave power

$$P_{wave} = g\,w\rho\lambda\left(\frac{h_w^2}{16}\right) \times \sqrt{\frac{g}{2\pi\lambda}} = \frac{w\rho\,g^2 h_w^2}{32\pi f} \tag{12.5}$$

This equation leads to the following estimate of the wave power, which can be converted into electrical power: A wave of 1-km width, 5-m height, and 50-m wavelength has a water power capacity of 130 MW. Even with 2% conversion efficiency, it can generate 2.6 MW of electrical power per kilometer of coastline, sufficient to power approximately 1000 homes in a coastal community in developed countries and several thousand homes in developing countries.

An estimated 2 to 3 million MW of wave power is breaking up on the shores of the world. At favorable coastlines, the power density could be 30 to 50 MW/km. At this rate, all the energy requirement of the world can be met by wave energy. Thus,

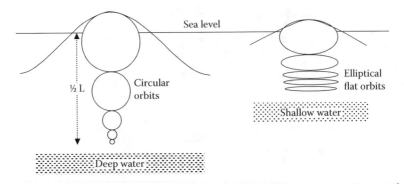

**FIGURE 12.3**    Orbits of water particles in deep and shallow waters.

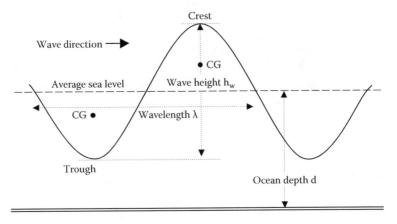

**FIGURE 12.4**    Ideal sinusoidal representation of ocean wave. CG = center of gravity.

the exploitable energy from the ocean waves is much greater than that from the tidal or marine currents.

As indicated by Equation (12.5), the power in the wave depends on the frequency, which is a random variable. The frequency distribution of the ocean waves is shown in Figure 12.5, which shows that the energy comes from waves at a frequency from 0.1 to 1.0 Hz, with most energy coming from 0.3-Hz waves. The actual wave may have a long wave superimposed on a short wave from a different direction. Multiple single-frequency analyses may be superimposed for an approximate estimate of the total power from multiple-frequency waves.

## 12.2   POWER GENERATION SCHEMES

Although such enormous energy is available in the ocean waves, the extraction of this energy has not been fully developed as yet. Converting the wave energy into electricity is particularly difficult because of complications arising in capturing the energy at a wide range of amplitudes and frequencies that characterize the wave motion. This variable-voltage and variable-frequency power is converted

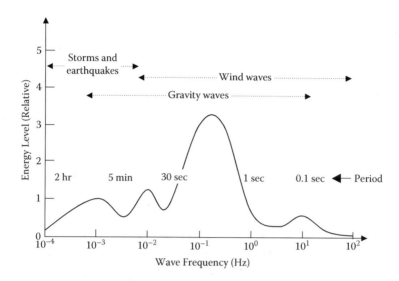

**FIGURE 12.5** Energy level versus frequency distribution of ocean waves.

into fixed voltage and fixed frequency for local use or for feeding to the grid using the traditional variable-frequency converters in the reverse as shown in Figure 12.6.

A number of companies in the United States, Europe, Canada, and around the world are now engaged in extracting ocean energy to generate electrical power. A dedicated facility for testing electrical power generation from waves has been established in Blyth, United Kingdom, where a 70 × 18 -m testing tank is converted from a dry dock with wave-generating machines. The few alternative technologies for converting wave power to electrical power are presented next.

## 12.2.1 TURBINE GENERATOR

The turbine generator scheme uses the circulating water particles in the waves to create local currents to drive a turbine. A prototype built and connected to the Danish grid in 2002 is working well, even with irregular waves as small as 20 cm. A combined Wells turbine and Darrieus rotor is used to capture the wave energy from both up-and-down and back-and-forward currents. It works on the principle of lift in a circulating flow using hydrofoil blades.

## 12.2.2 FLOAT GENERATOR

In this most promising scheme, a float on the wave drives an electrical generator. Figure 12.7 shows a wave energy farm using multiple units of PowerBuoy® developed by Ocean Power Technologies, Inc. of Princeton, New Jersey. Each buoy is anchored to the seabed and can generate power with wave heights between 1.5 and 7 m (4.9 to 22.9 ft). The PowerBuoy model PB150 (Figure 12.8) offers a sustained maximum peak-rated output of 150 kW and a power factor of 0.9 lagging to 0.9 leading.

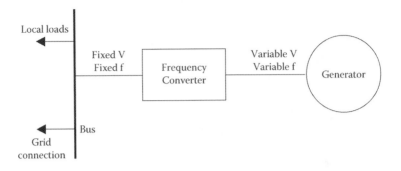

**FIGURE 12.6**   Power system architecture for extracting ocean wave energy.

**FIGURE 12.7**   Wave energy farm with array of floats. (With permission from Ocean Power Technologies, Inc.)

A 10-MW power plant would occupy about 30 acres (0.125 km²) of ocean space. Typical capacity factors range between 30% and 50%, depending on location. Ocean Power Technologies, Inc. has developed an Underwater Substation Pod (USP)—also shown in Figure 12.8 with the tower—which can collect the electrical output from up to 10 offshore power generators (e.g., wave, tidal, and offshore wind) into one common interconnection point for economic undersea power transmission and data to the shore. It is designed to step up low voltages generated by offshore devices to medium voltage (11–15 kV) compatible with onshore electrical distribution.

The electrical generator is made of a permanent magnet reciprocating in a coil that works like a linear alternator. In the PowerBuoy, an electrical coil is secured to the heaving buoy inside and surrounds the magnetic shaft, which is anchored to the sea floor (Figure 12.9). When the wave causes the coil to move up and down relative to the fixed magnetic shaft, a voltage is induced that powers the connected load. Each buoy is presently designed for 150 kW, but the technology can be scaled up or down to suit a variety of energy needs. The University of Oregon estimated that a fleet of about 200 buoys at 150 kW could power the business district of downtown Portland.

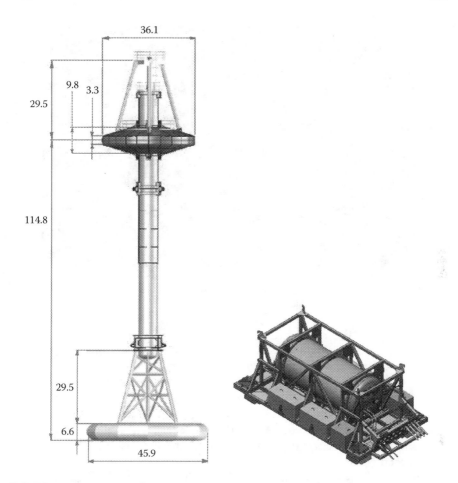

**FIGURE 12.8**  PowerBuoy® tower construction and PowerPod™ for power and data transmission to shore. (With permission from Ocean Power Technologies Inc., Princeton, N.J.)

Recent examples of this technology are the 40-kW PowerBuoys installed in Hawaii and New Jersey shores and a 1.39-MW wave farm off the northern coast of Spain. The project is a joint venture with the Spanish utility Iberdrola SA. A larger demonstration wave farm of up to 5-MW capacity is also planned for installation in U.K. waters.

### 12.2.3  PIEZOELECTRIC GENERATOR

The piezoelectric crystal produces electrical voltage under mechanical stress. It has been used for many years for measuring mechanical stress and acceleration. The glow-as-you-walk sneakers kids wear use this voltage with light-emitting diodes. The piezoelectric phenomenon can be deployed to generate electrical power using the mechanical power of ocean waves. Initially developed by Ocean Power Technologies, it has not been commercially viable as yet due to various material and design limitations. In this scheme, piezoelectric polymer sheets convert the

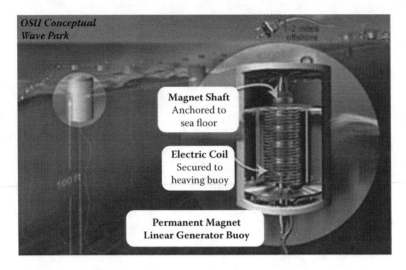

**FIGURE 12.9** Wave power generation with permanent magnet linear alternator on float. (Photo credit: University of Oregon.)

wave energy directly into electricity. The piezoelectric material, when subjected to a mechanical stress $\sigma_{xy}$ over an area, generates an electrical field (voltage gradient) in the transverse direction (Figure 12.10), which is given by

$$E_z = \alpha \times \sigma_{xy} \text{ volts/m} \tag{12.6}$$

where $\alpha$ = the piezoelectric coupling coefficient, defined as the ratio of the electrical field across the thickness to the longitudinal mechanical stress.

The coefficient $\alpha$ of most commonly used piezoelectric materials is less than 0.5 V/m per N/m². Low-cost piezoelectric polymer sheets developed by companies like AMP Incorporated have a higher $\alpha$ that can convert ocean wave power directly into electrical power. Such a hydropiezoelectric generator (HPEG) has potential applications in providing power for remote islands, coastal communities, lighthouses, offshore platforms, buoys, and so on.

The HPEG uses many piezoelectric modules, each made of multiple piezoelectric polymer sheets laminated together. The modules are hung from a raft in the ocean and anchored to the ocean floor (Figure 12.11). The sheets stretch when a wave lifts the raft, and the resulting strain generates electrical power. The power electronics circuit converts the low-frequency power into dc and then to standard 60-Hz or 50-Hz power. This concept differs from other ocean power schemes, which generally involve large moving parts and rotating or reciprocating electrical generators.

The electrical model of the piezoelectric power generator is shown in Figure 12.12a. It represents the piezoelectric material as a charge generator with an internal static capacitance $C_i$ in series with resistance $R_i$ and inductance $L_i$ representing the internal voltage drops under load. The parallel shunt resistance $R_{sh}$ represents the power leakage to the ground. Under a sudden dc stress, the HPEG would generate a charge $C_i$, which would eventually decay to zero in five time constants via $R_{sh}$ (relatively large).

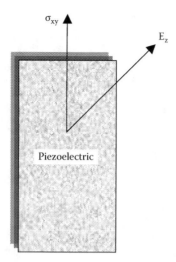

**FIGURE 12.10**  Piezoelectric phenomenon generating electrical voltage under mechanical stress.

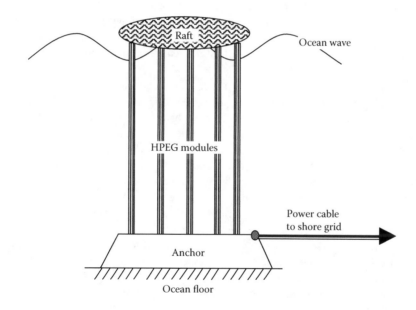

**FIGURE 12.11**  Hydropiezoelectric generator construction features.

The electrical model of Figure 12.12a can be simplified as shown in Figure 12.12b with the following reasoning: The HPEG, when stressed, generates a charge $Q_i$, which in turn results in voltage $V_i$ across the capacitance $C_i$. This voltage under alternating mechanical stress may be represented by an alternating voltage source $V_i \sin \omega t$.

Since the ocean wave frequency is low—generally a fraction of 1 Hz—the inductance $L_i$ can be ignored for all low frequencies. Moreover, the shunt resistance $R_{sh}$

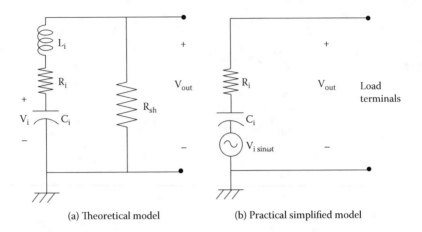

(a) Theoretical model                    (b) Practical simplified model

**FIGURE 12.12**    Electrical circuit model of hydropiezoelectric generator.

may be ignored for practical HPEG designs. Under a low-frequency strain $\delta$, a sustained alternating voltage $V_i$ proportional to the strain rate $\delta$ is internally generated, which appears at the terminal under no load, that is, $V_i = K_i\delta$. The maximum allowable strain rate $\delta_{max}$ depends on the yield and fatigue strength of the piezoelectric sheets. The yield strength is typically in a range of a few percent. The values of the circuit parameters shown in Figure 12.12b can be determined from the open-circuit and short-circuit tests. The load resistance (not shown), when connected to the output terminals, would absorb the electrical power generated in the HPEG.

Since the wave power is in use-it or lose-it mode, one objective of the HPEG design is to transfer the maximum possible power to the load when operating under load. This requires both mechanical and electrical load matching as explained further here. For mechanical load matching, the restraining spring constant of the piezoelectric sheets must be optimum, such that the deflection of the sheets under wave is exactly equal to the wave height, no more or no less. If the sheets were infinitely soft, they will deflect without any stress, resulting in zero power generation. On the other hand, if the sheets were infinitely stiff, they would not deflect at all, and the wave will ride over them, not delivering its energy to the sheets and subsequently wasting the wave energy. The principle of mechanical load matching is illustrated by Figure 12.13, which plots the wave energy capture versus spring constant of the piezoelectric sheets. The zero-energy capture at the two extremes on the $x$ axis infers that there exists an optimum spring constant that would give the maximum energy capture from the waves.

For electrical load matching, the electrical load impedance must be equal to the internal (Thevenin) impedance of the HPEG. For this purpose, a transformer is generally required between the HPEG and the load as shown in Figure 12.14. By the Thevenin theorem, the maximum power would transfer when the internal resistance and the load resistance as reflected on the source side are equal. For this, the optimum electrical design should have

$$R_i = n^2 R_L \tag{12.7}$$

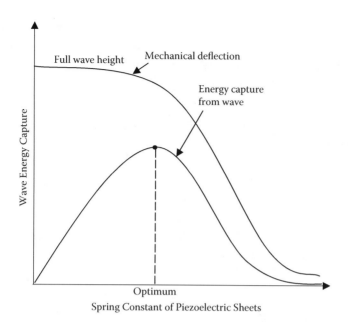

**FIGURE 12.13**   Mechanical load matching for maximum energy capture from wave.

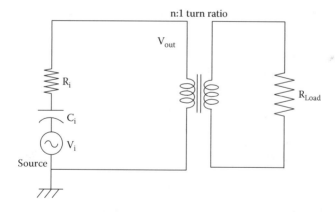

**FIGURE 12.14**   Electrical load matching for maximum power transfer to load.

where $n$ = turn ratio of the transformer = (Source side turns ÷ Load side turns).

Even with such load matching—electrical and mechanical—the extremely low frequency of the ocean waves compared to the conventional power frequencies (60 or 50 Hz) poses a design challenge in extracting the maximum power. One practical design requirement of the HPEG power system is to convert the fractional frequency power into 60 Hz for consumers. The ac-dc-ac frequency converter would serve this purpose, except that the energy storage requirement during the off time of the switching period is much greater than that found in the traditional frequency converter used in the power electronics motor drives. This is again due to the low frequency of the ocean wave power. The high energy density supercapacitor in a full-wave rectifier circuit is one potential candidate for the HPEG, but such designs need to be developed on an economical scale.

The electromechanical considerations of the HPEG system would show that only a couple of percent of the mechanical energy of the wave could be converted into electrical energy. The remaining energy does not dissipate in the internal resistance but remains in the waves. For this reason, one can build an ocean wave energy farm stacking the piezoelectric sheets in multiple columns along the wave propagation. This helps minimize the offshore energy farm spread for beachgoers.

## QUESTIONS

*Question 12.1*: Explain the use of piezoelectric crystal in measuring mechanical stress and acceleration.

*Question 12.2*: In developing HPEG, briefly explain the two load-matching requirements for maximum power generation.

*Question 12.3*: Identify the frequency band of ocean waves carrying the most wave energy.

*Question 12.4*: Identify the most challenging aspects of power electronics design for capturing and converting ocean wave power into fixed-frequency and fixed-voltage power.

## FURTHER READING

Bedard, R., et al. 2005. *Ocean Wave Power Feasibility Demonstration Project.* EPRI Report No. E21 Global WP009-US. EPRI: Palo Alto, CA.

Brooke, J. 2003. *Wave Energy Conversion.* Oxford, UK: Elsevier.

Cruz, J. 2010. *Ocean Wave Energy: Current Status and Future Perspectives (Green Energy and Technology.* Berlin: Springer.

Gross, N. 1994. Jolt of electrical juice from choppy seas. *Business Week,* November 28, p. 149.

Kaligh, A., and O.C. Onar. 2009. *Energy Harvesting.* Boca Raton, FL: CRC Press/Taylor & Francis.

Mario, C. 1995. Ocean power envisions a piezoelectric energy farm. *U.S. Business Journal,* January 18, P11.

Melnyk, M., and R. Andersen. 2009. *Offshore Power: Building Renewable Energy Projects in U.S. Waters.* Tulsa, OK: PennWell.

Myers, J.J., C.H. Holm, and R.F. McAllister. 1968. *Handbook of Ocean and Underwater Engineering.* New York: McGraw-Hill.

Patel, M.R. 2003. Piezoelectric conversion of ocean wave energy. *AIAA Proceedings of the First International Energy Conversion Engineering Conference.* Paper No. 6073. August 2003. Portsmouth,

Patel, M.R. 2004. Optimized design of piezoelectric conversion of ocean wave energy. *AIAA Proceedings of the Second International Energy Conversion Engineering Conference.* Paper No. 5703. August 2004. Providence, RI.

Taylor, G.W., et al. 1985. *Piezoelectricity.* New York: Gordon and Breach.

Temeev, et al. 1999. Natural-artificial power-industrial system based on wave energy conversion. *AIAA Proceedings of the 34th Intersociety Energy Conversion Engineering Conference.* Paper No. 1-2556. August 1999. Vaneouver, Canada.

# 13 Marine Current Power

The uneven solar heating at different latitudes of the earth results in a pressure difference, which generates wind in the atmosphere and marine currents in the ocean. The tidal currents, on the other hand, result from enormous movement of water due to gravity forces of the sun and the moon on the oceans.

There are two types of marine currents: (a) surface currents (surface circulation) in the upper 400-m layer of the ocean, which makes up about 10% of all water in the ocean, and (b) deep-water currents (thermohaline circulation) in the other 90% of the ocean water, moving around the ocean basins by density-driven forces and gravity. Since the density difference is a function of differential temperature and salinity, water sinks into the deep ocean basins at high latitudes where the temperatures are colder, causing the density to increase. The marine currents are influenced by two types of forces:

*Primary forces:* These result from solar heating, wind, gravity, and coriolis and start the water moving.

*Secondary forces:* These determine where the currents flow. The water expansion under solar heating is about 8 cm higher near the equator than in the midlatitudes. The resulting slope causes the water to flow down the slope from the equator to the Tropics of Cancer and Capricorn.

The persistent surface currents are found in the oceans all around the globe (Figure 13.1), but the interest in harvesting electrical energy from them has emerged only recently. For example, the Florida current (Figure 13.2) is a well-defined component of the Gulf Stream system. The inner edge of the system is less than 10 miles from Miami and Fort Lauderdale, with a marine current speed of about 2 m/s—with seasonal variations—within a few miles from the coast. Its short distance from the shore and good persistent current speed make the Florida coast viable for installing and operating marine current energy farms.

The principle of converting the kinetic energy of the marine current into electrical energy is similar to that of the wind turbine, except that the marine current turbine works in water (Figure 13.3). The analysis of electrical power generation in both the marine current and the wind is the same, except for the fluid density and speed differences. Compared to air, water is 800 times denser, but the marine current speed is about one-half that of wind. This gives the theoretical marine current power density per square meter of the blade-swept area approximately 100 times greater than in wind, as indicated in the next section. Moreover, marine currents have an additional advantage over the wind: They are persistent and more predictable.

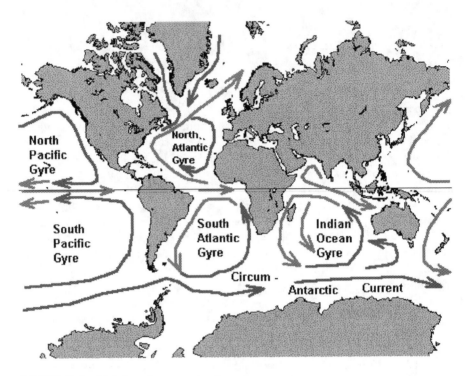

**FIGURE 13.1**   Global marine current patterns.

## 13.1   SPEED AND POWER RELATIONS

The kinetic energy in fluid (water or air) with mass $M$ moving at speed $V$ is given by

$$Kinetic\ Energy = \frac{1}{2} M\ V^2 \tag{13.1}$$

The power in moving fluid is the flow rate of kinetic energy per second, that is,

$$Power = \frac{1}{2} \times (Mass\ flow\ rate\ per\ second) \times V^2 \tag{13.2}$$

If we let $P$ = mechanical power in moving fluid (watts), $\rho$ = fluid density (kg/m³), $A$ = area swept by moving fluid (m²), and $V$ = velocity of fluid (m/s), then the moving fluid has the volumetric flow rate of $A \times V$ m³/sec, mass flow rate of $\rho \times A \times V$ kg/sec, and the mechanical power in the moving fluid is given by

$$P = \frac{1}{2}(\rho\,AV)V^2 = \frac{1}{2}\rho\,V^3 \times A\ \text{watts} \tag{13.3}$$

Two potential power-generating sites are compared in terms of the *specific power* expressed in watts per square meter of circular blade area swept by the moving fluid.

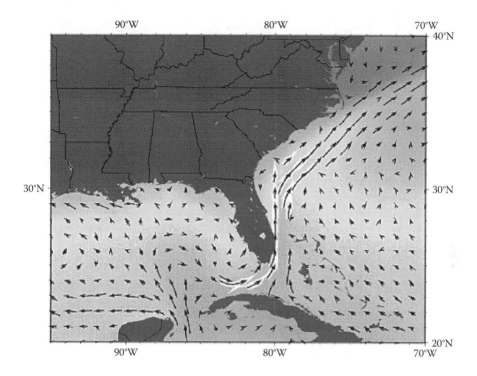

**FIGURE 13.2**    Florida current in Gulf Stream system has about 2-m/s flow close to the coast.

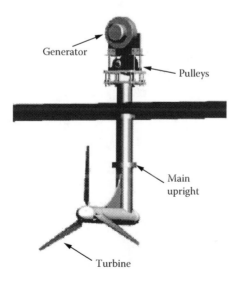

**FIGURE 13.3**    Marine current turbine for electrical power generation. (From W. Batten, L. Blunden, and A. Bahaj, 2007, Marine Current Turbine, University of Southampton, UK.)

It is also referred to as the *power density* of the site and is given by the following expression:

$$Power\,density\,of\,site = \frac{1}{2}\rho V^3 \text{ watts/m}^2 \text{ of moving fluid} \qquad (13.4)$$

### 13.1.1 Turbine Power

Not all, but a good fraction of the power of moving fluid can be extracted by blades driven by the fluid. We consider the marine current turbine or wind turbine of blade diameter $D$ placed in fluid of density $\rho$, with incoming speed $V$ and exit speed $V_o$ past the turbine as shown in Figure 13.4. The power extracted by the rotor blades of such a turbine is the difference between the upstream and the downstream fluid powers. Using Equation (13.2), we obtain the power output of the rotor shaft delivered to the electrical generator:

$$P_o = \frac{1}{2}(Mass\,flow\,rate\,per\,second)\{V^2 - V_o^2\} \quad \text{watts} \qquad (13.5)$$

where

$P_o =$ rotor shaft output power = mechanical power extracted by turbine blades
$V =$ upstream fluid velocity at entrance of rotor blades
$V_o =$ downstream fluid velocity at exit of rotor blades

We leave details regarding blade aerodynamics for many excellent books on the subject and take here a macroscopic view of the fluid flow around the blade. Macroscopically, the fluid velocity has a discontinuous drop from $V$ to $V_o$ at the *plane of the rotor blades*, with an average speed of ½($V + V_o$). Multiplying the fluid density with the average velocity therefore gives the mass flow rate of fluid through the rotating blades as follows:

$$Mass\,flow\,rate\,per\,second = \rho A\frac{V + Vo}{2} \qquad (13.6)$$

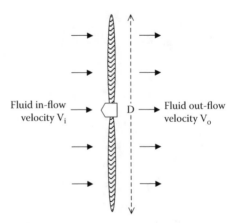

**FIGURE 13.4**    Horizontal axis turbine in moving fluid (air or water).

Substituting Equation (13.6) in Equation (13.5) gives the mechanical power captured by the rotor and delivered to the electrical generator, which is given by

$$P_o = \frac{1}{2}\left(\rho A \frac{(V+V_o)}{2}\right) \times \{V^2 - V_o^2\}$$  (13.7)

This expression is algebraically rearranged in the following form for further analysis:

$$P_o = \frac{1}{2}\rho A V^3 \frac{\left(1+\dfrac{V_o}{V}\right)\left[1-\left(\dfrac{V_o}{V}\right)^2\right]}{2}$$  (13.8)

The power output of the blades is customarily expressed as a fraction $C_p$ of the upstream fluid power, that is,

$$P_o = \frac{1}{2}\rho V^3 A C_p$$  (13.9)

where

$$C_p = \frac{\left(1+\dfrac{V_o}{V}\right)\left[1-\left(\dfrac{V_o}{V}\right)^2\right]}{2}$$  (13.10)

Comparing Equation (13.3) with equation (13.9), we observe that $C_p$ is the fraction of the upstream fluid power that is extracted by the rotor blades and fed to the electrical generator. The remaining power is discharged or wasted in the downstream fluid. Therefore, the factor $C_p$ is called the *power coefficient* of the rotor or the *rotor efficiency*.

For a given upstream fluid speed, Equation (13.10) clearly shows that the value of $C_p$ depends on the ratio of the downstream to the upstream fluid speeds, that is, the $(V_o/V)$ ratio. The plot of the output power versus $(V_o/V)$ in Figure 13.5 shows that $C_p$ has a single maximum value of 0.59 when the $(V_o/V)$ ratio is one-third. Therefore, the maximum power is extracted from the moving fluid when the downstream fluid speed equals one-third of the upstream speed. Under this condition,

$$P_{max} = \frac{1}{2}\rho V^3 A \times 0.59$$  (13.11)

Thus, the theoretical maximum value of $C_p$ is 0.59. The $C_p$ is often expressed as a function of the rotor tip-speed ratio (TSR) as shown in Figure 13.6, where the *Tip speed ratio = (Linear speed of rotor's outermost tip ÷ upstream fluid speed).* The aerodynamic analysis of the fluid flow around the moving blade with a given pitch angle establishes the relation between the rotor tip speed and the fluid speed. As seen in Figure 13.6, the maximum achievable $C_p$ in practical designs ranges between 0.4 and 0.5 for modern high-speed two- or three-blade turbines. If we take 0.5 as the

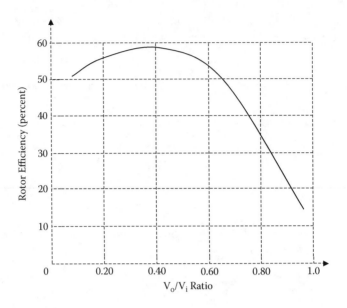

**FIGURE 13.5**   Rotor efficiency versus $V_o/V$ ratio has a single maximum.

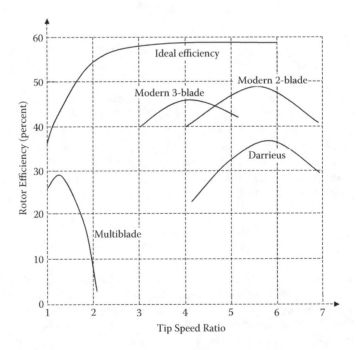

**FIGURE 13.6**   Ratio of rotor efficiency versus tip speed for two- and three-blade rotors.

practical maximum rotor efficiency, the maximum power output of the fluid turbine becomes a simple expression as follows:

$$P_{max} = \frac{1}{4} \cdot \rho \cdot V^3 \text{ W/m}^2 \text{ of swept area} \qquad (13.12)$$

The marine and tidal currents exceeding 3 m/s present an attractive source of energy that can be economically exploited. For example, a marine current of 3 m/s (6 knots) represents the impinging mechanical power of 14 kW/m² of the blade-swept area. A minimum current of about 1.25 m/s bringing in 1-kW/m² mechanical power is required for a practical power plant that may be economically viable. Thus, a potential site must have a sustained current exceeding 1.25 m/s at sufficient depth to install a turbine of required diameter for a desired electrical power-generating capacity.

The fluid turbine efficiently intercepts the fluid energy flowing through the entire swept area even though it has only two or three thin blades with solidity between 5% and 10%. The solidity is defined as the ratio of the solid area to the swept area of the blades. The modern two- blade turbine has a low solidity ratio and is cost effective as it requires less blade material to sweep large areas and yet produces the same or even greater power with the same diameter three-blade rotor.

## 13.1.2 Hub Height and Power Potential

The shear at the ground surface causes the fluid speed to increase with height in accordance with the following expression:

$$V_2 = V_1 \cdot \left( \frac{H_2}{H_1} \right)^{\alpha} \qquad (13.13)$$

where
  $V_1 =$  fluid speed measured at the reference height $H_1$
  $V_2 =$  fluid speed estimated at height $H_2$
  $\alpha =$  riverbed (or ground) surface friction coefficient

The friction coefficient $\alpha$ is low for smooth terrain and high for rough terrain. It is noteworthy that the offshore wind tower, being in low-$\alpha$ terrain, always sees a higher wind speed at a given height and is less sensitive to the tower height. For average terrain on land, the average value of $\alpha$ for wind speed is about 1/7. The 1/7 power relation applies for river currents also, as validated by data measured by Verdnant Power Corporation in the East River in New York City.

The power generation potential of a marine current site primarily depends on the current speed cube, which is also influenced by the blade hub height from the floor in case of the river current. Therefore, placing the marine current turbine higher from the floor gives an advantage in the annual energy capture. For example, if the current speed is, say, 2 m/s at 4 m height, then it would be $(10/4)^{1/7} = 1.14$ times greater at

10 m height, and the power generation will be $1.14^3 = 1.48$, which is 48% greater at 10 m height than at 4 m from the riverbed.

### 13.1.3 Rotor-Swept Area

As seen in Equation (13.11), the output power of the fluid turbine varies linearly with the rotor-swept area. For the horizontal axis turbine shown in Figure 13.4, the rotor-swept area is given by

$$A = \frac{\pi}{4}(rotor\,diameter)^2 \tag{13.14}$$

The conventional horizontal axis turbine must be continuously oriented perpendicular to the fluid flow to get the maximum power. With the vertical axis machine shown in Figure 13.7 (called the Darrieus turbine after its inventor) has no such requirement because it is always perpendicular to the flow. Determination of the swept area of the Darrieus turbine is complex as it involves elliptical integrals. However, approximating the blade shape as a parabola leads to the following simple expression for the swept area for the vertical axis turbine:

$$A = \frac{2}{3} \cdot (maximum\,rotor\,width\,at\,the\,center) \cdot (height\,of\,the\,rotor) \tag{13.15}$$

The Darrieus turbine is not well developed yet. The marine current turbines that are being installed at present are of the horizontal axis design that is well proven in many wind farms around the world.

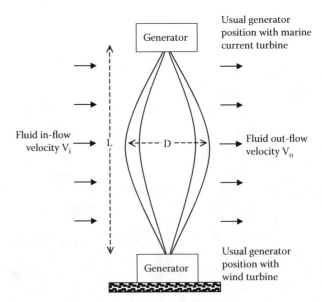

**FIGURE 13.7**    Vertical axis Darrieus rotor in moving fluid needs no directional orientation.

### 13.1.4 Maximum Power Capture

The fixed-speed turbine has a limitation on energy capture. When the available current speed is high, it must change the pitch to keep its speed and frequency constant. This invariably wastes the available energy in high currents, which comes in use-it or lose-it mode. The variable-speed turbine runs at high speed in high current and at low speed in low current. In the variable-speed power system (Figure 13.8), the variable-frequency output power of the generator (induction or synchronous) is first converted into dc and then converted into 60-Hz or 50-Hz fixed-frequency power before distributing to the users or feeding to the grid. The cost of power electronics converters (ac-dc rectifier and dc-ac inverter) is typically paid back within a few years by higher energy capture every year over the life of the turbine (20 to 30 years). The initial capital cost of the converters may be partially offset by the fact that the system may not need the main frequency (60-Hz or 50-Hz) power transformer, which is generally required for a fixed-speed system. The necessary voltage step-up may be incorporated in the converter design.

The dc link converter has a built-in advantage of incorporating large-scale energy storage in batteries connected to the dc link. Such energy storage is often required for a wind farm to improve power availability during low wind periods. Moreover, it can support an additional type of power generation from the dc link, thus serving a diverse group of users, some using dc and some using ac.

## 13.2 PRESENT DEVELOPMENTS

The use of ocean waves and marine currents, although not fully developed yet, is receiving increasing attention for renewable energy around the world. The European Commission funded a study that estimated 48 TWh/year exploitable tidal current energy potential at more than 100 locations around Europe. Similar studies around the world have shown the renewable ocean energy potentials of 70 TWh/year in the Sibulu Passage in the Philippines and 37 TWh/year on the Chinese coasts.

Successful prototypes for marine current power generation have been demonstrated using the conventional horizontal axis turbine, although a Darrieus turbine is also possible. After a few successful demonstrations, the work is under way for

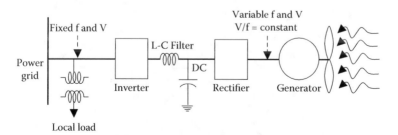

**FIGURE 13.8** Variable-frequency marine current power generation system for maximum energy capture.

two commercial grid-connected power generators, one rated at 300 kW with a horizontal axis turbine in the United Kingdom and another 250-kW vertical axis turbine in Canada. A 300-kW tidal current turbine has been installed 1 km offshore of Devon in the United Kingdom; this was funded by the United Kingdom, Germany, and the European Commission. The machine was built by the Marine Current Turbine Company and partly funded by the electrical power company of London. It is mounted on steel pipes set into the seabed. Its 11-m diameter blades generate 300 kW of power under tidal currents of 2.7 m/s. The top of the whole installation is a few meters above the water surface. Another design is also being developed that does not require attachment to the seabed.

Marine Current Turbines company has successfully developed the power generator SeaGen™ in various ratings (Figure 13.9). Using this technology, the company completed the first installation phase of the 1.2-MW SeaGen Tidal System in the fast-flowing waters of Strangford Narrows off the coast of Northern Ireland. It was the largest grid-connected tidal stream system in 2010.

The maximum marine current speed tends to be near the surface, so the turbine rotor must intercept as much of the depth as possible, beginning from the

**FIGURE 13.9** A marine current tidal power-generating tower. (With permission from Marine Currents Turbine Ltd., Bristol, UK.)

near-surface area. The rotor can be secured in position several ways, similar to those used in offshore oil drilling and mining platforms.

The U.K. Department of Trade and Industry in the recent past has provided £50 million for marine research for harnessing wave and tidal current power. A 2.25-MW marine current power plant (the largest in 2009) was operating in Agucadoura, Portugal. Scotland is leading the way in this research, with the first such plant in operation on the island of Islay. The European Marine Energy Center in Orkney, Scotland, offers the development, testing, and monitoring of grid-connected ocean energy conversion systems.

The government of Scotland in 2010 announced the Saltire Prize competition for £10 million (US$16 million). The award is for demonstrating in Scottish waters a commercially viable wave or tidal current energy technology with a minimum output of 100 GWh over a 2-year period using only the ocean power. More than 100 potential competitors showed interest, 30 of them from the United States. Scotland is aiming for 50% of their electrical energy from renewable sources by 2050. The country has a potential to generate 25% of Europe's total tidal energy and 10% of wave energy.

The government of Taiwan is considering large-scale ocean current power generation using the strong Kuroshio current off the eastern coast of Taiwan. The project will start with a 5-MW marine turbine on a trial basis, with the goal of testing both the power generation efficiency and the related technologies. Based on the surveys done by National Taiwan University, the sea area of some 6,000 km$^2$ between the eastern county of Taitung and the outlying Green Island in the Pacific Ocean appears to meet all the requirements. The maximum potential capacity there exceeds 1500 billion kWh per year, compared to Taiwan's current annual demand of about 100 billion kWh. The government thinks it is possible that Taiwan's existing coal power plants could be retired, while the nuclear power generators could be used as a backup system, thereby resulting in a great reduction in Taiwan's total carbon dioxide emissions. A site 25 km$^2$ located in the shallow, high-current zone could support the deployment of a thousand 1-MW marine turbines to form an ocean energy park that will have output equal to one large nuclear power plant.

In the United States, a tidal energy project is being considered in New York City's East River. The first turbine that was installed in the waterways that connect the Long Island Sound with the city's harbor failed in the powerful currents, but the work continued. In 2008, Verdant Power LLC demonstrated grid-connected tidal energy turbines delivering energy to New York City and collected significant environmental data that showed no evidence of increased fish injury or mortality. With the U.S. Federal Energy Regulatory Commission (FERC) license, the Roosevelt Island Tidal Energy (RITE) Project, proposed for installation in the East Channel of the East River in New York City, would use the natural tidal currents of the East River to generate up to 1 MW power in 2012 by turbine generator units anchored on the riverbed. It will have an annual generation of 2.40 GWh after the completion of phase 3.

Since tidal currents vary in magnitude and direction over time, tidal stream time series data over at least 1 year are required to predict the energy potential of a tidal site. A tidal stream atlas is an early source of field data on tidal streams until actual flow meters are installed and data recorded over 1 year. The tidal power plant also requires consideration of operating the turbine in fixed or yawed orientation modes relative to the flow. Such operation will have an impact on the annual yield and the cost of energy production. Some proposed marine current turbines are designed for a fixed orientation to the flow but can invert the blades to operate the turbine in the reverse direction. However, some tidal power sites can have a swing of 20–30° away from 180° when the flow reverses. Use of vertical axis or yawing horizontal axis turbines at such sites would have no effect on the energy extracted by tidal generators.

The turbine design speed and orientation depend primarily on maximizing the energy capture and secondarily on the installation and maintenance costs. Often, designing a power plant at a lower current speed generating 15% to 20% less energy may be more profitable overall due to the lower costs of designing the structure for lower power and thrust. Nevertheless, full economic life-cycle costing is required to justify one design over another.

## QUESTIONS

*Question 13.1*: The marine currents at site A have 30% higher speed than that at site B. Compare the electrical energy capture potential at site A with that at site B.

*Question 13.2*: In a river current, which is a better location for a water turbine, near the surface, in the middle, or near the riverbed? Why?

*Question 13.3*: What is the advantage of the vertical axis Darrieus turbine over the traditional horizontal axis turbine?

*Question 13.4*: Compare the end-to-end electric propulsion of ships with the marine current power generation scheme.

## FURTHER READING

Bahaj, A.S., and L.E. Myers. 2003. Fundamentals applicable to the utilization of marine current turbines for energy production. *Renewable Energy,* 28(14), 2205–2211.

Batten, W.M.J., A.S. Bahaj, A.F. Molland, and J.R. Chaplin. 2006. Hydrodynamics of marine current turbines. *Renewable Energy,* 31(2), 249–256.

Batten, W.M.J., A.S. Bahaj, A.F. Molland, and J.R. Chaplin. 2007. Experimentally validated numerical method for the hydrodynamic design of horizontal axis tidal turbines. *Ocean Engineering,* 34(7), 1013–1026.

Blunden, L.S., and A.S. Bahaj. 2007. Tidal energy resource assessment for tidal stream generators. *Proceedings of the Institution of Mechanical Engineers, Part A: Journal of Power and Energy,* 221(2), 137–146.

Burton, T., D. Sharpe N. Jenkins, and E. Bossanyi. 2000. *Wind Energy Handbook.* New York: John Wiley & Sons.

Charlier, R.H., and C.W. Finkl. 2010. *Ocean Energy: Tide and Tidal Power.* Berlin: Springer.

# 14 Offshore Wind Power

Wind power is the cheapest source of new electrical power at present, cheaper than coal, nuclear, or photovoltaic power or any other source—conventional or renewable. That is why new wind power plants on land, off shore, and far shore are being built to add new megawatt generation capacity at an annual growth rate of 20–40% depending on the part of the world.

Germany plans to close its nuclear reactors by 2022 after the Fukushima disaster in Japan raised serious safety concerns. The government will promote energy from wind turbines and solar panels with the additional cost financed by higher energy bills. It plans to boost the renewable energy output from 17% in 2010 to 35% in 2020. Wind turbines contributed 6.2% of the total power in Germany in 2010, while nuclear power accounted for 23%. Germany had an onshore wind power generation capacity of 27.2 GW in 2010 and additional capacity on many large offshore wind farms.

Offshore wind farms fall in two broad categories, near shore (<6 km) and far shore (>6 km). Most existing European farms are less than 10 km off shore in water less than 10 m deep on average, but new larger ones are installed far from the shore (Figure 14.1). Public opinion favors farshore installations even though they cost more. In addition to being out of sight, the wind speed and the resulting energy yield improve with distance from the shore, as seen in Figure 14.2 for a turbine on a tower 80 m tall. With wind coming from the land, the energy yield can be 25–30% higher at 5 to 6 km away from the shore, whereas it is 12–15% lower 5 to 6 km inland. Similar gains are seen with wind coming from the sea or along the shore. Thus, the difference between 5 km offshore and 5 km inland can be 40–50% in energy yield. Ten kilometers from the shore, wind speed is typically 1 m/s (10–15%) higher, and that gives 40–50% higher energy yield.

Early offshore wind farms installed in Europe are listed in Table 14.1. The early projects were relatively small in shallow or sheltered waters. At present, more than 100 offshore wind farms are being planned worldwide, with concentration in northern Europe. In the 4-year period of 2004–2007, new offshore installations in Europe alone amounted to adding more than 4000 MW capacity. Three wind farms in the United Kingdom—one off the northwestern coast and two off the eastern coast—were installed by 2010. One of them has 38 turbines at 2 MW on towers that are 60 m high and 3 km off the coast of Great Yarmouth at a cost of US$110 million. Soon after that, a large 500-MW farm took shape off the Irish coast.

Wind farms placed in operation between 2004 and 2011 are listed in Table 14.2, whereas Table 14.3 lists the largest 9 offshore wind farms in construction for 2012 completion. Table 14.4 lists the 10 largest proposed wind farms in Europe, where Dogger Bank in the United Kingdom, with 9000-MW capacity, is equivalent to nine average-size thermal or nuclear power plants.

**FIGURE 14.1** Large offshore wind farm in Denmark. (With permission from Vestas Wind Systems, Denmark.)

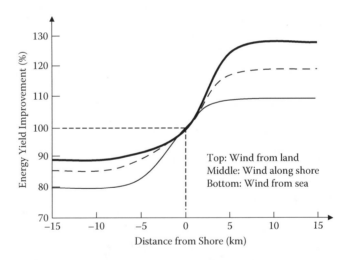

**FIGURE 14.2** Energy yield improvement with distance from shore for turbine at 80-m hub height. (Adapted from Bjerregaard, H. *Renewable Energy World*, James & James, London, March–April 2004, p. 102.)

## 14.1 WIND POWER THEORY

The theory of extracting wind power and converting it to electrical power is the same as that presented in Chapter 13, Section 13.1 for marine current, except that the density of air is 1.225 kg/m$^3$ versus 1027 kg/m$^3$ for seawater, and wind speed is 5 to 10 m/s versus 1 to 3 m/s for marine currents. The power output that can be extracted from wind is given by

## TABLE 14.1
## Earlier Offshore Wind Farms in Europe

| Location[a,b] | Country | Online | MW | No. | Tower Power Rating |
|---|---|---|---|---|---|
| Vindeby | Denmark | 1991 | 5.0 | 11 | Bonus 450 kW |
| Lely (Ijsselmeer) (close onshore) | Holland | 1994 | 2.0 | 4 | NedWind 500 kW |
| Tunø Knob | Denmark | 1995 | 5.0 | 10 | Vestas 500 kW |
| Irene Vorrink (close to shore) | Denmark | 1996 | 16.8 | 28 | Nordtank 600 kW |
| Dronten (Ijsselmeer) (close onshore) | Holland | 1996 | 11.4 | 19 | Nordtank 600 kW |
| Gotland (Bockstigen) | Sweden | 1997 | 2.75 | 5 | Wind World 550 kW |
| Blyth (offshore) | United Kingdom | 2000 | 3.8 | 2 | Vestas 2 MW |
| Middelgrunden, Copenhagen (cooperative) | Denmark | 2001 | 40 | 20 | 2 MW (mixed) |
| Uttgrunden, Kalmar Sound | Sweden | 2001 | 15 | 7 | Enron 1.5 MW |
| Horns Rev | Denmark | 2002 | 160 | 80 | Vestas 2 MW |
| Samso | Denmark | 2003 | 23 | 10 | Bonus 2.3 MW |
| Frederikshavn (close to shore) | Denmark | 2003 | 5.3 | 2 | Vestas 3 MW + Bonus 2.2 MW |
| Nysted, Rodsand | Sweden | 2003 | 158 | 72 | Bonus 2.2 MW |

[a] Minimum distance to shore 0 to 14 km with an average of 7 km.
[b] Water depth varies from 0 to 20 m with an average of 8 m.
*Source:* British Wind Energy Association.

$$P_o = \tfrac{1}{2}\rho\, V^3 A \times C_p \tag{14.1}$$

where $\rho$ = air density, $A$ = swept area of turbine blades, $V$ = wind speed, and $C_p$ = rotor efficiency, having a typical value of 0.40 and theoretical maximum of 0.59. We note that the electrical power output of the wind turbine varies with the wind speed cubed. The unrestricted wind speed in farshore ocean waters is higher by 15–30%. Therefore, the wind turbines installed on tall towers in ocean water can produce 1.5 to 2.2 times more electrical energy every year compared to an identical land-based turbine at the same height. Such high annual energy yield can more than offset the higher installation cost in ocean waters. For this reason, many large wind farms have been installed in European oceans, and more are under construction around the world.

The wind speed varies over the year with the Weibull or Rayleigh probability distribution function. Since the power is a function of the wind speed cubed, the

**TABLE 14.2**

**Wind Farms Placed in Operation between 2004 and 2011**

| Capacity (MW) | Country | Turbines × Model | Completed |
|---|---|---|---|
| 300 | U.K. | 100 × Vestas V90–3 MW | 2010 |
| 209 | Denmark | 91 × Siemens 2.3–93 | 2009 |
| 207 | Denmark | 90 × Siemens 2.3–93 | 2010 |
| 194 | U.K. | 54 × Siemens 3.6–107 | 2008 |
| 180 | U.K. | 60 × Vestas V90–3 MW | 2010 |
| 172 | U.K. | 48 × Siemens 3.6–107 | 2010 |
| 165 | Belgium | 55 × Vestas V90–3 MW | 2010 |
| 120 | Netherlands | 60 × Vestas V80–2 MW | 2008 |
| 110 | Sweden | 48 × Siemens 2.3 | 2007 |
| 108 | Netherlands | 36 × Vestas V90–3 MW | 2006 |
| 102 | China | 34 × Sinovel SL3000/90 | 2010 |
| 90 | U.K. | 30 × Vestas V90–3 MW | 2005 |
| 90 | U.K. | 30 × Vestas V90–3 MW | 2006 |
| 90 | U.K. | 25 × Siemens 3.6–107 | 2007 |
| 90 | U.K. | 25 × Siemens 3.6–107 | 2009 |
| 60 | U.K. | 30 × Vestas V80–2 MW | 2004 |
| 60 | Germany | 6 × REpower 5M, 6 × AREVA Wind–5M | 2009 |
| 48 | Germany | 21 × Siemens 2.3–93 | 2011 |
| 32 | China | 2 × 3 MW, 2 × 2.5 MW, 6 × 2 MW, 6 × 1.5 MW | 2010 |
| 30 | Finland | 10 × WinWinD 3 MW | 2008 |
| 30 | Belgium | 6 × REpower 5 MW | 2008 |

increase in the energy capture in high wind is much more than the decrease in low wind. The end result is that the annual energy capture in kilowatt-hours turns out to be twice that calculated using the average speed. Assuming $C_p = 0.40$, the yearly average power generation is given by

$$P_{annual.average} = 2 \times 0.4 \times \tfrac{1}{2}\rho V^3 A \times C_p \qquad (14.2)$$

**TABLE 14.3**

**Nine Largest Offshore Wind Farms Currently in Construction for 2012 Completion**

| Wind Farm | Capacity (MW) | Country | Turbines × Model |
|---|---|---|---|
| London Array (Phase I) | 630 | U.K. | 175 × Siemens 3.6–120 |
| Greater Gabbard | 504 | U.K. | 140 × Siemens 3.6–107 |
| Bard 1 | 400 | Germany | 80 × BARD 5.0 |
| Sheringham Shoal | 315 | U.K. | 88 × Siemens 3.6–107 |
| Lincs | 270 | U.K. | 75 × 3.6 MW |
| Walney Phase 2 | 184 | U.K. | 51 × Siemens 3.6 |
| Walney Phase 1 | 184 | U.K. | 51 × Siemens 3.6 |
| Ormonde | 150 | U.K. | 30 × REpower 5M |
| Tricase | 90 | Italy | 38 × 2.4 MW |

**TABLE 14.4**

**Ten Large Proposed Wind Farms**

| Wind Farm | Capacity (MW) | Country |
|---|---|---|
| Dogger Bank | 9,000 | U.K. |
| Norfolk Bank | 7,200 | U.K. |
| Irish Sea | 4,200 | U.K. |
| Hornsea | 4,000 | U.K. |
| Firth of Forth | 3,500 | U.K. |
| Bristol Channel | 1500 | U.K. |
| Moray Firth | 1300 | U.K. |
| Triton Knoll | 1200 | U.K. |
| Codling | 1100 | Ireland |
| He Dreiht | 600 | Germany |

As for the electrical generator, the conventional induction generator has been the most widely used machine in the past and continues to be. However, some newer installations may use one of two generators discussed next.

The doubly fed induction generator is the least-cost variable-speed generator within its operating range since it requires a frequency converter rated for only 20–30% of the stator power rating. On the other hand, the conventional induction generator requires the frequency converter rated for 100% of the stator power. The doubly fed induction generator, however, is not a viable option for a direct-drive wind turbine with the generator running at low speed.

The permanent magnet synchronous generator has higher power density and hence offers a lighter weight option on the tower. Even then, this machine is increasingly used in new wind farms at present with variable-speed drives. It can be designed for a direct drive at 30–100 rpm with a large number of poles on a large-diameter generator, eliminating the gearbox altogether. However, it may require a large magnetic core for the generator and for the step-up transformer due to low-frequency power generation. Another disadvantage is the lack of field control.

## 14.2   OFFSHORE ADVANTAGES

Offshore and farshore wind farms cost more to erect and operate due to difficult access and the continuous need for support from the shore. However, benefits from the high wind speed compared to that of land-based wind farms exceed the additional cost. The major advantages of the offshore wind power plants are as follows:

- They produce significantly greater energy depending on the distance from the shore.
- Tall towers can be used in water, which further improves the annual energy yield as the wind speed is greater at higher hub heights.
- They can be out of sight from costal property owners and beachgoers.
- Land-based wind turbines are restricted from running downwind because of the low-frequency noise caused by the tower shadow. Offshore systems do not have this restriction. This could lead to lighter weight rotors, less-costly yaw drives, and reduced turbine loads.
- Wind is less turbulent at sea and hence generates better quality power with fewer power pulsations, which also extend blade life.

The negative sides of being farther out in the sea are (a) deeper seabed, (b) higher construction and operating costs, and (c) higher grid connection cost, all of which partially offset the higher energy yield, but still leave a net positive advantage for offshore installations.

## 14.3   POWER TRANSMISSION TO SHORE

Wind turbines in a typical offshore wind farm are connected in the radial configuration shown in Figure 14.3. The number of turbines in one radial arm determines the feeder cable capacity. This configuration is less reliable as a damaged cable would disable all the turbines connected to it. A ring configuration may be preferred for reliability. In either configuration, the cable is buried 1 to 2 m deep in offshore ground to protect it from ship anchors and sea currents. This may not provide adequate protection near busy ports, where anchors of large ships can go 10 to 12 m deep into the ground. It is not practical to bury cables at such depths. Good reliability can be achieved only by having redundant cables on separate routes. Getting permission for

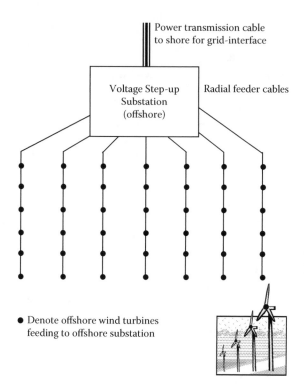

**FIGURE 14.3** Radial layout of the wind farm turbines.

one cable route is difficult enough, so most offshore wind farms accept the reliability achievable with one cable.

Offshore wind power is transmitted to the shore by cables running underwater. Most individual wind generators produce 690 V ac power, which is collected at one substation, as shown in Figure 14.3. Without an electrical substation in water next to the tower to step up voltage, the feeder cable voltage must be the same as the generator voltage. In large offshore wind farms, this would significantly increase the transmission loss. To eliminate such loss, a transformer is installed under each tower to step up the generator voltage to the substation level. The substation collecting power from farther turbines requires a high voltage (35 kV or even higher) to minimize the power loss.

Burying cables and pipes underwater requires underwater earthmoving equipment, many of which are made by Soil Machine Dynamics of the United Kingdom. Umbilical lines provide power up to 1 MW to pump seawater to create a 3-m deep fluidized trench behind the vehicle; the cable is buried while the vehicle moves between towers. For a rock seabed, a chain cutter is used to cut about a 1.5-m vertical slot below the vehicle.

The high-voltage offshore system not only improves energy efficiency but also adds high capital cost. Since the energy cost at the mouth of the wind farm is much lower than that at the customer's meter, an economic trade exists for using a low distribution voltage, which would be less efficient but also low in capital cost. The

160-MW Danish Hornes Rev wind farm uses 36 kV for the collecting substation offshore. It is then stepped up to 150 kV for feeding a 15-km ac cable to the shore. The offshore substation is built as a tripod structure. The 20 × 28 m steel building is placed 14 m above the mean sea level. The platform accommodates the 36/150-kV step-up transformer and the switchgear on sides, emergency diesel generator, instrumentation and controls, staff and service facilities, helipad, crane, and a boat. Larger wind farms (above 300-MW capacity) may require power transmission to shore at higher voltage in the 150- to 350-kV range. However, the cost of the transformer and switchgear increases with the voltage level. Such high-voltage offshore substations are new developments, although the oil and gas industry has been using 135-kV installations for a long time.

A 300-MW farshore wind farm has been built by the DEME Group of companies in Belgium even farther from shore—some 38.7 km away. The project is built by engineers in various marine disciplines with state-of-the-art material and construction techniques. The wind turbines are linked to each other by 33-kV cables, and a 150-kV high-voltage sea cable connects the wind farm to the existing grid on land. The cable weighs 81 kg per meter. The cable was first pulled through the pipe installed under the dunes. Then, it followed a 38.7-km route from the coastline to the wind farm. The cable layer pulled a large sea plow behind it, and the plow buried the high-voltage cable at a depth of about 2 m. That way the cable is protected from ships' anchors, fishing nets, and other threats. The high-voltage cable route had to overcome a number of obstacles, such as crossing the telecommunication lines and other power lines in the dredged trench.

Instead of using electrical cables, it is possible to produce hydrogen offshore using wind power and transmit onshore by pipelines. Such a proposal has been considered in Germany, where no tax is levied on hydrogen produced by offshore wind farms. However, initial studies indicated that it would be less expensive to transmit the electrical power to the shore than to use it for producing hydrogen and transmitting it to the shore.

## 14.4   AC VERSUS DC CABLE

The offshore wind farm requires an electrical link—ac or dc—to the shore grid. The ac link requires synchronous operation with the grid. A fault in the wind farm has a direct impact on the grid and vice versa. Fast voltage control is needed to maintain system stability under all possible faults in the grid and the wind farm. Most wind farms to date transmit power to shore using ac cables. However, the ac link in water has high capacitive VAR, which must be compensated by the offshore wind farm. Such VAR compensation adds cost, power loss, and complexity, particularly when the ac link is long in farshore installations. For large megawatt ratings and long transmission distances, ac has the following disadvantages:

- Power loss increases significantly.
- kVAR compensation is required at both ends of the cable by using a synchronous condenser or static power electronics kVAR compensators.

- Cable capacity may be limited. For example, the cross-linked polyethylene (XLPE) insulated cable is limited to 200 MW with one 150-kV three-phase cable. A 1000-MW wind farm would require five such cables in parallel, with one more for reliability.

The high-voltage dc (HVDC) cable can eliminate these disadvantages. Moreover, it significantly reduces the fault current contribution to the onshore network. HVDC cable in the ocean is an established technology. The first such 100-km, 20-MW, 100-kV dc submarine cable was installed between the Swedish island of Gotland and Swedish mainland in 1954. Since then, many HVDC links have been built around the world—in the United States, Canada, Japan, New Zealand, Brazil, and Paraguay—for power ratings up to 6300 MW and voltages up to ±600 kV.

The use of HVDC requires the ac-dc and then dc-ac conversions at two ends of the transmission cable. This can use conventional load-commutated or voltage source converter (VSC)-based power electronics technology. Either one adds significant capital cost, operating cost, and power loss. The capital cost can be several times that with conventional ac cable. It is estimated that the traditional HVDC transmission to the shore may not be economical for power ratings below 400–500 MW and distances shorter than 50 km. However, the new VSC technology offered by ABB (HVDC Lite™) and by Siemens (HVDC Plus™) are economical. Both are based on pulse width modulated (PWM) design using insulated gate bipolar transistors (IGBTs), as opposed to the line-commutated converter using thyristors. In the VSC, the current can be switched off without costly commutating circuits, and the active and reactive powers can be controlled independently. The VSC is bipolar—the dc circuit is not connected to ground—so it uses a two-conductor cable. The first new VSC link, a 70-km, 50-MW, ±80-kV dc link was built for voltage support for a large wind farm installed south of Gotland, Sweden. Many other VSC links are being built on shore and off shore around the world today.

The high-voltage dc link becomes economical over long distance with the following advantages:

- No limitation of the transmission distance
- No synchronous stability issue to manage
- More megawatt power transmitted per kilogram of conductor
- Fast control of power flow and direction
- Possible use of ground return, saving one conductor
- Absence of capacitive charging current (no VARs)
- Ability to limit short-circuit current using a current-limiting series reactor with zero reactance during steady-state dc operation but high reactance during fast-rising fault current

Two main architecture options for the HVDC link are: (a) load commutated converters, known as LCC, and (b) forced-commutated converters, typically using gate turnoff (GTO) thyristors or IGBTs, such as in the LightHVDC™ system.

As for cable insulation, XLPE can be used for ac as well as dc power. It gives a flexibility of converting the shore connection from ac to dc if and when the offshore wind farm gradually expands from a pilot phase to the full capacity. The present cost of 150-mm², 35-kV class cable for grid connection runs about $100 per meter length. The voltage drop and power loss would be about 0.15% per kilometer distance from the wind farm to the grid connection on land.

## 14.5  OFFSHORE FOUNDATIONS

The global average land elevation is 840 m, and the highest elevation is 8840 m at the top of Mount Everest. The global average ocean depth is 3800 m, and the maximum depth is 11,524 m in the Mindano Trench in the Pacific Ocean. Figure 14.4 is a cross section of the ocean floor, depicting the general terminology of the shore and the continental shelf. About 8% of the ocean floor is shallower than 1,000 m, and 5% is shallower than 200 m, where the majority of offshore oil and gas platforms are located. Most offshore wind farms in Europe have been installed in water less than 15 to 20 m in depth.

The tower foundation design depends on water depth, wave height, and seabed type. The foundation for an offshore turbine and its installation is a significant cost, about 20–30% of the turbine cost. The height and consequently the depth of installation are key factors in choosing the farm location. Turbines installed on land generally have a tower height equal to the rotor diameter to overcome the wind shear from ground obstacles. However, the offshore tower height can be 70% of the rotor diameter due to the low shear effect of the water.

The most favorable water depth for an offshore wind farm is 2 to 30 m. Depths of less than 2 m are not accessible by boat, and depths greater then 30 m make the foundation too expensive. In this range, three possible foundation designs (Figure 14.5)

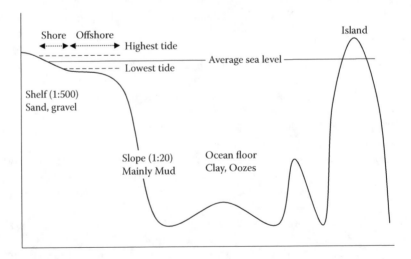

**FIGURE 14.4**  Ocean floor terminology: shore, offshore, and farshore.

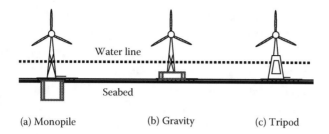

FIGURE 14.5 Tower foundation types for offshore wind turbines.

are (a) monopile for depths of 5 to 20 m, (b) gravity for depths of 2 to 10 m, and (c) tripod or jacket for depths of 15 to 30 m.

### 14.5.1 MONOPILE FOUNDATION

The monopile (Figure 14.5a) is the most common and cost-effective foundation in water 5 to 20 m deep. It is driven in the seabed and consists of the cylindrical steel pile being driven into the seabed up to a depth of 1.1 times the water depth depending on the seabed conditions. It does not need seabed preparation, and erosion is not a problem. However, boulders and some layers of bedrock may require drilling or blasting, which would increase the installation cost. Monopile construction is limited to water depths of 20 m due to stability considerations. A monopile foundation including installation for a 1.5-MW turbine in water 15 m deep would cost about $600,000 in 2010.

### 14.5.2 GRAVITY FOUNDATION

The gravity foundation (Figure 14.5b) is made of concrete and steel that sits on the seabed. It is often used in bridges and low-depth turbine installations. It is less expensive in shallow water (2 to 10 m deep) but is extremely expensive in water deeper than 15 m and hence is not used there.

### 14.5.3 TRIPOD FOUNDATION

The tripod foundation (Figure 14.5c) is suitable for water 15 to 30 m deep. It utilizes a lightweight, three-legged steel jacket to support the foundation. The jacket is anchored to the seafloor by piles driven into each leg. Each pile is driven up to 20–25 m in the seafloor depending on seabed conditions. Boulders in the pile area are the only concern, and they may be blasted or drilled if necessary. The tripod foundation cannot be used in depths less than 10 m due to the legs possibly interfering with ships. However, in water deeper than 10 m, the tripod is very effective, as tested in oil rig foundations. Preparation of the seafloor is not necessary, and erosion is not a concern with this installation. For a 1.5-MW turbine utilizing a tripod foundation at a water depth of 15 m, the Danish Wind Industry Association in 2005 estimated a foundation cost of about $400,000, including the installation.

### 14.5.4 OTHER FOUNDATIONS

Other foundations under development for the offshore towers include (a) lightweight foundations guyed for stability, (b) floating foundations with the turbine, tower, and foundations in one piece, and (c) bucket-type foundation.

The bucket-type foundation (Figure 14.6) has been successfully designed and used to support 3-MW turbines. The 135-ton structure has the shape of an upturned bucket sucked into the seabed through a vacuum process. Early indications are that such a design makes the fabrication, handling, installation, and subsequent removal easier and less expensive.

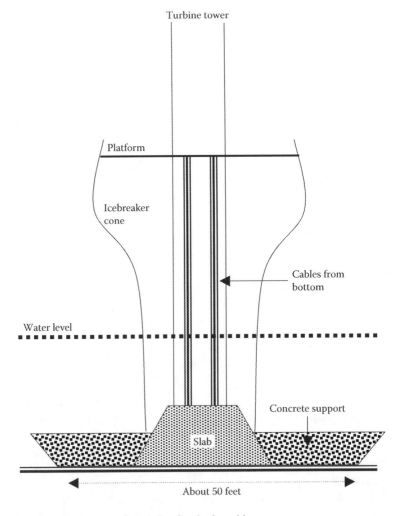

**FIGURE 14.6**   Bucket-type foundation for single turbine.

Perhaps it is more economical—in terms of manpower, equipment, and time—to install the turbine and tower as a single unit (Figure 14.7). The foundation is placed first, then the turbine, tower, and rotor are lifted and attached to the foundation in that sequence. A jack uplift barge is the most likely ship to be used for installing the foundations as it can carry the foundations and drive the piles.

### 14.5.5 Other Foundation Considerations

The foundation cost, however, varies with the seabed and water conditions. Approximate costs of foundation installation have been compiled by the Danish Wind Industry Association. It compared the foundation cost, including installation, against the water depth up to 25 m average. This comparison did not include the cost of installing the turbine tower, nacelle, and rotor. Typical foundation cost at present is about $50 per kilowatt on land and $300 per kilowatt off shore in water 10 m deep, increasing by roughly 2% per meter of additional water depth as shown in Figure 14.8.

The difficulties of remoteness from the onshore facilities and crews to construct and maintain offshore installations are being addressed by developing the offshore and farshore wind turbine sites with integrated assembly cranes, helicopter pads, and emergency shelters capable to sustain crews for several days of harsh weather offshore.

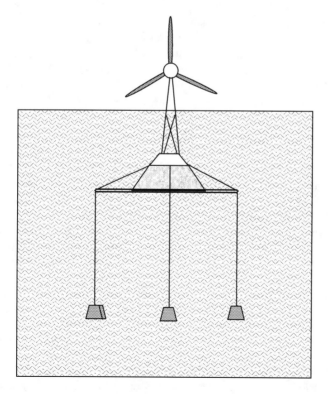

**FIGURE 14.7** Vertically moored foundation with single turbine.

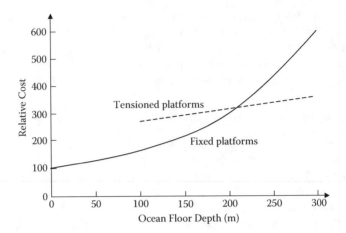

**FIGURE 14.8**   Relative cost of offshore structures versus water depth.

## 14.6   FARSHORE FOUNDATIONS

The current offshore and farshore wind tower installations have been limited to water depths of 30 m. This technology may gradually extend to deeper water farther from the shore, but eventually floating wind turbines may be the most economical way for harnessing wind energy in the coastal waters beyond the view of densely populated urban load centers. Worldwide, the deepwater wind resources have been estimated to be extremely abundant, with U.S. potential ranking second only to China.

About 30% of the world's oil and gas comes from offshore drilling rigs that can go up to 2.5 km deep from the floating platform. The equipment and the crew of 15 to 20 are served by support boats (cheetahs) and helicopters. The power required during installation often comes from the shore via a cable running parallel to the oil pipeline taking oil to the shore. The deployment economics of current offshore oil rigs, however, may not be applicable to the floating wind turbine platforms. For deepwater wind turbines, the floating structure may replace the conventional pile-driven monopiles or concrete bases commonly used to support shallow-water and land-based wind turbines. The floating structure must provide enough buoyancy to support the turbine weight and to limit pitch, roll, and heave motions.

The deepwater farshore wind turbine economics primarily depends on the additional cost of the floating structure and the power cable to the shore. The added cost is offset by higher farshore wind speed and close proximity to a large load center on land near the shore. In addition, the following differences may result in some reduction of high cost:

- Extremely high safety margin on oil platforms for stability, spill prevention, and safety of permanent personnel may be somewhat lax on unmanned wind platforms.
- Wind platform water depths could be typically less than 750 ft (150 m), whereas oil platforms are deployed in depths from 1500 ft (450 m) to 8000 ft (2400 m).

- Wind platforms can be mass produced to bring the economy of scale.
- Submerging wind platforms minimizes the structure exposed to wave loading (oil platforms must maximize above-water deck/payload area).

Some farshore structures that are being investigated by National Renewable Energy Lab (NREL) and others are (a) single turbine on a single tower, (b) multiple smaller turbines on a single tower and platform, and (c) multiple turbines on a common pontoon-type floating platform (Figure 14.9) to share the anchor cost and provide wave stability.

Because turbine spacing may be poorly optimized in both multiple-turbine concepts, the floating structure is required to either yaw with wind direction changes or compromise energy production when the wind shifts off the prevailing direction. A single-turbine floating structure is a more viable option in some cases.

## 14.7  INSTALLATIONS AND MAINTENANCE

Generally, the offshore wind farm costs 50–100% more to install compared to that on land. Even on land, the investment cost breakdown is highly site specific. The cost varies considerably with the location, particularly for remote sites where the cost of grid connection may be high. The cost breakdown of wind farms installed on land and offshore is summarized in Table 14.5.

The offshore project cost varies over a wide range between €1000 to €2000 per kilowatt, and the energy production cost varies between €0.04 and €0.06 per

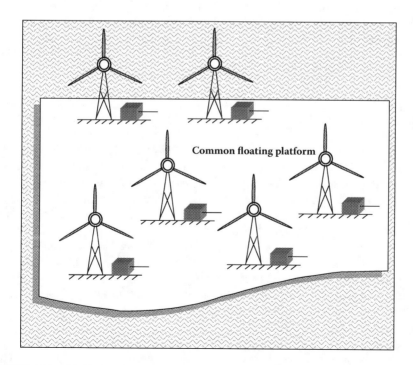

**FIGURE 14.9**   Multiple turbines floating on common platform.

**TABLE 14.5**

**Typical Cost Breakdown for Large Land-Based and Offshore Wind Farms**

| Cost Element | Cost on Land (% of Total) | Cost Offshore (% of Total) |
| --- | --- | --- |
| Wind turbines | 70 | 50 |
| Foundations | 6 | 16 |
| Internal connections | 7 | 5 |
| Grid connection | 7 | 18 |
| Engineering | 3 | 4 |
| All others | 7 | 7 |

kilowatt-hour. The primary cost parameters are the distance from the shore, water depth, and wind speed. The foundation and interconnect costs can exceed the wind turbine cost. The operation and maintenance costs are also difficult to predict with high confidence. Thus, possibilities of large cost variations pose an investment risk and make project financing difficult unless the risk can be adequately insured. However, the offshore wind farm industry is relatively new for both the financing companies and the insurance companies. Therefore, potential investors need adequate and reliable data to determine how best to maximize the profitability and minimize the investment risk. The data collection includes stationary data towers pile-driven into the seabed at the proposed site.

Maintenance of the wind turbine and foundation is important for longevity for keeping the energy cost down. Regular scheduled inspections are necessary, particularly because offshore towers do not receive incidental periodic visits. No federal regulations exist for inspecting the equipment in offshore wind farms. The Naval Facilities Engineering Command has outlined procedures based on good engineering practices for inspecting offshore structures. Although not mandatory, implementation would reduce the maintenance cost in the long run. The period of inspections must be planned in accordance with the weather conditions and the wave height.

Wind farms at present are built to operate for about 20 years with minimal maintenance. Visual inspections of the turbine and tower should be conducted periodically during the operating life of the farm. Visual inspections are done to search for large physical signs of deterioration, damage, and the like. The procedures for inspecting the tower follow those laid out for inspecting the underwater foundation.

Inspections of the foundation and the tower at the splash line are important as this part of the structure is exposed to corrosion due to galvanic action. A level 1 inspection is conducted annually for the underwater structures. A level 2 inspection is conducted biannually for the underwater structures, and a level 3 inspection occurs every 6 years. The procedures and equipment used to carry out these inspections are listed in the underwater inspection criteria published by the Naval Facilities Engineering Service Center.

Generally, visual inspections allow detection and documentation of most forms of deterioration of steel structures. Some types of corrosion may not be detected by visual inspections. For example, inside a steel pipe piling, anaerobic bacterial corrosion caused by sulfate-reducing bacteria is especially difficult to detect by visual inspection. Fatigue distress can be recognized by a series of small hairline fractures perpendicular to the line of stress, but these are difficult to locate by visual inspection. A cathodic protection system must be closely monitored both visually and electrically for wear of anodes, disconnected wires, damaged anode suspension system, or low voltage. A nondestructive test plan may be developed and tailored to any specific area of concern at a particular wind farm site.

Serving the offshore wind farm in rough sea conditions may be difficult and costly. Boats with water jets are one viable option often used due to their good maneuverability and station-keeping characteristics.

## 14.8   FORCES ON STRUCTURE

Offshore platforms, along with undersea pipelines for transportation to shore, have been built around the world since the early 1950s to drill for oil and gas. Offshore mining is also being developed now, and offshore wind farms are the latest addition to such structures installed to provide means of exploiting ocean energy resources and transporting them to the shore.

The offshore structure must withstand mechanical forces exerted by the ocean waves, currents, wind, storms, and ice. The wave force is the most dominant. The structure must absorb, reflect, and dissipate the wave energy without degradation in performance over a long life.

The weight and cost of a fixed platform that will withstand the forces resulting from the waves, currents, and wind increase exponentially with the water depth. The offshore cable or pipelines carrying the cable must withstand forces due to the inertia, drag, lift, and friction between the floor and the pipe. The water drag and lift forces for an underwater structure in somewhat streamlined water current can be derived from classical hydrodynamic considerations. They depend on the water velocity near the floor, which follows the 1/7 power law as stated in Chapter 13, Equation (13.13).

## 14.9   MATERIALS AND CORROSION

The ocean shore and shelf material generally consists of sand and gravel coming from land via rivers and blown in by wind. The ocean water density of seawater at atmospheric pressure and 10°C is 1027 kg/m$^3$. It contains 35 g of salt per kilogram of seawater, which is expressed as the salinity of 35 ppt (parts per thousand). The salinity is measured by measuring the electrical conductivity of seawater since the two are related. The seawater salt composition in percentage is shown in Figure 14.10. It is 55% chlorine, 31% sodium, and 14% all others.

Offshore equipment is necessarily made to withstand the marine environment. The equipment has marine-grade material, components, seals, and coatings to protect from the corrosive environment. The corrosion is caused by the electrochemical reaction in which a metal anode corrodes by oxidation and a cathode undergoes a reduction reaction. The seawater works as an electrolyte for the transfer of ions and electrons between the two electrodes. The corrosion rate must be accounted for in the design. Table 14.6 lists the corrosion rates of commonly used metals in seawater.

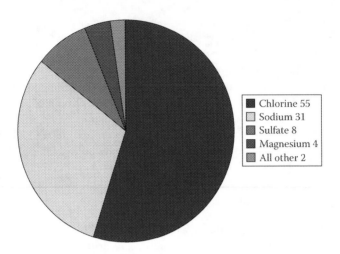

**FIGURE 14.10**    Salt composition of ocean water.

**TABLE 14.6**
**Corrosion Rate of Various Materials in Seawater**

| Material | Average Corrosion Rate (mm/year) |
| --- | --- |
| Titanium | None |
| Stainless steel, Nichrome | <2.5 |
| Nickel and nickel-copper alloys | <25 |
| Copper | 10–75 |
| Aluminum alloys | 25–50 |
| Cast iron | 25–75 |
| Carbon steel | 100–175 |

Two types of cathodic protection are widely used for corrosion protection of materials submerged in seawater. The impressed current system gives more permanent protection but requires electrical power. The galvanic protection employs aluminum, magnesium, or zinc anodes attached to the steel structure in seawater. Under the cathodic protection principle, a metal receiving electrons becomes a cathode, which can no longer corrode. The zinc anode is most widely used as a sacrificial material to protect steel hulls of ships. When depleted, it is replaced for continued protection. Zinc provides about 1000 Ah charge transfer capacity per kilogram. The sacrificial anode design follows the basic electrical circuit principles.

Table 14.7 lists materials commonly used in seawater structures along with their mechanical strengths and relative costs. The marine environment is harsh enough to cause rapid and sometimes unanticipated degradations of the equipment. For example, all 80 of the 2-MW turbines at Denmark's Horns Rev—one of the largest wind farms built in 2002—were moved back to the shore within a couple of years after installation. This was to repair or modify the generators and transformers at a considerable expense

**TABLE 14.7**
**Mechanical Strength of Structural Materials Commonly Used in Seawater**

| Material | Yield Strength (ksi) | Ultimate Strength (ksi) | Elongation (%) | Relative Cost |
|---|---|---|---|---|
| Mild steels | 30–50 | 50–80 | 15–20 | 1 |
| Medium-strength steels | 80–180 | 100–200 | 15–20 | 3–5 |
| High-strength steel | 200–250 | 250–300 | 12.15 | 10–15 |
| Stainless steels 302 and 316 | 35–45 | 80–90 | 55–60 | 5–10 |
| Aluminum alloys | 30–70 | 45–75 | 10–15 | 7–8 |
| Titanium alloys | 100–200 | 125–225 | 5–15 | 20 |

of removal and reinstallation of the turbine. It was believed that the equipment design and material would not have withstood the harsh marine environment for long. Similar problems have been encountered in other offshore farms.

Since land-based wind turbines are highly visible on tall towers, they need to be pleasing in appearance for public acceptance. The paint coatings must last over the design life of 20 to 25 years. A typical coating system on the land-based wind turbine is a three-coat system. It consists of 50–80 μm epoxy-zinc-rich primer, 100–150 μm epoxy midcoat, and 50–80 μm polyurethane topcoat. Protection for an offshore wind turbine in a highly corrosive marine environment is a different matter. It is defined in the EN-ISO-12944 standard, which is necessarily different for parts under the water and for the tower parts above water.

## 14.10   OFFSHORE WIND POWER TRENDS

The largest offshore wind farm in 2002 was the 160-MW Horns Rev installed in Danish waters at a total cost of €268 million. It was built in the harsh environment of the North Sea, has eighty 2-MW Vestas turbines on 62-m towers. The mean wind speed is 9.7 m/s. The site is 17 km offshore, where the water depth is 6.5 to 22.5 m and the seabed is firm sand and gravel. The wind farm layout is rectangular, in rows 560 m apart. The electrical energy yield is estimate to exceed 600 GWh/year.

Since then, many countries in Europe, Asia, and the Americas have been building large wind farms and evaluating new proposals every year. Winergy of Long Island has applied for 12-GW capacity off the U.S. eastern coast, and Australia's Tasmanian coast is being seriously explored for large wind farms. In 2009, the world's largest offshore wind farm was launched into full power generation by DONG Energy, a partially state-owned utility company in Northern Europe, which undertook the project in 2007. The wind farm, Horns Rev 2—which followed Horns Rev—has 91 wind turbines and one transformer platform installed 30 km to 35-km off the western coast of Jutland Denmark in the North Sea for 209-MW capacity costing about $1 billion. The electricity produced by the project could provide enough electricity for 200,000 Danish households. DONG Energy estimated that its offshore capacity will be expanded to 1200 MW by 2020. MAKE Constantans estimated that the European offshore wind farms will grow at a 45% annual rate in the coming years.

Germany's Borkum West has the planning permission for construction atleast 12-km away from the shore at a water depth of about 30 m. It will initially have twelve 5-MW turbines in a pilot plant 45 km north of Borkum Island in the North Sea, connected to the onshore transformer station via 100-kV ac cable over a 115-km distance. After 3 years of pilot operation, further developments may proceed to full 1000-MW capacity at a cost of €1.5 billion.

The U.S. eastern coast, thousands of miles long, is more promising than many European countries with coastlines in hundreds of miles. Among the main attractions along the U.S. northeastern coast are strong, steady winds; shallow waters; low wave height; and a growing regional market for the renewable energy. The southern New England shore is viewed as a potential Saudi Arabia of wind energy. It is the most heavily populated urban region with heavy transmission constraints. Power produced elsewhere cannot be easily transmitted for use in the cities. Although there are good wind regimes along the coast, both in and out of water, the wind speed dramatically drops off in just 30 to 60 km inland. For these reasons, developers are considering large wind farms offshore of the northeastern United States.

A recent report by the U.S. Department of Interior estimated a 900,000-MW potential in the nation's outer continental shelf (OCS), close to the total installed capacity in the United States. Since then, interest in offshore wind power has grown rapidly in the Northeast. As the United States aims to reduce its carbon footprint, the U.S. Department of Energy and the American Wind Energy Association (AWEA) estimated that a goal of 20% wind energy generation in America by the year 2030 is required to reduce $CO_2$ emissions from the nation's electric sector by 25%, or about 825 million metric tons. The AWEA estimated that approximately 85,000 people were employed by the wind energy industry in 2008, amounting to an increase of 143% from 2007 levels. A target of 20% wind energy by 2030 could potentially support more than 800,000 jobs in areas related to wind energy.

The offshore wind projects are generally large, often costing $100 to $300 million. For reducing the cost per megawatt capacity, wind turbines built for offshore applications are larger than those used onshore. The 2-MW turbine of Vesta and 3.6-MW turbine of GE are just two examples in wide use today. Even larger turbines in the 5- to 7-MW range are being developed for offshore installations, such as a 7-MW turbine developed by Vestas Wind Systems of Denmark. At present, a 300-MW onshore wind farm would probably use 150 to 200 wind turbines, while an offshore wind farm of the same capacity would probably use fewer than 100 turbines. The project construction cost decreases with increasing turbine size because fewer barges, ships, and cranes are required per megawatt installed.

Along the U.S. northeastern coast, a proposed 420-MW project, Cape Wind, is located 8 km off Hyannis village in a 40-km² shallow area in Nantucket Sound. It consists of one hundred thirty 3.6-MW turbines placed 0.5 km apart and connected to the New England power grid by two 130-kV submarine cables to an electrical substation on Cape Cod. Although this project holds great wind power potential, it has met with both legislative and environmental opposition.

A study funded by New York State found that about 5200-MW capacity could be installed in a 800-km² band 5 to 10 km off the south shore of Long Island. About 2500-MW capacity could be installed even by placing the wind turbines in a smaller

400-km² band where the water depth is typically 15 m or less. The Long Island Power Authority invited proposals from interested developers and environmental groups for initial 100-MW capacity. The cost estimates were $150 to $180 million for 100-MW wind turbines and $40 to $70 million for the grid-interconnecting substation. The energy cost estimate was $0.06 to $0.09 per kilowatt-hour, about one-half of what the local consumers are paying. The initial 100-MW capacity would use thirty-five 3-MW turbines 5 to 10 km offshore. The rotor would be 50 m in diameter on an 80-m hub above the water surface, with the rotor tip reaching out 130 m above the surface of the water. Jones Beach has been mentioned as a possible location, along with Montauk. Further detailed studies are under way for specific site locations. The permitting process, due to multiple oversight entities, may take a long time.

## 14.11   ENVIRONMENTAL REGULATIONS

Since installations in water for energy farming are still emerging, many regulations have yet to be streamlined by various government agencies. Under the U.S. National Environmental Policy Act, an environmental impact statement for the entire project must be conducted and submitted for approval before a private interest is allowed to erect any structure on the OCS lands. The environment study must include the Migratory Bird Treaty Act, the Marine Mammal Protection Act, the Endangered Species Act, and the National Environmental Policy Act. It must also include the initial data-gathering structure and the work to be carried out offshore (e.g., localized disturbance of the seabed and possible pollution). Analytical simulation programs developed by British Maritime Technology to evaluate the environment impact of offshore installations and their waste, such as drill cuttings, lubricants, and other chemicals, may be of help in the United States as well.

It has not been investigated yet whether many large wind farms extracting more than a third of the wind energy from the region could affect the local weather and ocean wave patterns. As for the shipping traffic, individual turbines could be spaced such that small vessels could sail through the array. However, the proposed site may interfere with established routes of large ships or pleasure vessels. The wind turbines and the underground cable could interfere with fishing areas or even migrating/mating whale habitats.

The environmental issues involved with the placement of offshore wind farms include the impact on the avian population and underwater ecosystems. This is especially a factor when known birds of endangered and threatened species reside or migrate through the area. Avian collisions with the turbines resulting in death are a major concern along with the disruption of feeding, nesting, and migrating habits. Although avian deaths by the blades had been seen on a large scale in early wind farms at the Altamont Pass location, it has not been experienced offshore.

## 14.12   INTERNATIONAL SAFETY REGULATION

According to the European Wind Energy Association (EWEA), 9.3 GW of new wind power capacity was installed in the European Union during 2010, with offshore wind power installations growing 51% from 582 MW in 2009. With further projects in the pipeline from the United Kingdom, Germany, the Netherlands, and most recently France, Europe may continue to be the largest offshore wind energy-producing region in the world for years to come.

Until 2010, all offshore wind farms constructed have been relatively small and positioned relatively close to the shore, with transfer time of about 1 hour. Future offshore wind farms will result in wind farms positioned significantly farther in the sea and with more wind turbines that are larger. That will result in significant operational challenges for the transfer vessels. Transporting a wind turbine installation crew farther offshore will require a change in the current regulations. More important, it will also be required that the vessels are designed to operate comfortably and safely in the rougher sea conditions likely to be encountered and longer transit times in the farshore wind farms. The new regulations will address these issues to minimize seasickness among the personnel and to provide safe transfer.

Wave height is the main safety concern with turbine access. The higher wave height poses more danger. Most access systems at present can operate at up to 1.5-m wave height, but installations in 2- to 3-m wave height may soon be required with acceptable safety regulations in rougher seas. This may require sophisticated motion control and dynamic positioning.

Although offshore wind power developers are establishing key strategies early in the project to meet the new challenges of farshore installations, a wider collaboration will be needed to mitigate the future safety risks of farshore wind project installation and operation. The Global Wind Organization (GWO) is a collaboration between some of the main European wind developers and the utilities that is taking the lead in establishing the basic safety standards and regulations as a part of training for offshore wind power operations.

## QUESTIONS

*Question 14.1*: What is the energy density (kW/m$^2$) in 10-m/s wind? What is the maximum percentage that can be extracted to drive the electrical generator?

*Question 14.2*: If site B has 20% higher wind speed compared to site A, how much more energy would site B yield per year?

*Question 14.3*: Why does the offshore wind tower yield higher wind energy capture per year?

*Question 14.4*: What is the conversion efficiency range of wind to electrical power that can be practically achieved?

*Question 14.5*: Discuss the ac versus dc transmission of power from a farshore wind farm to the shore, including the high-voltage dc link.

*Question 14.6*: Identify alternative offshore foundation schemes.

*Question 14.7*: List the reasons why offshore and farshore wind farms are gaining momentum around the world.

*Question 14.8*: Discuss the positive and negative environmental impacts of wind energy farms (a) on land, (b) offshore, and (c) farshore.

*Question 14.9*: If you have worked on an offshore construction site, share your work experience with others in the class.

## FURTHER READING

Ackerman, T. 2000. Transmission systems for offshore wind farms. *IEEE Power Engineering Review*, December, pp. 23–28.

Cockburn, C.L., S. Stevens, and E. Dudson. 2011. *Accessing the Farshore Wind Farm*. London: Royal Institution of Naval Architects.

Danish Wind Industry Association: Frederiksberg, DK-1970. 2003. *Wind Turbine Offshore Foundations*.

Halfpenny, A. 2000. *Dynamic Analysis of Both On- and Offshore Wind-Turbines in the Frequency Domain*. Ph.D. thesis, University College London.

Henderson, A.R., and M.H. Patel. 2003. On the modeling of a floating offshore wind turbine. *Wind Energy Journal, 6*, 53–86.

Henderson, A.R., R. Leutz, and T. Fujii. 2002. Potential for floating offshore wind energy in Japanese waters. *Proceedings of the 12th International Offshore and Polar Engineering Conference,* May 26–31. ISBN 1-880653-58-3; ISSN 1098-6189.

Krohn, S. 2002. *Offshore Wind Energy: Full Speed Ahead*. Danish Wind Industry Association Report. Frederiksberg, DK-1970.

Musial, W., S. Butterfield, and A. Boone. 2003. *Feasibility of Floating Platforms Systems for Wind Turbine*. National Renewable Energy Laboratory, Golden, CO. Report No. NREL/CP-500-34874.

Patel, M. R. 2006. *Wind and Solar Power Systems*. Boca Raton, FL: CRC Press/Taylor & Francis.

Randall, R.E. 1997. *Elements of Ocean Engineering*. Jersey City, NJ: Society of Naval Architects and Marine Engineers.

# Part D

## System Integration Aspects

The topics covered in Part D of this book—energy storage and reliability—relate to the aspects required for the overall integration of an electrical power system. For example, energy storage is required in all practical power system designs to provide (a) peak power beyond the generator capability, (b) power for critical and essential loads until the emergency generator is brought online after the main power failure, (c) starting power to bring the system up from cold, (d) compressed air for many pneumatic equipments, and so on. Large-scale energy storage can eliminate the emergency or redundant generator altogether, changing the overall system architecture.

Another aspect of system integration is the overall system reliability. The system—electrical or mechanical—is made of numerous piece-parts and subassemblies (components), more so in power electronics components using numerous semiconducting devices (diodes, thyristors, transistors, etc.). Numerous components connected in series make the system less reliable. The classical statement in reliability engineering is that *the system is less reliable than its weakest link*. The system engineer designs the system with reliability that is required by the contractual specification or that matches or exceeds the level that is acceptable by industry norms.

# 15 Large-Scale Energy Storage

In any practical power system of high power rating, some form of large-scale energy storage is required for various purposes. On ships, this requirement is traditionally met by the electrochemical battery and compressed air. Regardless of how the energy is stored, the process of first storing and then using the stored energy will result in some losses both ways. The goodness of the energy storage device is best measured in terms of the round-trip energy efficiency, which can be taken as the figure of merit in comparing various alternative energy storage technologies. It is defined as

$$Round\,trip\,energy\,efficiency = \frac{energy\,delivered\,to\,loads\,during\,discharge}{energy\,injected\,to\,restore\,initial\,charge\,level} \quad (15.1)$$

Five alternatives for large-scale energy storage are discussed next: electrochemical battery, supercapacitor, rotating flywheel, superconducting coil, and compressed air.

## 15.1 ELECTROCHEMICAL BATTERY

Among the basic energy storage devices, electrochemical batteries are the most mature and hence widely used technology. The energy stored in the battery that is made of N cells connected in series—each with cell voltage of $V_{cell}$ and ampere-hour capacity of $Ah_{cell}$—is given by

$$Battery\ energy\ storage = N \times V_{cell} \times Ah_{cell} \ \ watt\text{-}hours \quad (15.2)$$

There are many types of batteries available today, each having different electrochemistry. The selection of the optimum electrochemistry for a given application depends on the desired performance and cost parameters. Table 15.1 lists relative performance of various types of batteries presently available. The design method and working details for various batteries can be obtained from a companion book, *Shipboard Electrical Power Systems*, or from other specialty books on batteries.

The battery is made of numerous elementary cells, each of small capacity, connected in series-parallel combination to obtain the desired voltage and ampere-hour ratings. It is generally used for small- and medium-scale energy storage. The battery, being highly modular in design (i.e., built from numerous cells), has no fundamental technological size limitation. Until recently, submarines were diesel-electric, by which the diesel engine uses the air drawn from the surface through a snorkel to generate electrical power that is stored in large batteries. The battery power is then used to propel the ship during submerged operations. The battery can be designed

**TABLE 15.1**

**Life and Cost Comparison of Various Batteries**

| Electrochemistry | Cycle Life in Full Discharge Cycles | Calendar Life in Years | Self Discharge per Month at 25°C (%) | Relative Cost ($/kWh) |
|---|---|---|---|---|
| Lead-acid | 500–1000 | 5–8 | 3–5 | 200–300 |
| Nickel-cadmium | 1000–2000 | 10–15 | 20–30 | 1500 |
| Nickel-metal hydride | 1000–2000 | 8–10 | 20–30 | 400–600 |
| Lithium-ion | 1500–2000 | 8–10 | 5–10 | 500–800 |
| Lithium-polymer | 1000–1500 | n/a | 1–2 | >2000 |

by stacking numerous cells together for large-scale energy storage. For example, the largest 40-MW peak power battery in the world was built and commissioned in 2003 at a cost of $30 million. The system uses 14,000 sealed NiCd cells manufactured from recycled cadmium by Saft Corporation at a total cell cost of $10 million. The cells will be recycled again in 2023 after the 20-year calendar life of the system. The battery system is operated by the Golden Valley Electric Association in Fairbanks for an Alaskan utility company. The spinning energy reserve of the battery provides continuous voltage support to the utility and lowers the blackout possibility.

The demand for energy storage is large enough that other technologies with specific niches are developed to compete with the battery with better performance at a lower cost. The supercapacitor and rotating flywheel are discussed next, among others.

## 15.2 SUPERCAPACITOR

The energy stored in the capacitor is $\frac{1}{2}CV^2$ joules, where $V$ = voltage, and $C$ = capacitance, which is given by

$$C = \frac{plate\,area \times permittivity\,of\,insulation\,between\,plates}{distance\,between\,positive\,and\,negative\,plates} \quad (15.3)$$

This indicates that the capacitor can store more energy if the plate area is increased and the distance between the positive and negative plate is reduced without the voltage breakdown of the insulation in-between. The supercapacitor—also known as an *ultracapacitor* and an *electrical double-layer capacitor*—stores energy like any other capacitor, but its energy storage density in terms of watt-hours per kilogram can be as high as 100 times that for the traditional electrolyte capacitor. As such, the supercapacitors are not competing with other capacitors but with the battery in industrial and automotive applications. For that reason, they are made in cells in the voltage range matching with that of battery cells.

In the supercapacitor, the electrodes are made of highly porous activated carbon to provide a much larger surface area (Figure 15.1), and the double layer formed by the ionic separation results in a very small distance between the positive and negative charges, both factors contributing to very high capacitance. Supercapacitors up to 3000 farads are available in low-voltage ratings of typically 2.5 V. They can be connected in series-parallel combinations as required for the desired voltage and current ratings of the system.

Table 15.2 compares the performance of the supercapacitor with the lithium-ion battery. It shows that the supercapacitor is not volumetrically efficient and is more expensive than the battery, but it has many other advantages over the battery. For comparison, the watt-hour per kilogram density of the supercapacitor is around 5 to 8 Wh/kg, versus 100–130 Wh/kg for the lithium-ion battery. But, the supercapacitor can deliver much higher power density in watts per kilogram, and its cycle life in terms of charge-discharge cycles is much higher (in millions) versus cycles in the thousands for the battery. The lithium-ion battery can deliver around 250 W/kg of power, whereas the supercapacitor can deliver 10,000 W/kg of power without

**FIGURE 15.1**   High surface area and close distance of graphene in supercapacitor results in high capacitance.

## TABLE 15.2
## Performance Comparison between Supercapacitor and Lithium-Ion Battery

| Function | Supercapacitor[a] | Lithium-Ion Battery |
|---|---|---|
| Charge time | 1–10 seconds | 10–60 minutes |
| Cycle life | 1 million or 30,000h | 500 and higher |
| Cell voltage | 2.3 to 2.75 V | 3.6 to 3.7V |
| Specific energy (Wh/kg) | 5 (typical) | 100–200 |
| Specific power (W/kg) | Up to 10,000 | 1000 to 3000 |
| Cost per Wh | $20 (typical) | $2 (typical) |
| Service life (in vehicle) | 10 to 15 years | 5 to 10 years |
| Charge temperature | −40 to 65°C (−40 to 149°F) | 0 to 45°C (32° to 113°F) |
| Discharge temperature | −40 to 65°C (−40 to 149°F) | −20 to 60°C (−4 to 140°F) |

[a]   Adapted from Maxwell Technologies Incorporated.

performance degradation. Moreover, the supercapacitor does not degrade performance in cold weather. These considerations make the supercapacitor the preferred choice in applications requiring a large amount of energy storage to be delivered in high-power bursts repetitively.

The most significant advantage of the supercapacitor over the battery is its ability to be charged and discharged quickly and repetitively without performance degradation as in the battery. For this reason, the battery and supercapacitor are often used in conjunction with each other. The supercapacitor supplies the high-energy bursts since it can be charged and discharged quickly, whereas the battery supplies the bulk of slow energy since it can store and deliver a larger amount of energy slowly over a longer period.

The supercapacitor is important for the next generation of fuel cell vehicles to provide fast transient power during acceleration. It can also absorb energy during regenerative braking, whereas the fuel cell is nonregenerative in operation. For its much longer cycle life compared to the battery, the supercapacitor may also replace the lithium-ion, NiMH, and lead-acid batteries currently used in the hybrid electric vehicles.

## 15.3   ROTATING FLYWHEEL

The flywheel stores kinetic energy in rotating inertia. The kinetic energy can be converted to electrical energy with high efficiency. Flywheel energy storage is an old concept, which has now become commercially viable due to advances made in high-strength, lightweight fiber composite rotors and the magnetic bearings that operate at high speeds up to 100,000 rpm. The flywheel energy storage systems have been developed for a variety of potential applications and are expected to make more inroads in the near future. Some commercially available uninterruptible power supplies now use the flywheel instead of the battery for energy storage. The round-trip energy conversion efficiency of a large flywheel system can be around 90%, which is much higher than around 70% with the battery.

The flywheel can replace the battery in some applications, with the following benefits:

- Higher round-trip energy efficiency
- Higher energy density, hence lower mass and volume
- Deeper depth of discharge, up to 90%
- Longer cycle life that is insensitive to the depth of discharge
- Insensitivity to temperature, hence lower thermal system mass and cost
- Easy charge/discharge control as the stored energy depends on the wheel speed alone
- No capacity or voltage degradations over its lifetime
- Much lower trickle charge requirement
- Flexibility in design with electrical machine for desired voltage and current
- Improved quality of power as the electrical machine is stiffer than the battery
- Higher peak power capability without overheating concerns
- Easy power management, as the state of charge is simply measured by the speed
- Environmental friendliness compared to the battery, which has toxic elements in construction

These benefits may make the flywheel the least-cost energy storage alternative per kilowatt-hour delivered over the operating life. The following potential applications of flywheel energy storage in navy ships are being explored at present.

- Electrical start of gas turbines, which presently use compressed air pneumatic starts. Each start may need up to 2.5 kWh, so a 25-kWh flywheel system can provide 10 consecutive starts.
- Power continuity while reconfiguring the electrical system of the ship. Reconfigurable ship technology is aggressively developed for new Navy ships to meet the changing needs of the ship over its service life.
- Launch power in the burst mode for the weapons onboard.

Considerable development efforts are under way around the world for high-speed flywheels to store large amounts of energy. The present goal of these developments is to achieve five times the energy density of the currently available rechargeable batteries. This goal is achievable with the following enabling technologies, which are already in place in their component forms: (a) high-strength fibers having ultimate tensile strength of over 1 million pounds per square inch, (b) advances made in designing and manufacturing fiber-epoxy composites, and (c) high-speed magnetic bearings, which eliminate friction, vibration, and noise.

The flywheel system is made of a fiber-epoxy composite rotor, supported on magnetic bearings, rotating in a vacuum, and mechanically coupled with an electrical machine that can work as the motor or the generator. Two counterrotating wheels are placed side by side where gyroscopic effects must be eliminated, such as on a ship, in a city transit bus, on a train, or in an automobile.

The energy stored in a flywheel having the mass moment of inertia $J$ rotating at an angular speed $\omega$ is $(1/2)J\omega^2$. The centrifugal force in the rotor material of density $\rho$ at radius $r$ is given by $\rho \times (r\omega)^2$, which is supported by the hoop stress in the rotor rim. Since the linear velocity $V = 2\pi \times (r\omega)$, the maximum centrifugal stress in the rotor is proportional to the square of the outer tip velocity. The allowable stress in the material imposes an upper limit on the rotor tip speed. Therefore, a smaller rotor can run at a higher speed and vice versa. The thin-rim-type rotor has a high ratio of inertia to weight and stores more energy per kilogram. For this reason, the rotor in all practical flywheel system designs is in the thin-rim configuration. For such a rotor with inner radius $R_i$ and outer radius $R_o$, it can be shown that the maximum energy that can be stored for an allowable rotor tip velocity $V_{tip}$ is as follows, where $K_1$ is the proportionality constant:

$$Maximum\,energy\,storage\,E_{max} = K_1 \times V_{tip}^2 \left[ 1 + \left( \frac{R_i}{R_o} \right)^2 \right] \qquad (15.4)$$

A figure of merit for the flywheel design is its specific energy, defined as its energy density in watt-hours per kilogram. Since the $R_i/R_o$ ratio is always less than unity, Equation (15.4) indicates that the thin-rim flywheel with $R_i/R_o$ ratio approaching unity results in a high specific energy for a given allowable stress limit. Higher ultimate strength of the material results in higher specific energy. Lower material density results in a lower centrifugal stress, which leads to a higher allowable speed and specific energy. The maximum energy storage therefore can also be expressed as

$$E_{max} = K_2 \frac{\sigma_{max.design}}{\rho} \qquad (15.5)$$

where $K_2$ = a proportionality constant, $\sigma_{max.design}$ = maximum allowable hoop stress in the design, and $\rho$ = density of the rotor material.

A good flywheel design, therefore, has a high $\sigma_{max.design}/\rho$ ratio to achieve high specific energy. It also has a high $E_{young}/\rho$ ratio for rigidity, where $E_{young}$ = Young's modulus of elasticity.

The metallic flywheel has a low specific energy because of a low $\sigma_{max}/\rho$ ratio, whereas high-strength polymer fibers—such as graphite, silica, and boron, having much higher $\sigma_{max.design}/\rho$ ratio—can store an order of magnitude higher energy per unit weight. Figure 15.2 compares the specific energy of various metallic and polymer fiber composite rotors and the battery. The actual realizable watt-hours per kilogram of the flywheel system would be about 50% of those shown in the figure when other components are added, such as the electrical machine, bearings, enclosure, and so on. In addition to a high specific energy, the composite rotor has a safe mode of failure as it disintegrates to a fluff rather than fragments as in the metal flywheel.

Figure 15.3 shows the mass and specific energy versus $\sigma_{ult}/\rho$ ratio of the material for a prototype 10-kWh flywheel, where $\sigma_{ult}$ = ultimate breakdown tensile strength

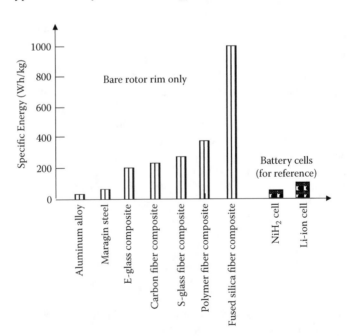

**FIGURE 15.2** Theoretical maximum specific energy of rim flywheel of various materials.

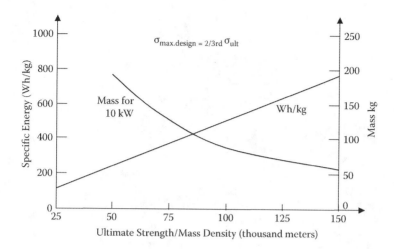

**FIGURE 15.3**  Specific energy and mass of 10-kW flywheel versus $\sigma_{ult}/\rho$ ratio.

of the rotor fiber composite. It is noteworthy that the weight decreases inversely and the specific energy increases linearly with the $\sigma_{ult}/\rho$ ratio. In this prototype design, the maximum allowable design stress $\sigma_{max.design}$ was kept at two-thirds of the ultimate breakdown tensile strength $\sigma_{ult}$.

Figure 15.4 shows a rotor design that was developed at the Oakridge National Laboratory. The fiber-epoxy composite rotor is made of two rings. The outer ring is made of high-strength graphite and the inner ring of a low-cost glass fiber. The hub is made of a single piece of aluminum in the radial spoke form. Such construction is cost effective because it uses the costly material only where it is needed for strength—in the outer ring—where the centrifugal force is high, resulting in high hoop stress.

The complete flywheel energy storage system requires the following components:

- High-speed rotor attached to the shaft via a strong hub
- Bearings with good lubrication system or with magnetic suspension in high-speed rotors
- One machine that works as a motor during charging and as a generator during discharging the energy
- Power electronics to drive the motor and to condition the generator power
- Control electronics for controlling the magnetic bearings and other functions

Conventional bearings are used for speeds up to a few tens of thousands rpm. Speeds approaching 100,000 rpm are possible only by using magnetic bearings, which support the rotor by magnetic repulsion and attraction. The mechanical contact is eliminated, thus eliminating the friction. Running the rotor in a vacuum eliminates the aerodynamic drag (windage).

The magnetic bearing comes in a variety of configurations using permanent magnets and dynamic current actuators to achieve the required restraints. A rigid

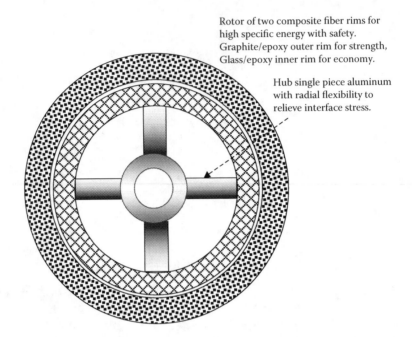

Rotor of two composite fiber rims for
high specific energy with safety.
Graphite/epoxy outer rim for strength,
Glass/epoxy inner rim for economy.

Hub single piece aluminum
with radial flexibility to
relieve interface stress.

**FIGURE 15.4**    Two-rim flywheel design for low cost and high energy density. (From Oakridge National Laboratory.)

body—like a rotor—can have six degrees of freedom. The bearings retain the rotor in five degrees of freedom, leaving one free for rotation. The servo control coils provide active control to maintain the shaft stability by providing restoring forces as needed to maintain the shaft in the centered position. Various position and velocity sensors are used in an active feedback loop. The electric current variation in the actuator coils compels the shaft to remain centered in position with desired clearances. Small flux pulsation as the rotor rotates around the discrete actuator coils produces a small electromagnetic loss in the metallic parts. This loss, however, is negligible compared to the friction loss in the conventional bearings.

Figure 15.5 shows the flywheel developed by NASA as an alternative energy storage candidate for the International Space Station instead of the battery. It forms volume-efficient packaging. High-speed magnetic bearings were used to achieve high watt-hours per kilogram. The electromechanical energy conversion in both directions is achieved with one electrical machine, which works as the motor for spinning up the rotor for energy charge and as the generator while decelerating the rotor for a discharge. Two types of electrical machines can be used with the flywheel: the synchronous machine with variable frequency converter or the permanent magnet brushless dc machine.

The machine voltage varies over a wide range with speed. The power electronic converters provide an interface between the widely varying machine voltage and the fixed bus voltage. It is possible to design a discharge converter and a charge converter with input voltage varying over a range of 1 to 3. This allows the machine speed to vary over the same range. That is, the low rotor speed can be one-third of the full speed. Since the energy storage is proportional to the speed squared, the flywheel

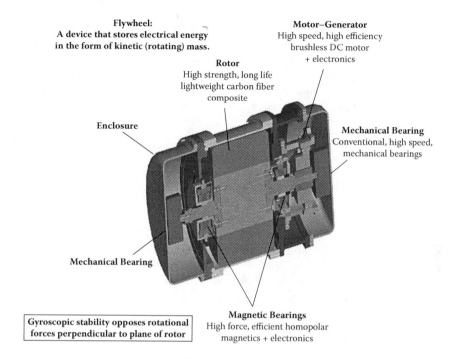

**Flywheel:**
A device that stores electrical energy
in the form of kinetic (rotating) mass.

**Motor–Generator**
High speed, high efficiency
brushless DC motor
+ electronics

**Rotor**
High strength, long life
lightweight carbon fiber
composite

**Enclosure**

**Mechanical Bearing**
Conventional, high speed,
mechanical bearings

**Mechanical Bearing**

Gyroscopic stability opposes rotational
forces perpendicular to plane of rotor

**Magnetic Bearings**
High force, efficient homopolar
magnetics + electronics

**FIGURE 15.5** NASA flywheel module in volume-efficient packaging for space applications. (From Gieh, E.B., E. W. Gholdson, B. Manners, and R. A. Delventhal Electrical power systems of the International Space Station. NASA Report 210209, NASA Glenn Reserch Center, Cleveland, OH.)

state of charge at low speed can be as low as 0.10. That means 90% of the flywheel energy can be discharged with no hardship on the power electronics or other components of the system.

As for the number of charge–discharge cycles the flywheel can withstand, the fatigue life of the composite rotor is the limiting factor. Experience indicates that the polymer fiber composites in general have a longer fatigue life than solid metals. A properly designed flywheel therefore can last much longer than a battery and can discharge to a much deeper level. Flywheels made of composite rotors have been fabricated and tested to demonstrate more than 10,000 cycles of full charge and discharge cycles. This is an order of magnitude more than any electrochemical battery can deliver at present.

## 15.4  SUPERCONDUCTING COIL

The energy stored in the magnetic field of a coil is given by

$$Coil\,energy\,storage = \frac{1}{2}\frac{B^2}{\mu_o}\ joules/m^3 = \frac{1}{2}LI^2\,joules \qquad (15.6)$$

where $B$ = magnetic field density in the inner air of the coil (Wb/m$^2$), $\mu_o$ = magnetic permeability of air = $4\pi\ 10^{-7}$ (H/m), $L$ = inductance of the coil (H), and $I$ = coil current (A). The coil carrying current produces the required magnetic field. The superconducting coil can carry current 100–150 times greater than the conventional copper coil, giving energy storage density greater by several orders of magnitudes in practical designs.

Old superconductors worked at liquid helium temperature. The new superconductors operate at a relatively much higher (liquid nitrogen) temperature and hence are called high-temperature superconductors. Since their discovery, they have opened up many new commercially viable applications due to their much lower cryogenic cooling requirement.

The theory and operation of the superconducting energy storage was discussed in Chapter 9; Figure 9.5 described the main components in a typical large-scale superconducting energy storage system. In that figure, the superconducting coil is charged by an ac-to-dc converter in the coil power supply. Once fully charged, the converter continues providing small voltage needed to overcome losses in the room temperature parts of the circuit components. This keeps a constant dc current flowing in the superconducting coil. In the storage mode, the current is circulated through a normally shorted coil. The coil is opened to discharge its stored energy into a load. The number of times the coil can be charged and discharged is limitless in theory, whereas the battery can be charged and discharged no more than a few thousand times, after which it must be replaced.

## 15.5 COMPRESSED AIR

Compressed air is common on ships for pneumatic equipment. On a large scale, it can store excess energy of a power plant (thermal, nuclear, wind, or photovoltaic) and supply energy when needed during lean periods or peak demands. The compressed air energy storage in an electrical power system consists of an (a) air compressor, (b) expansion turbine, (c) electric motor-generator, and (d) overhead storage tank or an underground cavern.

The compressed air stores energy in a pressure-volume relation. If $P$ and $V$ represent the air pressure and volume, respectively, and if the air compression from pressure $P_1$ to $P_2$ follows the gas law $PV^n = $ constant, then the work required during this compression is the energy stored in the compressed air, which is given by

$$Energy\, stored = \frac{n(P_2 V_2 - P_1 V_1)}{n-1} \tag{15.7}$$

The temperature at the end of the compression is given by the following:

$$\frac{T_2}{T_1} = \left(\frac{P_2}{P_1}\right)^{\frac{n-1}{n}} \tag{15.8}$$

The energy stored is less with a smaller value of $n$. The isentropic value of $n$ for air is 1.4. Under normal working conditions, $n$ is about 1.3. When air at an elevated temperature after constant-volume pressurization cools, a part of the pressure is lost with a corresponding decrease in the stored energy.

The stored energy can be used to drive pneumatic equipment. On a large scale, it can also be used to generate electrical power by venting the compressed air through an expansion turbine that drives a generator. One million cubic feet of air stored initially at 600 psi provide an energy storage capacity enough to deliver about 0.25 million kWh of electrical energy to the loads. The compressed air system may work under a constant-volume or a constant-pressure configuration.

In the constant-volume compression, the compressed air is stored in pressure tanks, mine caverns, depleted oil or gas fields, or abandoned mines. Such a system, however, has a disadvantage. The air pressure reduces as the compressed air is depleted from the storage, and the electrical power output decreases with the decreasing pressure.

In the constant-pressure compression, the air storage may be in an aboveground variable-volume tank or an underground aquifer. A variable-volume tank maintains a constant pressure by a weight on the tank cover. If an aquifer is used, the pressure remains approximately constant while the storage volume increases because of water displacement in the surrounding rock formation. During electrical generation, the water displacement of the compressed air causes a decrease of only a few percent in the storage pressure, keeping the electrical generation rate essentially constant.

The operating energy cost would include cooling the compressed air to dissipate the heat of compression. Otherwise, the air temperature may be significantly high to shrink the storage capacity and adversely affect the rock wall of the mine. Energy is also lost to the cooling effect of expansion when the energy is released.

The energy storage efficiency of the compressed air storage system is a function of a series of component efficiencies, such as the compressor efficiency, motor-generator efficiency, heat losses, and compressed air leakage. The overall round-trip energy efficiency of about 50% can be achieved with practical designs.

At a large scale suitable for a power grid application, the air is compressed during off-peak hours and used to satisfy peak demand during the day. Since a large volume is required in such a large system, abandoned mines are typically used as the air receiver. Low-pressure, high-volume air compression can be used in some areas. A low-pressure air system would have a low leakage rate, but the electromechanical energy conversion efficiency is another matter.

The natural gas industry has developed much of the technology that may be applied to a variety of compressed air energy storage systems. It stores massive volumes of compressed natural gas in naturally occurring caverns in the earth's bedrock that have been flushed of rock salt. Some of these caverns may measure up to 1 mile in diameter by 6 miles vertical height. Depending on the depth of the cavern dome below the ground surface, the caverns may hold pressure anywhere between 1000 and 3000 psi.

Compressed air storage is one option that may be applied to short-distance marine services or ship propulsion near ports for environmental reasons. It has successfully been used to power an early generation of mining locomotives and is being developed for a large-scale energy storage applications. Some of the large cylindrical-shaped natural gas storage tanks can hold over 750-bar (10,000-psi) pressure. Many large marine vessels can accommodate such large-volume tanks. Compressed air-powered ships may operate to and from terminals located near a large compressed air energy storage systems. A compressed air-powered ship may carry a complement of combustible fuel that would be used to superheat the compressed air immediately upstream of the engine inlet. During operation, compressed air would transfer from a high-pressure tank to a lower-pressure running tank that would remain at constant pressure for the duration of each voyage. Heat from the onboard thermal energy storage system would preheat the air as it flows from the constant-pressure tank.

Some positive-displacement engines may be able to directly drive the propeller, while some continuous-flow engines may drive an electrical generator to provide propulsion power to Azipods. The size of the ship, volume of the onboard storage tanks, and the available onboard thermal energy would determine the operating range of a ship propelled by compressed air.

The technology needed to develop a marine compressed air propulsion system already exists and is at work in some applications. It may be suitable for ferry services and tugboats that operate short-distance voyages.

## QUESTIONS

*Question 15.1*: Identify five energy storage technologies and their suitability in various applications.

*Question 15.2*: What is the best figure of merit for comparing various energy storage alternatives?

*Question 15.3*: What makes the superconducting coil the most dense energy storage system in terms of the watt-hours per kilogram and watt-hours per liter?

*Question 15.4*: List five advantages of storing energy in a flywheel versus an electrochemistry battery.

*Question 15.5*: Much pneumatic equipment is used on ships. Indicate which equipment can be replaced with an equivalent electrical product and identify the potential advantage or disadvantage.

## FURTHER READING

Barnes, F.S. 2100. *Large Energy Storage Systems.* Boca Raton, FL: CRC Press/Taylor & Francis.

Nazri, G.A., and Pistoia, G. 2009. *Lithium Batteries Science and Technology.* Dordrecht, the Netherlands: Kluwer.

Patel, M.R. 2006. *Wind and Solar Power Systems.* Boca Raton, FL: CRC Press/Taylor & Francis.

Reddy, T.B., and D. Linden. 2010. *Linden's Handbook of Batteries.* New York: McGraw-Hill.

Patel, M.R. 2011. *Shipboard Electrical Power Systems.* Boca Raton, FL: CRC Press/Taylor & Francis.

# 16 System Reliability Fundamentals

The economic success of any system depends on its availability for the intended use. The overall availability has two components: reliability and maintainability. System reliability is defined as the probability of numerous components working together to deliver the specified performance over the mission duration. Failure of any one component in a chain of many components renders the assembly useless. The failure rate of a given component is derived from tests on a large number of identical components under actual operating conditions. Sometimes in practice actual failure rate data may be available for only a few components to make any meaningful statistical inferences with high confidence. This may make the reliability estimate difficult, sometimes debatable or even questionable. However, the estimate is extremely useful, at least on a relative basis, in identifying and correcting weak links in the chain of reliability. The perspective of system reliability developed during the reliability analysis is perhaps more important than the final reliability number.

## 16.1  FAILURE MECHANISMS

The probability of failure of a component comes from four sources:

*Random failure*, which is just a matter of chance.

*Wear-out failure,* which is well understood and accounted for in the design. Components such as bearings, solenoid valves, frequently operating relays, and battery electrodes wear out in a predictable way.

*Design failure,* which occurs if operating stress exceeds the design limit under an anticipated or unanticipated disturbance. It is avoided by allowing adequate design margin.

*Manufacturing failure,* which depends on the consistency and sensitivity of manufacturing processes and the quality assurance procedure in place. Two manufacturing problems that can have an impact on the reliability of electrical power components are

- Nicked wire that may occur during removing the enamel insulation from wire strands. This degrades the wire both electrically and mechanically and may lead to a short or open circuit.
- Protruding screw in a component, puncturing the surrounding wire insulation and shorting to another part. It may also cause dielectric stress concentration, leading to corona or even arcing under a voltage spike.

## 16.2    AGING OF POWER ELECTRONICS DEVICES

Power electronics devices have a much longer life than electrical machines, for which insulation aging under heat and vibration limits the operating life. All power electronics devices have two power terminals (conducting channel) and a control terminal (triggering channel). The control signal applied at the control terminal determines whether current can flow between the power terminals. In metal-oxide semiconducting field effect transistor (MOSFET) and insulated gate bipolar transistor (IGBT) devices, although a layer of dielectric material electrically insulates the control terminals, the electric field applied across them alters the conductivity between the power terminals.

The power electronics devices do not fail in a conventional way that suddenly ends performance. They often reach the end of life when the intended performance is not fully met, although they continue operating with degraded performance. Power electronics aging mechanisms are as follows:

- Over time, charge carriers—electrons for the n-channel and holes for p-channel—stray out of the conducting channel and become trapped in the insulating channel at the control terminal. This gradually builds up sufficient charge in the insulated control channel, eventually increasing the signal strength needed to turn on the device. As a result, the device switching time gradually increases, eventually to a point that it does not perform fully as initially designed.
- The conducting joints between the device and the outside load circuit degrade due to gradual migration of the joint material from one terminal to the other, eventually leading to degraded performance.
- The control signal at the gate may sometimes create an electrically active defect in the gate dielectric. If the number of such defects exceeds a certain limit, these defects make a chain and cause a short circuit between the gate and the conductive channel, leading to an outright failure, as opposed to a gradual degradation of performance.

## 16.3    FAILURE RATE IN TIME

The component reliability estimate is based on its *failure rate in time* (FIT), also known as the *hazard rate*, which is quoted in number of failures in a million hours. The failure rate of any component varies with age and follows the well-known bathtub shape shown in Figure 16.1a. The initial high failure rate—the infant mortality rate—is caused by manufacturing defects. The rate decreases rapidly with time in use, particularly in electronic components. Therefore, newly manufactured components are placed in operation for some time in a process called burn-in to weed out infant failures. For power electronics components, most early failures occur in the first 12 hours of burn-in operation.

A low constant rate of failures for a long time in service makes the flat and stable part of the bathtub curve. In this region, the failures are random in nature and are treated probabilistically in engineering designs. The rising failure rate near the

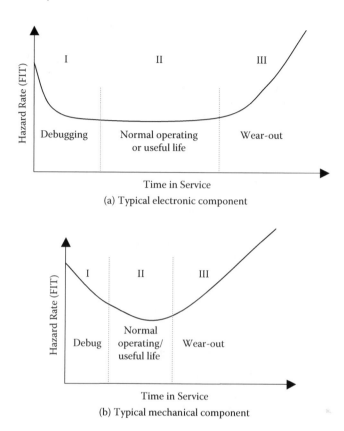

**FIGURE 16.1**   Component hazard rate versus time in service (reliability bathtub).

end of life is due to wear-out mechanisms that are generally well understood and accounted for by the design engineer.

The flat part of the curve is considered to be the useful life of the component. Therefore, only the random failure rate is used in developing the reliability estimate over the useful life. Electronic components with no moving parts have a relatively longer flat part of the tub. On the other hand, mechanical components with moving and physically wearing parts have a relatively shorter flat part, as shown in Figure 16.1b.

## 16.4   RANDOM FAILURES

If identical components of a large population are put to test simultaneously, the number of random failures in a given time duration is constant regardless of the beginning or end of life. In other words, under random (chance) failure, if the component has not failed at a given time, it is as good as new. However, under the accumulative nature of chance failures, the probability of the component still working decreases exponentially with lapse of time. The reliability is therefore an exponential function of time, known as the Poisson distribution function, expressed as

$$R(t) = e^{-\lambda t} \tag{16.1}$$

where $\lambda$ = failure rate, that is, the number of failures in unit time, also known as FIT and hazard rate $h(t)$. The probability of failure in time $t$, that is, unreliability $U$, is given by

$$U(t) = 1 - R(t) = 1 - e^{-\lambda t} \quad and \quad \frac{dU}{dt} = \lambda e^{-\lambda t} = f(t) \tag{16.2}$$

The probability of a component failing during a time interval $\Delta t$ at any time $t$ is $\Delta U = f(t) \times \Delta t$. For this reason, $f(t)$ is known as the *failure probability density function*. For a constant hazard rate $\lambda$ shown in Figure 16.2a, $f(t)$ is an exponentially decaying function as shown in Figure 16.2b. The component unreliability is the area under the $f(t)$ curve from time 0 to $t$, and the reliability is the shaded area from time $t$ to $\infty$.

The expected *mean time between failures* (MTBF) in a large population of a given component is

$$\text{MTBF} = \frac{1}{\lambda} \tag{16.3}$$

That is, if numerous identical parts were put to test under identical conditions, then the MTBF would be $1/\lambda$, always the same because $\lambda$ is constant. The probability of a component working at any time $t$ in terms of the MTBF is therefore

$$R(t) = e^{-\lambda t} = e^{\frac{-t}{MTBF}} \tag{16.4}$$

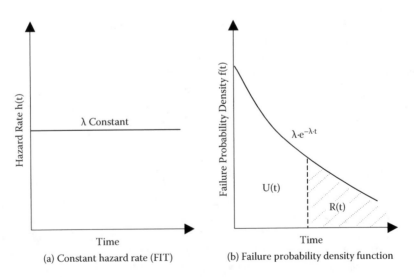

(a) Constant hazard rate (FIT)  (b) Failure probability density function

**FIGURE 16.2**   Constant hazard rate and failure probability density function versus time in service.

And the probability of a component not working at any time $t$ is

$$U(t) = 1 - R(t) = 1 - e^{-\lambda t} = 1 - e^{-t/MTBF} \tag{16.5}$$

## 16.5 FUNDAMENTAL THEOREMS OF RELIABILITY

The reliability of any complex system having components in series, parallel, or any combination thereof is determined by the following two fundamental theorems of reliability.

1. If $A$ and $B$ are two independent events, with their individual probability of occurrence $P(A)$ and $P(B)$, respectively, then

   Probability of both $A$ and $B$ occurring $= P(A) \times P(B)$ $\qquad$ (16.6)

2. When there are $n$ identical components each with reliability $R$, then the probability of *exactly* $k$ components working at any time $t$ is given by the binomial theorem,

$$P_k = \frac{n!}{(n-k)!k!} R^{n-k}(1-R)^k \tag{16.7}$$

Using the first theorem, the total reliability of a component under all failure modes is the product of the reliability under each mode separately. For example, the overall reliability under the four failure modes identified in Section 16.1 is

$$R_o = R_r \times R_w \times R_d \times R_m \tag{16.8}$$

where the four reliabilities on the right-hand side account for the random, wear-out, design, and manufacturing failures, respectively. The component is designed with an adequate design margin. The reliability under wear-out mode is made almost equal to unity by making the mean wear-out failure time much longer than the design lifetime. Eliminating potential failures using consistent manufacturing and quality assurance procedures leaves only the random failures to account for in the total reliability estimate and so is done in practice. The following section deals with only random failures.

## 16.6 SERIES-PARALLEL RELIABILITY

Two or more components are said to be working in series in the reliability sense if all components have to work for the whole assembly to work successfully. On the other hand, two or more components are said to be working in parallel in the reliability sense if only one component has to work for the assembly to work successfully. The series-parallel working of various components for reliability of an assembly is not to be confused with the series–parallel connection in electrical circuits. Electrically parallel components can be in series in the reliability sense and vice versa.

If two components having individual reliability $R_1$ and $R_2$ are working in series as shown in Figure 16.3a, then using Equation (16.6), reliability of the assembly is given by

$$R = R_1 \times R_2 \tag{16.9}$$

If two components having individual reliability $R_1$ and $R_2$ were working in parallel as shown in Figure 16.3b, then the assembly would not work only if both were not working. That is, $U = U_1 \times U_2$, and $R = 1 - U$, or

$$R = 1 - U_1 \times U_2 \tag{16.10}$$

For an assembly such as a battery, fuel cell, and solar array made of numerous identical series cells having equal reliability, Figure 16.4 depicts the assembly reliability versus number of series components. The rule of 72 is noteworthy here, which is as follows:

Reliability of assembly reduces to one-half when (Number of series component × Percentage unreliability of each component) = 72

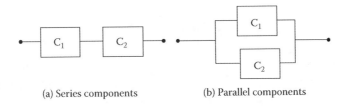

(a) Series components        (b) Parallel components

**FIGURE 16.3**    Assembly of two series and parallel components.

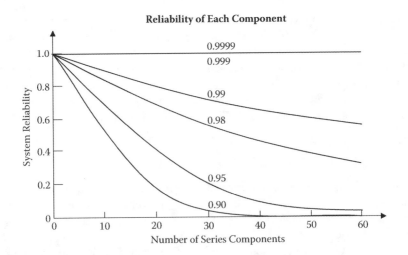

**FIGURE 16.4**    System reliability versus number of identical series components with given reliability of each component (Rule of 72).

**TABLE 16.1**

**Diminishing Rate of Return in System Reliability Using Parallel Units with Each Unit 80% Reliable**

| Number of Component | Overall System Reliability, $R_n = 1 - R_1{}^n$ | Incremental System Reliability | Improvement in Parallel Reliability over Single Component (%) |
|---|---|---|---|
| 1 | 0.800000 | Reference | Reference |
| 2 | 0.960000 | 0.160000 | 20.00 |
| 3 | 0.992000 | 0.032000 | 24.00 |
| 4 | 0.998400 | 0.006400 | 24.80 |
| 5 | 0.999680 | 0.001280 | 24.96 |
| 6 | 0.999936 | 0.000256 | 24.99 |

The assembly reliability increases with the number of parallel components, but with rapidly diminishing rate of return. This is seen in the last column of Table 16.1, which gives the percentage improvement in reliability by adding parallel components starting with a single unit. The reliability improvement over a single unit is given by $(R_n - R_1) R_1$, where $R_n$ is the assembly reliability with $n$ components in parallel, each having reliability of $R_1$. Using Equation (16.9), we have $R_n = 1 - R_1{}^n$. If each component has 80% reliability, then two components in parallel would improve the assembly reliability by 20%, boosting the assembly reliability to 96%. Three components in parallel would boost it to 99.2%, six components to 99.99%; it would take infinite many components in parallel to improve the assembly reliability to 100%. This clearly shows that using two components significantly improves the reliability, and placing more than three components in parallel is almost fruitless.

An instance of the series-parallel reliability principles presented is as follows:

Practical systems are made with some components in series and some in parallel. We calculate below the reliability of the overall system shown in Figure 16.5 to illustrate the general procedure. Dividing the system in subassemblies A, B, C, and D, we determine their unreliabilities and reliabilities as follows:

$$U_a = (1 - 0.90) \times (1 - 0.90) \times (1 - 0.90) = 0.001 \text{ and } R_a = 1 - 0.001 = 0.999$$

$$U_b = (1 - 0.80) \times (1 - 0.80) \times (1 - 0.80) = 0.008 \text{ and } R_b = 1 - 0.008 = 0.992$$

$$U_c = (1 - 0.992 \times 0.99) \times (1 - 0.97 \times 0.98) = 0.000885 \text{ and } R_c = 1 - 0.000885$$
$$= 0.999115$$

And the subassembly D, having only one part, has $R_d = 0.998$. The overall system reliability is therefore

$$R_s = R_a \times R_c \times R_d = 0.999 \times 0.999115 \times 0.998 = 0.99612$$

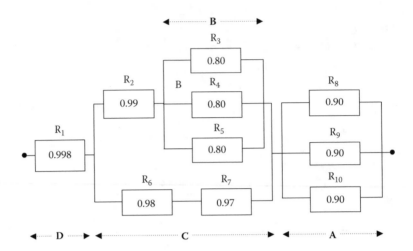

**FIGURE 16.5**   Assembly and subassemblies of series and parallel components.

The reliability program in some electrical power software tools calculates reliability indices and cost effect for alternative system designs based on the part failure rates, the series-parallel configuration of the assembly, active or dormant spare, time to repair, and cost impact of lost production. Many times, the part failure rate is not precisely known, making the final outcome of the study debatable. Even then, the process of going through such a study is more important than the final outcome—at least on a relative basis—to identify alternative ways of improving system reliability.

## 16.7   REDUNDANCIES

Redundancy is obtained by placing more units in parallel than necessary for the specified operation. Redundancy has a value only if the failure is instantaneously detected and acted on to switch over to the backup unit. It is therefore important to continuously monitor unit performance. Deviation in one or more performance parameters can be interpreted as failure. Needless to say, the failure detector must be of high reliability to match with the targeted reliability of the unit it monitors.

The redundant units can be operated in two ways shown in Figure 16.6: (a) active (hot) all the time or (b) dormant (cold standby) until the primary unit fails and then switched to the active state. Three types of redundancy are described next.

### 16.7.1   ACTIVE N FOR (N − 1) REQUIRED UNITS

If an assembly is made of $n$ identical units active all the time in parallel, where $(n − 1)$ units must work (i.e., any one may fail) for the assembly to perform the specified function, then the probability of the assembly to fail (i.e., all units to fail) is

$$U = U_1 \times U_1 \times \dots n \text{ times} = U_1^n$$

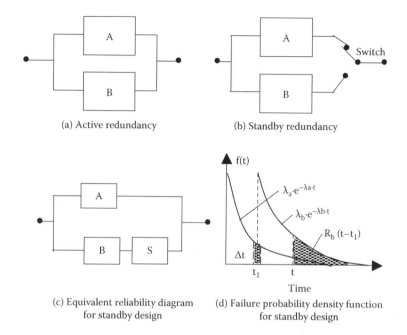

(a) Active redundancy

(b) Standby redundancy

(c) Equivalent reliability diagram
for standby design

(d) Failure probability density function
for standby design

**FIGURE 16.6** Active and dormant (cold standby) redundancies.

where $U_1$ is the probability of failure for each unit, assumed to be the same for all units. The reliability of the assembly is then

$$R = 1 - U = 1 - U_1{}^n \tag{16.11}$$

## 16.7.2 ACTIVE N FOR M REQUIRED UNITS

If a total of $n$ units are active in parallel and $m$ units ($m < n$) must work to meet the full-rated operation, the probability of exactly $k$ units working is given by the binomial distribution as per Equation (16.7),

$$P_k = \frac{n!}{(n-k)!k!} R_1{}^{n-k}(1-R_1)^k \tag{16.12}$$

where $R_1$ is the reliability of each unit. The probability of at least $m$ units working is the sum of all terms from $k = m, m + 1, m + 2, \ldots$ up to $k = n$, that is,

$$R = \sum_{k=m}^{k=n} \frac{n!}{(n-k)!k!} R_1{}^{n-k}(1-R_1)^k \tag{16.13}$$

### 16.7.3 Active *M* plus Dormant *D* Units

For an assembly with $(m + d)$ units in parallel, $m$ units active that must work, and $d$ units dormant as spares in the cold standby mode, any one of which can replace a failed active unit, the MTBF is given by

$$MTBF = \left(\frac{d+1}{m}\right) \times MTBF_1 \tag{16.14}$$

where $MTBF_1 = 1/\lambda_1$ for each unit.

For example, if there were 3 for 2 battery chargers (two active and one spare), the MTBF of the assembly would be $\{(1 + 1)/2\}\ MTBF_1, = MTBF_1$, that is, the same as that for each unit. If there were only two for two units with no spare, the MTBF of the assembly would be $\frac{1}{2}MTBF_1$. Here, all units are assumed to have identical failure rates, and the reliability of the switching element is assumed to be one. In reality, the dormant unit under no operating stress would have a much lower failure rate until switched on.

Two parallel units A and B can be assembled with both units active—each sharing half the load—as shown Figure 16.6a. An alternative working mode could be active A and backup B in dormant standby mode, with a switch at one end as seen in Figure 16.6b. The reliability is estimated with the switch in series with the backup unit as shown in Figure 16.6c. For the assembly to continue working after one unit fails, unit B and the switch both have to work at the time of switching. Figure 16.6d shows the assembly failure probability density functions, where unit B is switched on at time $t$. The reliability of the system at time $t$ is then determined by the shaded areas and diminishes as time progresses.

## 16.8 FAILURE RATE STATISTICS

The failure rate is derived from tests on a very large population of identical components. It is not practical to carry out such tests to end until all components in the population fail for it would take a very long time. Instead, testing is done for a predetermined duration, failed components are replaced as they fail, and the testing is terminated after a long enough time. This gives the results as intended because the failure rate is constant, and the unfailed components at any time are as good as brand new. For a given population, the failure rate in unit time under the test environment is then simply given by

$$\lambda = \frac{total\ number\ of\ failures}{total\ test\ duration} \tag{16.15}$$

Statistics on failure rates under particular operating conditions may be available from the component manufacturers. Various organizations are also active in compiling such information and publishing in various documents. For example, MIL-HDBK-217 compiles voluminous data on various components under various operating environments. A comprehensive review of these data showed that passive components—such as capacitors, resistors, diodes, and connectors—are the most reliable. The least reliable are the components with moving parts, such as slip rings, bearings, potentiometers, and relays. This is in line with what we expect. In electronics, traveling wave-tube amplifiers (TWTAs) used in communications networks are among the least-reliable components.

## 16.9 MIL-HDBK-217

Military handbook MIL-HDBK-217 establishes the uniform method of predicting reliability of military electronics parts, equipment, and systems that can be used for commercial systems as well. It lists failure rates of numerous parts under base thermal, electrical, and mechanical stresses. Any deviations from the specified operating base conditions would alter the failure rate as given by the following expression:

$$\lambda_a = \lambda_b \times \pi_p \qquad (16.16)$$

where

$\lambda_a$ = Failure rate under actual operating conditions in the design
$\lambda_b$ = Failure rate under base operating conditions specified in the handbook
$\pi_p$ = $\pi_1 \times \pi_2 \dots \pi_n$ = product of all modifying factors
$\pi_k$ = Factor to modify $\lambda_b$ for the environment, operating stress, construction differences, and so on for $k = 1, 2, \dots n$.

Major factors that modify part failure rates, particularly for electrical and electronics parts, are as follows:

$\pi_E$ = Environmental factor to account for factors other than temperature.
$\pi_Q$ = Quality factor to account for difference in quality level (class A, B, S, etc.). For example, this factor may be 1.0 for a part made under class B (commercial), but only 0.01 for the same part made under class S (space qualified).
$\pi_T$ = $K^{\Delta T}$ = Operating temperature deviation factor, where $\Delta T$ is the temperature deviation from the nominal rated value in degrees centigrade, and $K$ is a constant varying from 1.05 to 1.15 depending on part construction. With an average value of $K$ around 1.10, it is noteworthy that the failure rate factor doubles with every 10°C rise in the operating temperature and reduces to one-half for a 10°C decrease in the operating temperature.
$\pi_D$ = $(V_{op}/V_{rated})^a$ = Dielectric stress factor, where a = 2 to 4 near the rated voltage and 5 to 8 above the corona inception voltage. Obviously, the corona takes away life at a much faster rate. Therefore, corona in high-voltage equipment is avoided by proper design.
$\pi_V$ = Vibration factor, which depends on the fatigue life, which in turn depends on many factors, such as the vibration amplitude, frequency, stress concentration, fracture toughness of the material used in the part fabrication, and so on.

There are other modifying factors listed in the handbook, which must also be accounted for in the reliability estimates. The base failure rate $\lambda_b$ for some selected commercial and military power electronics parts before applying the modifying factors are reproduced from MIL-HDBK-217 in Table 16.2.

**TABLE 16.2**

**Base Failure Rate for Some Selected Electronic Parts**

| Component | Failure Rate $\lambda$ per $10^6$ Hours | | Component | Failure Rate $\lambda$ per $10^6$ Hours | |
|---|---|---|---|---|---|
| **Capacitors** | | | **Transistors** | | |
| Film | 78 | 26 | Signal, NPN | 71 | 14 |
| Tantalum | 200 | 63 | Signal, PNP | 100 | 20 |
| Electrolytic | 1600 | 480 | Power, NPN | 650 | 130 |
| **Resistors** | | | Thyristor, power | 1800 | 360 |
| Fixed, composition | 45 | 45 | ICs, digital £ 20 gates | 1050 | 18 |
| Fixed, wire wound | 100 | 100 | ICs, linear £ 32 Qs | 900 | 33 |
| Fixed, film | 3 | 3 | **Transformers** | | |
| Variable, cermetÔ | 2760 | 1380 | Signal | 10 | 3 |
| Variable, wire wound | 300 | 100 | Power | 82 | 32 |
| **Diodes** | | | Switches, thermal | 225 | 3 |
| Signal | 28 | 5.6 | Connectors, contact pairs | 32 | 4 |
| Power | 2160 | 432 | Fuses | 300 | 100 |

*Source:* MIL-HDBK-217F. Reliability Prediction of Electronic Equipment, Department of Defense, Washington, D.C.

## 16.10 PART COUNT METHOD OF RELIABILITY ESTIMATE

For a bid proposal and early design stage, assembly reliability can be estimated using a simple part count method outlined in MIL-HDBK-217 as follows: The information needed to apply this method is (a) generic part types and quantities, (b) part quality levels, and (c) operating environment of the unit. The general expression for assembly failure rate with this method is given by

$$\lambda_{eqp} = \sum_{i=1}^{i=m} N_i \cdot \lambda_{bi} \cdot \pi_{pi} \qquad (16.17)$$

where

$\lambda_{eqp}$ = Total equipment failure rate (failures per million hours)

$\lambda_{bi}$ = Base failure rate for the $i$th generic part (failures per million hours)

$\pi_{pi}$ = Product of all modifying factors for the $i$th generic part to account for the environment, operating stresses, and quality class difference

$N_i$ = Number of the $i$th generic parts used in the assembly, $i = 1, 2, 3, .... m$

$m$ = Number of different generic part categories used in the assembly

This method applies only if the assembly is used in one homogeneous environment. Otherwise, it is applied to portions of assembly (i.e., subassemblies) in each environment. Then, the failure rates of those portions are added to determine the total assembly failure rate.

## 16.11   DERATING FOR RELIABILITY

The failure rate of a component depends on operating stresses, such as the voltage, temperature, and so on. The electrical insulation at high temperature oxidizes, becomes brittle, and may crack, leading to failure (short circuit). The oxidation is a chemical degradation, which follows the Arrhenius exponential growth. Data on numerous equipment indicates that the failure rate doubles, or life is shortened to one-half, for every 7–10°C rise in the operating temperature. In the reverse, the life doubles for every 7–10°C reduction in the operating temperature. Similar degradation (wear) takes place above certain voltage, although it is not as well understood as that for the temperature. The rise in the failure rate with the operating stress level is not to be confused with the wear-out failure rate described in Section 16.2. It is to be seen as raising the flat part of the bathtub. The failure rate is still constant perunit time, although another constant at another operating stress level.

Derating is the reduction of electrical, thermal, and mechanical stress levels applied to a part to decrease the degradation rate and prolong the expected life. Lowering an operating stress on the component reduces the failure rate. Operating the component at lower than the nominally rated stress level is called *derating*. It is routinely used in engineering designs to decrease the failure rates in military and space-worthy designs. The derating in current is often done to lower the temperature. On the other hand, current derating in some active devices, such as transistors, may be to control the *di/dt* and *dv/dt* stresses, which can upset the semiconductor operation. Derating increases the margin of safety between the operating stress level and the actual failure level of the part. It provides added protection from system anomalies unforeseen by the design engineer.

## 16.12   QUICK ESTIMATE OF FAILURE RATE

Full-scale reliability testing is time consuming and expensive. A quick estimate of the reliability for screening a new proposed part for a potential application can be obtained by testing several units in parallel until the first one fails in normal or accelerated testing. If $T$ is the operating time to the first failure, obviously $T$ is not the MTBF since it is not derived from a large population. But, it can be used to roughly estimate the MTBF with different confidence levels as follows:

$$MTBF = \frac{1}{1-(1-C)^{\frac{1}{T}}} \qquad (16.18)$$

where $C$ is the desired confidence level, which is generally around 0.95 pu or 95% in the commercial world and could be 0.999999 pu or even much higher in defense and nuclear power equipment.

## 16.13   FAILURE MODES, EFFECTS, AND CRITICALITY ANALYSIS

The reliability analysis provides an estimate of the inherent reliability of the system. Failure modes, effects, and criticality analysis (FMECA), on the other hand, goes beyond that. Its primary purpose is to identify not only the failure modes but also their effects and criticality on mission success and safety. For example, criticality 1 means that if the component fails, it can cause injury to humans or cause the mission to fail completely. Obviously, such a component needs sufficiently high reliability. In other components, the probability of failure must be at the minimum possible level that is acceptable for the mission under design.

## QUESTIONS

*Question 16.1*: State the first theorem of reliability presented in this chapter and two simple examples of using it in everyday life.

*Question 16.2*: Identify four different ways a component can fail. Which of them is usually dealt with in reliability engineering, such as in this book?

*Question 16.3*: Differentiate the terms *series reliability* and *parallel reliability*.

*Question 16.4*: Two resistors are connected in parallel in an electrical circuit. In reliability analysis, when can they be considered in parallel or in series?

*Question 16.5*: How do various environmental factors influence the failure rate of components?

*Question 16.6*: What is the *derating* of a complement, and how is it used in practical engineering system designs?

## FURTHER READING

Billington, R., and R.N. Allan. 1984. *Reliability Evaluation of Power Systems*. New York: Plenum Press.

Brown, R.E. 2009. *Electric Power Distribution Reliability*. Boca Raton, FL: CRC Press/Taylor & Francis.

Ebeling, C.E. 1996. *Introduction to Reliability and Maintainibility*. New York: Mcgraw-Hill.

Rausand, M. 2003. *System Reliability Theory*. Newyork: John Wiley & Sons.

O'Connor, P. 2002. *Practical Reliability Engineering*. New York: John Wiley & Sons.

MIL-HDBK-217F. 1991. Reliability Prediction of Electronic Equipment. Department of Defense, Washington, D.C.

# INDEX

Printed in the United States
by Baker & Taylor Publisher Services